Minimizing Energy Consumption, Energy Poverty and Global and Local Climate Change in the Built Environment: Innovating to Zero

Minimizing Energy Consumption, Energy Poverty and Global and Local Climate Change in the Built Environment: Innovating to Zero

Causalities and Impacts in a Zero Concept World

Matthaios Santamouris
University of New South Wales, NSW, Sydney, Australia

Elsevier
Radarweg 29, PO Box 211, 1000 AE Amsterdam, Netherlands
The Boulevard, Langford Lane, Kidlington, Oxford OX5 1GB, United Kingdom
50 Hampshire Street, 5th Floor, Cambridge, MA 02139, United States

British Library Cataloguing-in-Publication Data
A catalogue record for this book is available from the British Library

Library of Congress Cataloging-in-Publication Data
A catalog record for this book is available from the Library of Congress

ISBN: 978-0-12-811417-9

For Information on all Elsevier publications
visit our website at https://www.elsevier.com/books-and-journals

Publisher: Candice Janco
Acquisition Editor: Amy Shapiro
Editorial Project Manager: Hilary Carr
Production Project Manager: Nilesh Kumar Shah
Cover Designer: Mark Rogers

Typeset by MPS Limited, Chennai, India

Dedication

To my Parents, My Maya and My Yiannos

Contents

Acknowledgments

The author is very grateful to the following people, who greatly helped with the preparation of this book:

The Faculty of Built Environment of UNSW, its Dean Prof. Helen Lochhead, and Prof. Catherine Bridge who have made the resources to edit the book and design the figures available. Special thanks to Erin Taylor who edited the book so nicely. A big thanks to my lovely students Jie Feng and Kai Gao who designed and typed the figures and tables of the book.

Thanks to everyone on the publishing team who helped me so much. Special thanks to Hilary Carr, the ever patient Publishing Manager.

Thanks to all the colleagues for permission to use copyright figures and other materials.

Finally, a big thank you to my family, colleagues, and friends for the invaluable help and support.

Mat Santamouris

Chapter 1

Introduction

THE BUILT ENVIRONMENT: CHARACTERISTICS, PROBLEMS, PROSPECTS AND FUTURE NEEDS

Building is one of the more dynamic economic sectors. Its main function and role is to protect human beings and ensure their quality of life. Buildings, and the construction sector in general, have an enormous potential to generate income and wealth and to create employment, but also to consume energy and resources and to produce pollution and waste. Additionally, it affects the local and global climate, while it is strongly associated with problems of poverty and vulnerability, at least for a large part of the population.

Buildings and construction are inherently dynamic. They act as one of the main promoters of major future development trends in technology and society, aiming to improve people's quality of life of and transform human societies. The building sector plays a very significant role in global economic activities. The sector is associated with several financial and commercial activities dealing with the management, construction, renovation and the extension of assets in the built environment, including buildings, open spaces and infrastructure. According to IHS Economics (2013), the total budget of the construction industry in 2013 exceeded USD$8.2 trillion, while forecasts for 2025 predict a total budget of close to USD$15 trillion (Global Construction Perspectives and Global Economics 2013). About USD$3 trillion are spent in the residential building sector, USD$2.7 trillion on infrastructure, and USD$2.5 trillion for commercial buildings (IHS Economics, 2013). Concerning the sources of investments, about 70% of the total budget is covered by the public sector and 26% by the private sector, while the rest is invested by official development assistance sources (BWI, 2006a, 2006b).

Buildings and construction generate almost 13% of global GDP, and this figure is expected to increase to 15% by 2020 (Global Construction Perspectives and Oxford Economics, 2013). About 35% of global construction output is in the so-called developing nations, and is expected to increase to 55% in 2020. The major players in the construction sector are countries with emerging economies such as China, Brazil and India, which are experiencing a very rapid increase in population, urbanization and the improvement of living standards (Global Construction Perspectives and Oxford Economics,

Minimizing Energy Consumption, Energy Poverty and Global and Local Climate Change in the Built Environment: Innovating to Zero. DOI: https://doi.org/10.1016/B978-0-12-811417-9.00001-5

2013). As well as their direct impact on and contribution to the global economy, buildings also contribute in indirect ways, influencing several industrial sectors through very well-defined growth linkages. The building sector presents a very high multiplier factor for the global economy. Industries that benefit from the construction sector include the manufacturing of components and materials, the production and exploration of energy and the development of equipment and machinery, as well as many other associated industries.

Parallel to the contribution of the building sector to the economy, the building sector is a very labor intensive sector, contributing a high number of jobs per investment. As mentioned by WIEGO (2015), the construction sector counts as the second most important employer in the world after agriculture. According to official statistical data, the construction sector employs more than 110 million people. Almost 75% of this is in developing countries where employment in the construction sector remains undeclared, and so the number may be much higher. It is characteristics that in India about 89% of the men and 97% of the women working in the building-construction sector are considered to be employees working under informal conditions (Pais, 2002). According to BWI (2006b), the total number of workers (both declared and undeclared) in the construction sector may exceeds180 million.

As mentioned, the construction sector presents an important multiplier factor for the global economy and the employment market. It is well accepted that for each new job in the construction sector many new jobs are also generated in the global economy. In the United States, for each new house built, almost 2.97 new jobs are generated (Emrath, 2015). About 1.76 jobs are directly associated with construction specifically, while the rest are created in the global economy in an indirect way. The industrial sectors that benefit the most from increased employment are manufacturing, wholesale and retail trade, warehousing and transportation real estate, services, and finance and insurance. The National Association of Home Builders (NAHB, 2015 has performed a study and estimated the impact of building 100 single family houses in a typical zone of the United States. They have estimated the impact of the construction activity (A), the effect of investment and the corresponding tax revenue (B), and the ongoing effect when the building is occupied (C). It is reported that during the three above phases, almost 463 new jobs are created, out of which only 185 are directly associated with the construction activity.

While the building and construction sector contributes highly to generating wealth and improving the quality of life of citizens, it is also associated with several negative impacts. Most of the problems created by the building sector are related to environment, energy, local and global climate change and human vulnerability. The main negative impacts and problems are:

a. Buildings are the highest consumer of energy, accounting for about 30%–40% of the total consumption on the planet (UNEP, 2012). Especially when the required energy for construction and demolition is counted, buildings are responsible for about 50% of the total energy consumption. The following chapters provide a detailed discussion about the energy consumption of the building sector and its characteristics. The global energy consumption by the building sector in 2010 was close to 23.7 PWh, while the International Energy Agency estimates that by 2040, consumption will increase tremendously in the developing countries and will reach 38.4 PWh (IEA, 2013).

b. The building sector is highly responsible for global and local climate change. According to UNEP (2012), the sector is liable for about 38% of the total greenhouses gas emissions. In parallel, buildings are responsible for the development of the urban heat island (UHI) phenomenon that increases the temperature in the dense part of cities. UHIs have significant impact on the amount of energy consumed for cooling, increase the concentration of harmful pollutants, deteriorate indoor and outdoor thermal comfort conditions, increase the ecological footprint of cities, and increase heat related mortality and morbidity (Santamouris, 2015).

c. Buildings have a serious impact on the global environment as they consume resources and produce pollution and waste. In parallel, buildings consume a very high quantity of raw materials and resources. They use almost three billion tons of raw materials per year, which represents almost 40%–50% of the total material use in the world (Hultgren, 2011). Additionally, the building sector consumes almost 12% of the total potable water in the world, and about 70% of the timber products, while is responsible for about 20%–25% of air pollutants, 70% of halocarbons, 25%–33% of the emissions of black carbon, 50% of landfill waste and 40% of the pollution of potable water (UNEP, 2012; BIMHow, 2015; IIASA, 2015).

d. There is a very serious deficit of adequate housing. The construction and building sector is facing the significant challenge of providing appropriate shelter for more than 1.6 billion people currently living in inadequate conditions in informal houses without proper sanitation (UN–Habitat, 2011). There are more 863 million people living in slums, and UN–Habitat (2013) forecasted that this number will increase to 1 billion by 2020. The demand for new urban dwellings is so high in the developing world that almost 20%–30% of new urban houses are associated with informal construction. Additionally, more than 100–150 million people in developed countries cannot afford the cost of energy and live under energy poverty conditions.

e. Tremendous urbanization in recent years, especially in the developing countries, challenges the capacity of the building and construction sector to satisfy the additional needs and requirements in housing, buildings and general infrastructure. According to the United Nations (2009), the urban population will exceed 6.3 billion of people by 2050. This is equivalent to an increase of close to 84% compared to 2009 levels. Forecasts predict that by 2020 there will be more than 527 cities where the population will exceed 1 million people.

Addressing the above challenges is a major issue for the building sector. A rich dialog is already taking place to identify the capacity of our societies to address these problems, while policy makers and civil society investigate which policies are the most appropriate to face the upcoming crisis. The design and implementation of policies, and of a concrete road map to address the challenges, requires a full and comprehensive knowledge and understanding of the characteristics and details of the problem. Under such conditions, it is possible to define specific qualitative and quantitative targets. Objectives and targets must be adapted to the needs and requirements of the local communities and should consider the specific characteristics of the problems in each area and community. Several factors control the capacity of the decision makers and in general of the societies to determine and implement adequate policies. Among other factors, the strength of the problem, the local economic conditions, the technological knowledge and competence, the development prospects, the existing business models and the availability of technological tools are the most important.

A book claiming to offer a roadmap for addressing all the challenges identified in the world is almost impossible and could be quite superficial. The present book concentrates on three major challenges and issues, focusing primarily on developed countries:

a. The political economic and technical context, the causes and the reasons defining and determining the strength and magnitude and the characteristics of the energy consumption in the building sector.
b. The specific impact of local and global climate change on the built environment, and vice versa.
c. The technological and social issues, characteristics and aspects of energy poverty.

These three problems seem to be among the most important for developed societies. They are highly interrelated, with strong synergies between them. The book aims to analyze the actual characteristics of the three problems, explore their multiple and complex synergies and trade offs, and address three fundamental questions regarding potential future developments (Santamouris, 2016):

a. What should the quantitative and qualitative future targets for each of the three problems in concern be?
b. What will be the major technological, macroeconomic and social forces and trends defining the progress and developments in the immediate future that will impact the economy and the sustainable advancement of societies? How can these mechanisms change the current state of affairs, defining a more proactive than a reactive agenda, and have a positive impact on the problem of energy consumption in buildings, local and global climate change, and energy poverty in the developed world?

c. How do we transform problems in opportunities? How can we increase the added value of technological, economic and social interventions to generate wealth and offer employment opportunities?

ANALYZING THE THREE MAIN PROBLEMS OF THE BUILT ENVIRONMENT: ENERGY CONSUMPTION AND ENVIRONMENTAL QUALITY, ENERGY POVERTY AND URBAN VULNERABILITY, AND LOCAL CLIMATE CHANGE

Energy Consumption in the Built Environment: Challenges and Opportunities

The amount of energy consumed in the built environment is extremely high. Forecasts of future energy demand show that it may soon increase considerably. The main drivers that define the evolution of energy consumption in the building sector are related to economic, technological, climatic, demographic and social issues. In particular:

- The expected tremendous increase of the planet's population, is believed to have a very important impact on future energy consumption, as 3–4 billion new citizens will be added to the global population.
- The expected increase in the ambient temperature caused by global and local climate change affects energy consumption in the built environment, increases energy demand for cooling and probably decreases energy requirements for heating purposes.
- The foreseen increase in global GDP and a corresponding increase in household income will result in a substantial rise in the surface area of dwellings and commercial buildings. In parallel, higher family income makes the use of air conditioning and other electrical equipment more affordable.
- Technological developments increase the energy efficiency of the equipment used in the built environment, while helping to considerably decreasing the energy consumption of buildings. Zero energy buildings and almost zero energy settlements are designed and built in most developed counties (Fig. 1.1). However, many questions remain regarding the higher cost of zero energy buildings and their impact of low income populations. Although the technological developments look promising, it is widely accepted that the expected technological improvements cannot compensate for the effects of the previously mentioned drivers.
- There is a big question mark about the future price of energy. Most analysts predict a significant increase in future cost, mainly due to considerable new investments in electricity generation and the need to decarbonize the energy system. Higher energy prices may reduce consumption, but will aggravate the problem of energy poverty.

FIGURE 1.1 The almost zero energy building NTL of the Cyprus Institute in Cyprus. Source: *Papanikolas et al. (2015).*

– Very intensive energy regulations aiming to decrease and optimize the use of energy consumption in the built environment have been adopted and implemented. New energy regulations, in combination with the application of advanced energy conservation technologies, have considerably decreased energy consumption for heating and lighting purposes.

Although the population increase may be negligible in developed countries, all other drivers determining future energy consumption will continue to regulate future energy consumption in the built environment. In particular, due to climate change and increased income, energy consumption for cooling purposes is expected to increase tremendously. The major challenge of the future energy policy in the developed world seems to be how to cover future cooling needs in the built environment using sustainable and green technologies and at a reasonable cost.

In the so-called developing world, the energy agenda seems to be completely different than in the developed countries. Overpopulation, urbanization and the need to improve living standards and supply most of the population with electricity skyrockets energy consumption and the need for additional electricity power plants. Soon energy consumption in just one single city of India, Mumbai, may be almost equal to the whole cooling consumption of the United States.

Such a huge energy increase has very important economic and environmental consequences as it increases the energy-related financial burden of the countries, requires tremendous investments to install additional energy infrastructures, and increases the emission of greenhouse gases and other harmful pollutants. The development of appropriate technologies based on local resources seems to be the main challenges for the developing world, especially the use of renewable energy technologies, the introduction of energy conservation measures and systems, the increased availability of energy sources for each one, the increase of energy equity, and the rationalization or even minimization of the cost of energy.

It is evident that energy in the built environment faces extremely serious future challenges, both in developed and developing countries. The main issue in question is how the challenges may be translated into opportunities that will generate wealth, create new employment, support social equity and boost the quality of life of the population.

Climate Change in the Urban Built Environment: Our New Big Problem

Urban overheating is a serious problem for the whole world. Although the UHI phenomenon has been known about for more than 100 years, rapid urbanization, combined with a continuous increase in the production and released of anthropogenic heat in cities, intensifies the magnitude of the phenomenon and aggravates the impact on energy, environment, comfort and health.

Scientific knowledge on the strength and the characteristics of urban overheating is continuously increasing. Experimental data are available for more than 400 cities worldwide, while new monitoring techniques based on the use of advanced measuring equipment have been developed and employed (Santamouris, 2015). In parallel, the impact of urban overheating on energy, health and environmental quality has been assessed in detail in many cities and countries (Baccini et al., 2008; Santamouris, 2014; Santamouris et al., 2015; Santamouris and Kolokotsa, 2015). All studies shown that the energy impact of urban overheating is quite high and may exceed 68 kWh per person and degree of temperature increase. In parallel, the average additional penalty on electricity power has been calculated as being close to 20 W per person and degree of temperature increase. Finally, all studies associating heat-related mortality with ambient temperature show that there is a threshold temperature over which the level of heat related mortality increases considerably. Fig. 1.2 summarizes the results of the above studies.

To counterbalance the impact of local climate change and urban overheating, mitigation technologies have been developed and implemented in several cities. The proposed mitigation technologies aim to decrease the strength of heat sources in the cities, such as solar radiation, and increase the potential of heat sinks, for example by using evaporative technologies. Technologies may be classified in the following technological groups:

1. Technologies aiming to decrease the surface temperature of the city. This is mainly achieved through the reduction of the amount of solar radiation absorbed and the increase of emission losses from the materials.

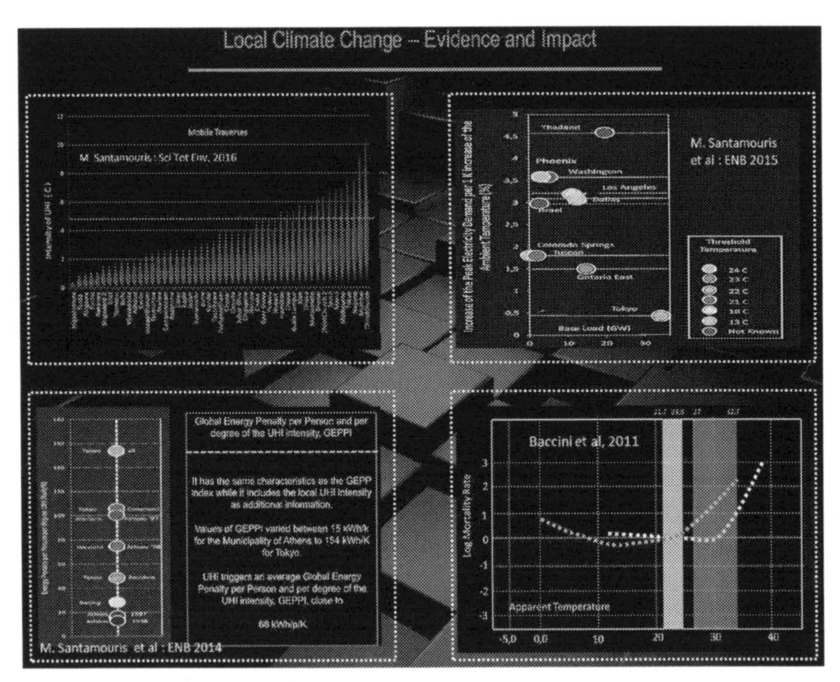

FIGURE 1.2 Impact of local climate change and urban overheating.

2. Greenery technologies aiming to increase evapotranspiration from trees and other greenery in the city.
3. Evaporative technologies aiming to decrease the ambient temperature through the rise of the latent losses at the city level.
4. Solar control technologies aiming to decrease the intensity of solar radiation at the earth's surface.
5. Radiative cooling technologies aiming to cool materials by increasing the radiative losses through the atmospheric window.
6. Ground cooling techniques aiming to dissipate the excess urban heat into the ground.

All of the above technologies have been extensively implemented in many large scale mitigation projects around the world. The cooling potential of the various technologies and their combinations has been analyzed by Santamouris et al. (2017). The potential decrease of the peak ambient temperature per technology and combination is shown in Fig. 1.3.

Although the reported results show that the ambient temperature in cities can decrease several degrees, the reduction of the average peak ambient temperature rarely exceeds 2.5–3.0°C.

As an example of the real potential of the various mitigation technologies to reduce the peak ambient temperature in cities, Table 1.1 estimates the

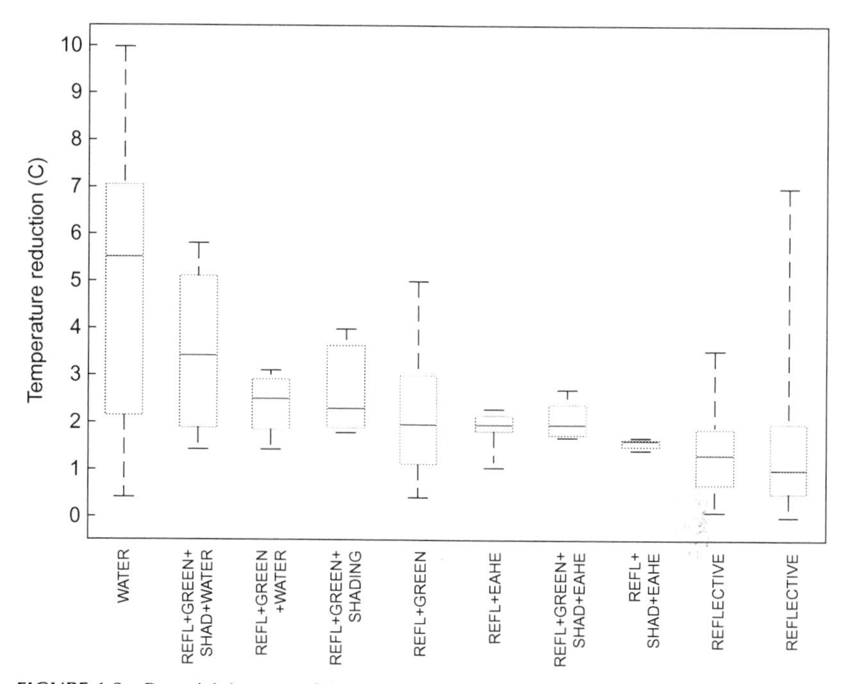

FIGURE 1.3 Potential decrease of the peak ambient temperature per mitigation technology and combination. Source: *Santamouris et al. (2017).*

potential of using several mitigation technologies for the city of Sydney. About twenty-one different mitigation scenarios are calculated.

The distribution of the ambient temperature in the zone of the city under consideration before the application of the mitigation technologies is given in Fig. 1.4. As shown, the ambient temperature in the area is found to vary from 24–29°C. In parallel, Figs. 1.5 to 1.9 report the calculated distribution of the ambient temperature when the mitigation scenarios are implemented. The whole study shows that the implementation of the specific mitigation technologies can result in a decrease of the peak ambient temperature by up to 2.5°C.

Although cooling technologies based on evaporative and evapotranspiration techniques are proven to be very efficient, their technological development is not impressive; indeed, almost the same level of technology has been available for the last 10 years. In contrast, progress in the development of new material technologies for mitigation purposes is quite impressive. Recent developments on the topic of advanced materials for mitigation are presented in Fig. 1.10 (Santamouris and Garshasbi, 2018).

The developed technologies are commercially available and are implemented in more than 200 large scale projects around the world. Monitoring

TABLE 1.1 Mitigation Scenarios and Their Characteristics Examined for the City of Sydney

Category	Scenario ID	Global Albedo	Albedo Streets	Albedo Pavements	Albedo Roofs	Greenery Pavements	Greenery Roofs
Base case	1	0.1				5%	
Global Albedo	2	0.2				5%	
	3	0.3				5%	
	4	0.5				5%	
	5	0.7				5%	
Albedo streets	6		0.2	0.1	0.1	5%	
	7		0.3	0.1	0.1	5%	
	8		0.5	0.1	0.1	5%	
	9		0.7	0.1	0.1	5%	
Albedo pavements	10		0.1	0.2	0.1	5%	
	11		0.1	0.3	0.1	5%	
	12		0.1	0.5	0.1	5%	
	13		0.1	0.7	0.1	5%	
Albedo roofs	14		0.1	0.1	0.2	5%	
	15		0.1	0.1	0.3	5%	
	16		0.1	0.1	0.5	5%	
	17		0.1	0.1	0.7	5%	
Greenery	18	0.1				20%	
	19	0.1				40%	
	20	0.1				60%	
Green roofs	21	0.1				5%	100%

Source: Santamouris et al. (2018).

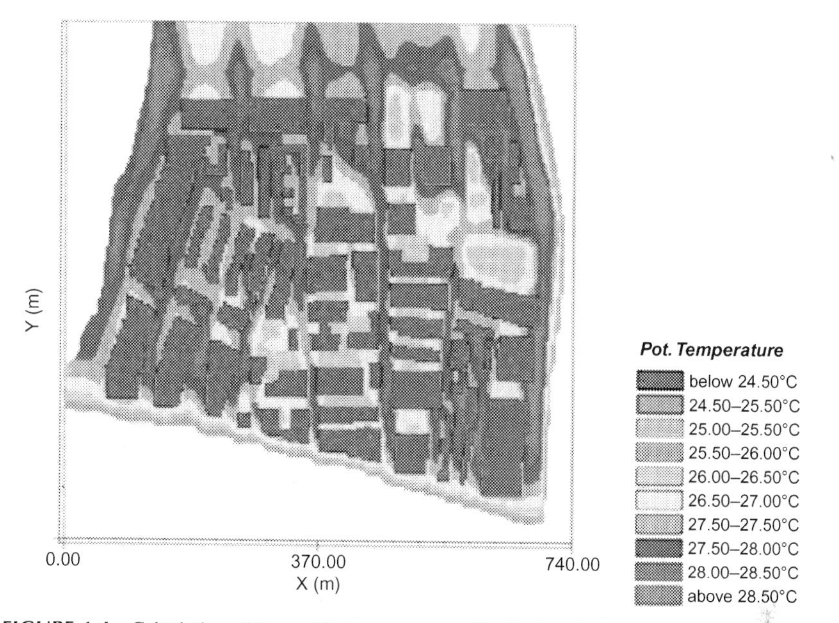

FIGURE 1.4 Calculation of the ambient temperature distribution in the considered urban zone of Sydney. Source: *Santamouris et al. (2018).*

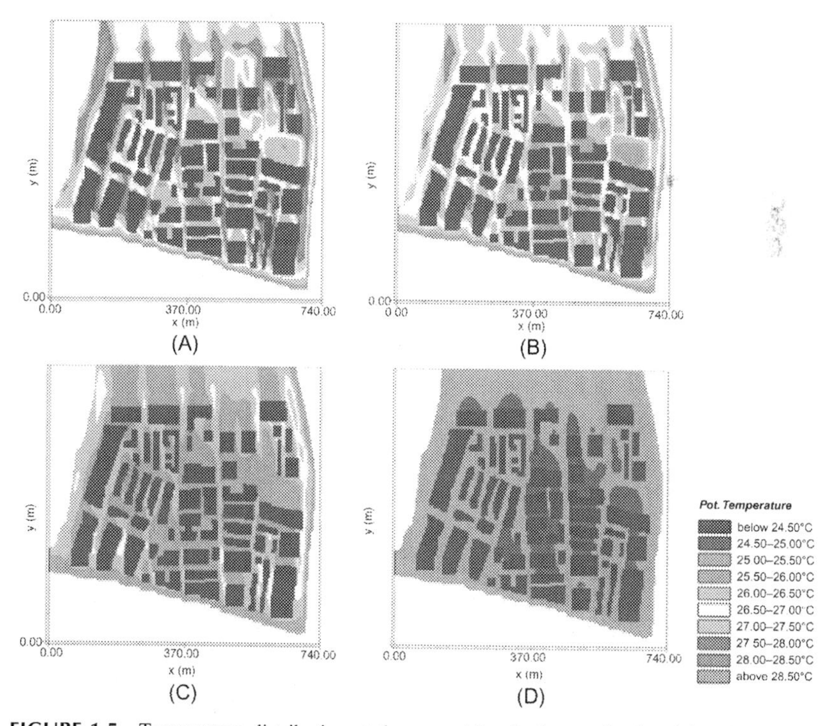

FIGURE 1.5 Temperature distribution at the ground level of scenarios involving the increase of the global albedo. (A) Scenario ID 2, (B) Scenario ID 3, (C) Scenario ID 4, (D) Scenario ID 5. Source: *Santamouris et al. (2018).*

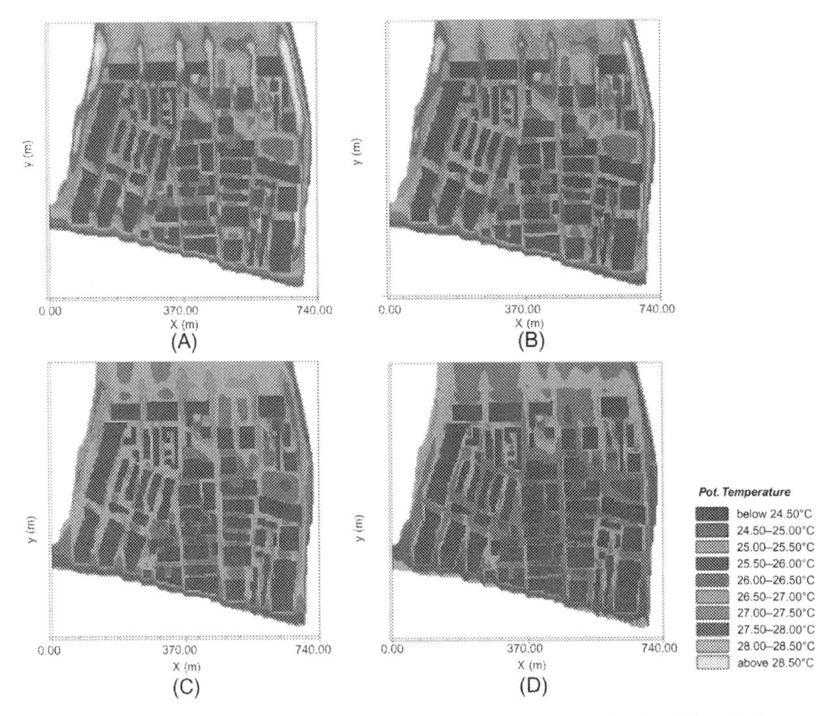

FIGURE 1.6 Temperature distribution at the ground level of scenarios involving the increase of the albedo of streets. (A) Scenario ID 6, (B) Scenario ID 7, (C) Scenario ID 8, (D) Scenario ID 9. Source: *Santamouris et al. (2018).*

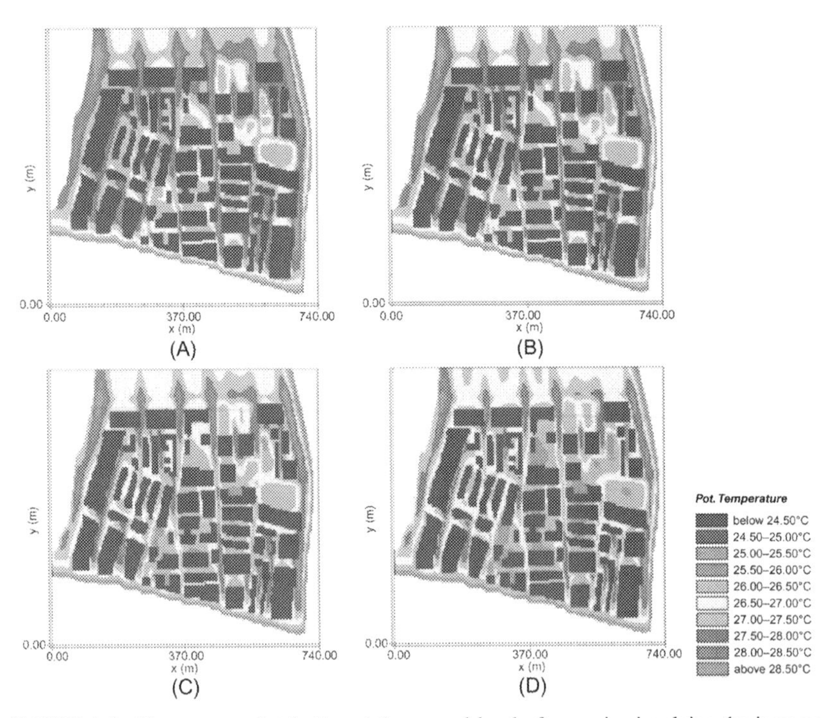

FIGURE 1.7 Temperature distribution at the ground level of scenarios involving the increase of the albedo of pavements. (A) Scenario ID 10, (B) Scenario ID 11, (C) Scenario ID 12, (D) Scenario ID 13. Source: *Santamouris et al. (2018).*

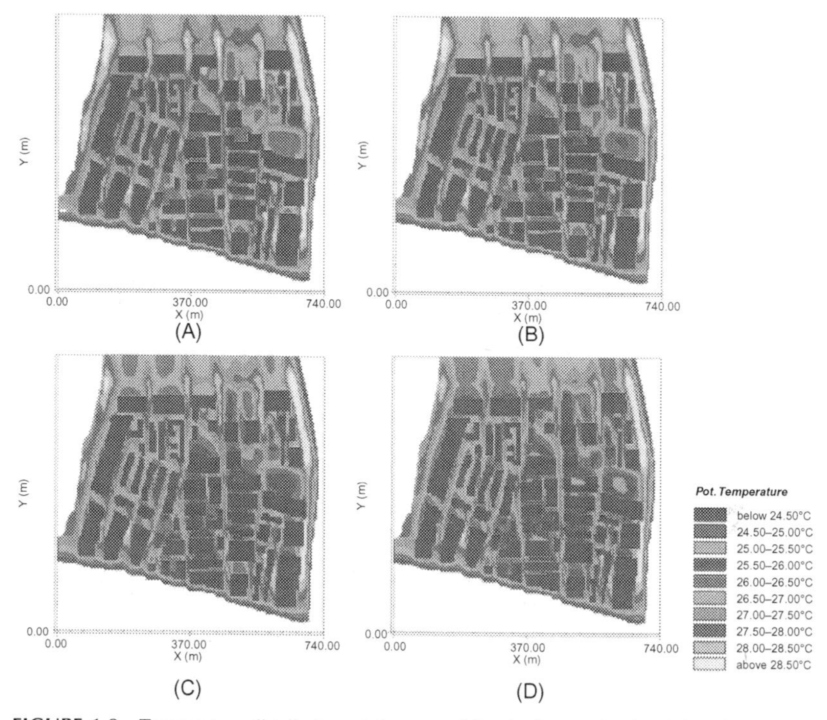

FIGURE 1.8 Temperature distribution at the ground level of scenarios involving the increase of the albedo of roofs. (A) Scenario ID 14, (B) Scenario ID 15, (C) Scenario ID 16, (D) Scenario ID 17. Source: *Santamouris et al. (2018).*

results have shown that the existing mitigation technologies are able to decrease the peak ambient temperature by up to 2.5–3 K. However, given than the magnitude of the urban heat island frequently exceeds 7°C–10°C, there is a profound need to develop advanced mitigation technologies and passive and active cooling techniques able to further decrease the peak ambient temperature of the built environment.

Fig. 1.11, presents a map of the existing non-compression cooling technologies displaying a significant potential to decrease the surface and ambient temperature of cities and provide cooling in the urban environment. The proposed mitigation technologies may be classified as passive or active. In principle, active technologies are based on the implementation of a thermodynamic cycle to provide heating and cooling, such as in a heat pump. Passive cooling technologies and materials used in the urban environment can provide cooling through natural and non-forced heat transfer to the environment, such as convection, emission, evaporation and conduction, without the use of any mechanical system.

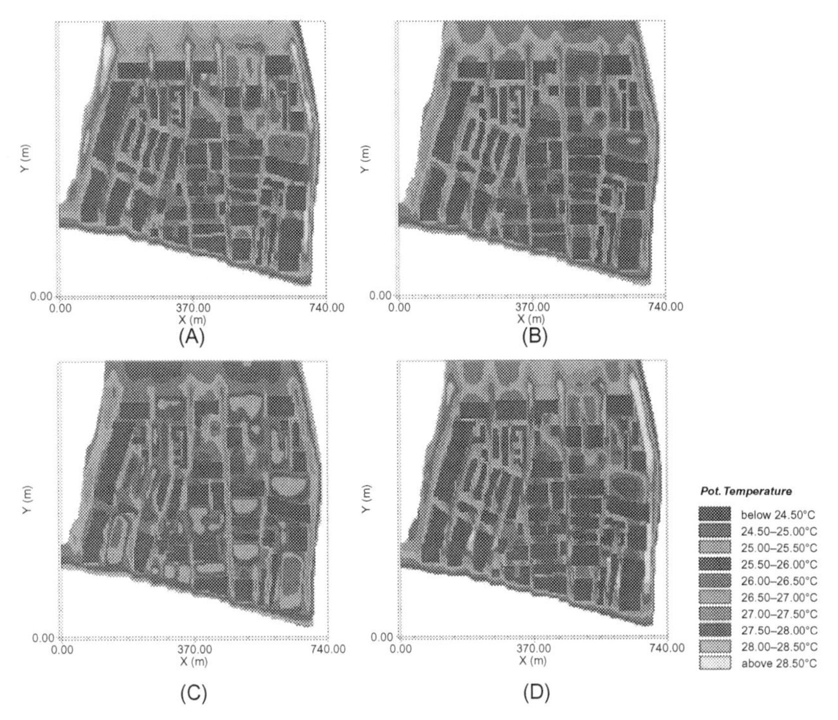

FIGURE 1.9 Temperature distribution at the ground level of scenarios involving an increase in greenery. (A) Increase in greenery at ground level to 20% (ID 18), (B) Increase in greenery at ground level to 40% (ID 19), (C) Increase in greenery at ground level to 60% (ID 20), (D) Implementation of green roofs on 100% of the area (ID 21). Source: *Santamouris et al. (2018).*

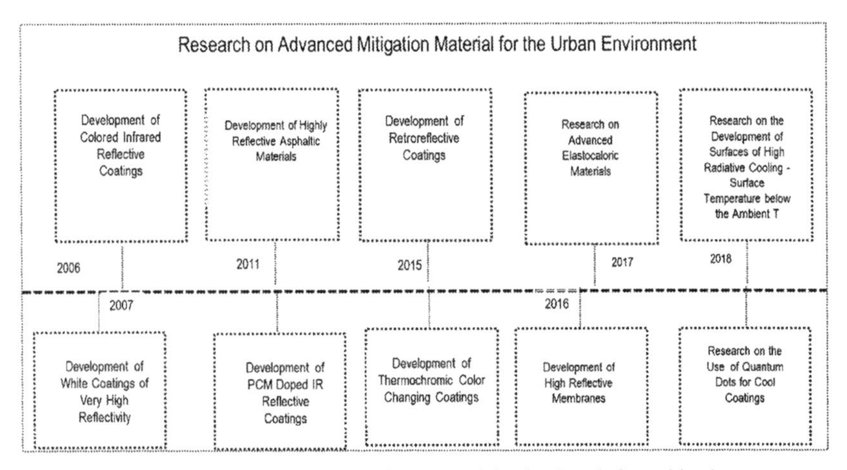

FIGURE 1.10 An overview of the various materials developed for mitigation purposes. Source: *Santamouris and Garshasbi, 2018.*

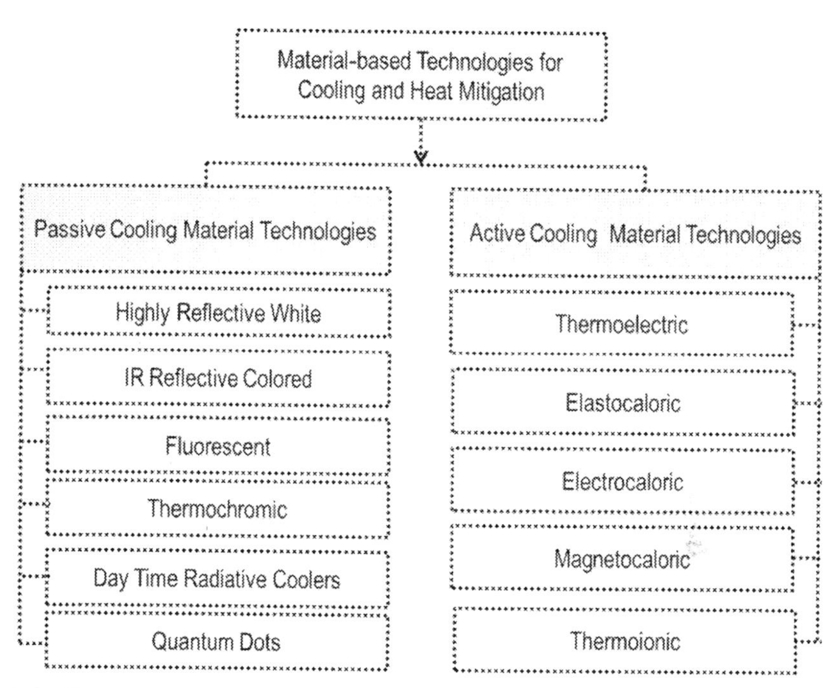

FIGURE 1.11 An overview of the prominent passive and active cooling technologies that are able to provide cooling in the built environment and mitigate the UHI phenomenon. Source: *Santamouris and Garshasbi (2018).*

Active non-compression cooling technologies are principally based on the use of caloric solid state materials. These materials can provide cooling once an external excitation is applied (Magnetic field, tension, pressure, voltage). Solid state materials are known for many years, but it is quite recently that thin film caloric materials have exhibited a substantial cooling capacity and potential. Among the various caloric materials, electrocaloric and elastocaloric materials seems to be the more appealing candidates as they demonstrate a higher potential to be used in applications related to urban mitigation and cooling.

The proposed passive cooling mitigation technologies, based on the use of urban mitigation materials, involve the use of white or colored reflective materials, presenting a high reflectivity in the whole or just in the infrared solar spectrum, thermochromic materials that are able to change their color and reflectivity as a function of their surface temperature, and fluorescent technologies that enhance the thermal losses of the material through the emission of a part of the absorbed light, as well as daytime and nighttime radiative cooling processes.

Most of the above-mentioned passive mitigation technologies contribute significantly to decreasing both surface and ambient temperatures in the urban built environment. The measured and expected decrease of the peak surface and ambient temperature caused by each of the considered passive technologies is reported in Fig. 1.12.

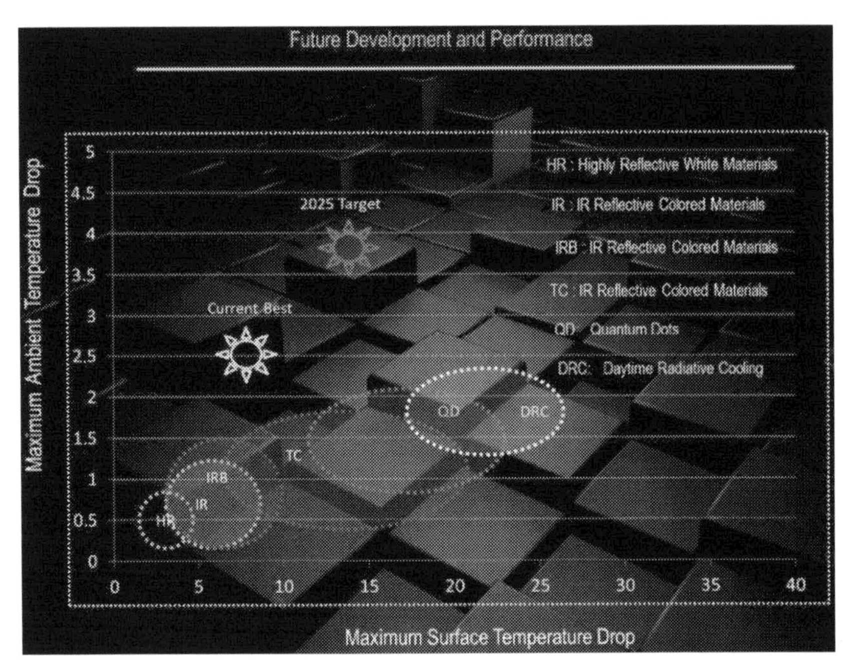

FIGURE 1.12 Measured and expected levels of the peak surface and ambient temperature in relation to the considered mitigation technologies. Source: *Santamouris and Garshasbi (2018).*

As mentioned, the impact of the various mitigation technologies on energy usage, peak electricity demand and health has been assessed in many large scale mitigation projects around the world. Almost all projects have concluded that the implementation of mitigation technologies can have a highly significant impact on energy, peak electricity demand and health. Detailed studies have been carried out for Sydney and Darwin in Australia, as well as many other places around the world. Fig. 1.13 presents the distribution of energy consumption for cooling a representative residential building in Western Sydney for the reference as well as for all the mitigation scenarios. As shown the use of the proposed mitigation scenarios can decrease the cooling load of a single residential building by up to 43%.

Greenery techniques could reduce the cooling load between 15%–29%, evaporative techniques between 13%–30%, the combination between greenery and evaporative systems between 18%–32%, the use of cool materials between 29%–41%, and the combination of the cool materials and of the evaporative systems between 29%–43%.

Considering the distribution of the residential buildings in Western Sydney, the total expected energy savings from the use of mitigation technologies is calculated between 800–1000 GWh/year, which is a tremendous energy conservation (Fig. 1.14).

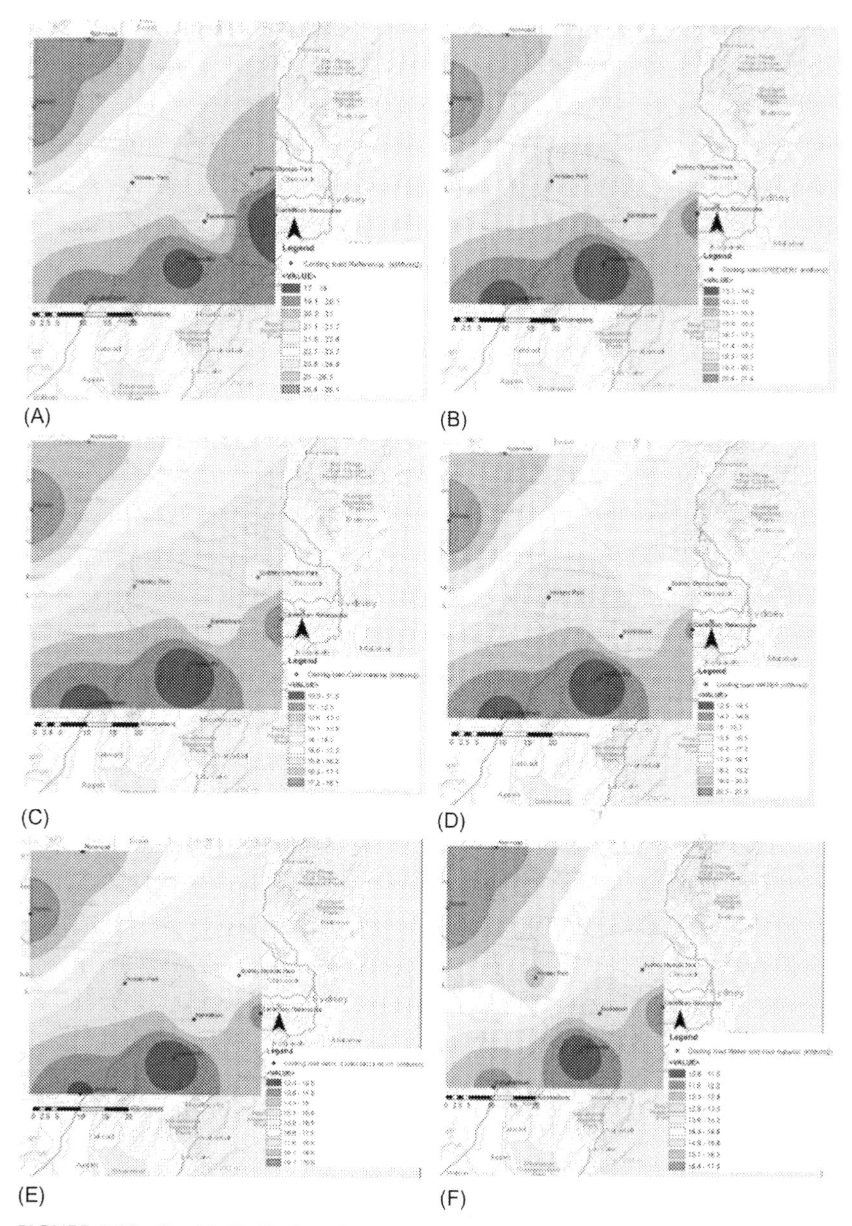

FIGURE 1.13 Spatial distribution of the cooling load of a representative residential building for the reference case (A) and each mitigation scenario, (B) Greenery, (C) Cool roofs & pavements, (D) Water, (E) Water technologies and greenery, (F) Water technologies and cool roofs and pavements.

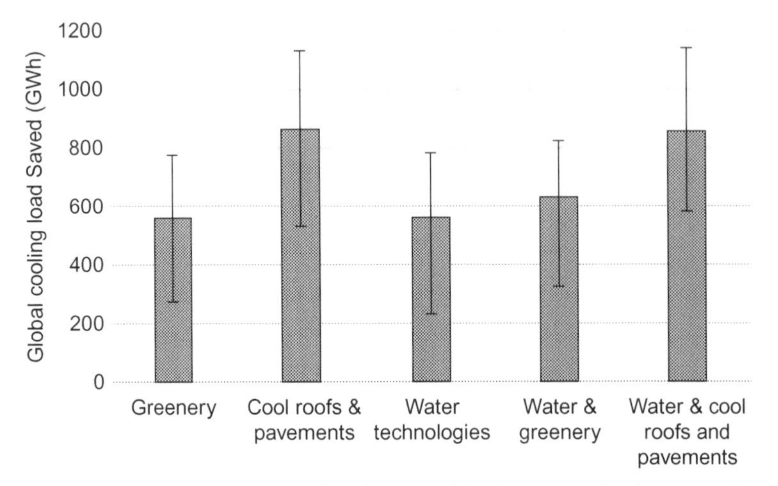

FIGURE 1.14 Global cooling load savings for each mitigation strategy for the area of Western Sydney.

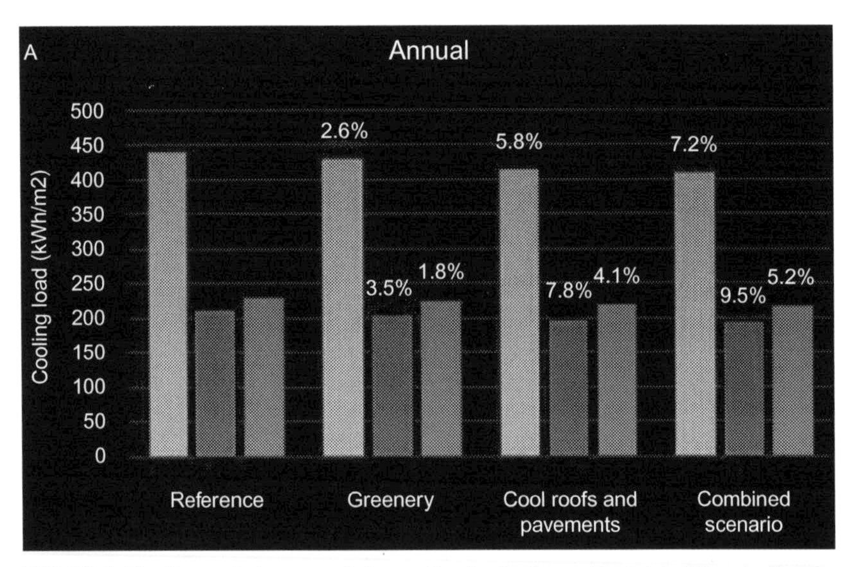

FIGURE 1.15 Expected decrease of the cooling load of a residential building in Darwin, Australia.

In a similar way, the energy conservation expected to arise from the application of urban mitigation technologies in the tropical city of Darwin are calculated as being between 2%–10% (Fig. 1.15). The considerably lower contribution compared to Sydney is due to of the very high latent load in tropical cities, which mitigation technologies cannot substantially reduce.

Apart from their very significant energy benefits, mitigation technologies help to significantly reduce peak electricity demand in cities. Fig. 1.16 shows

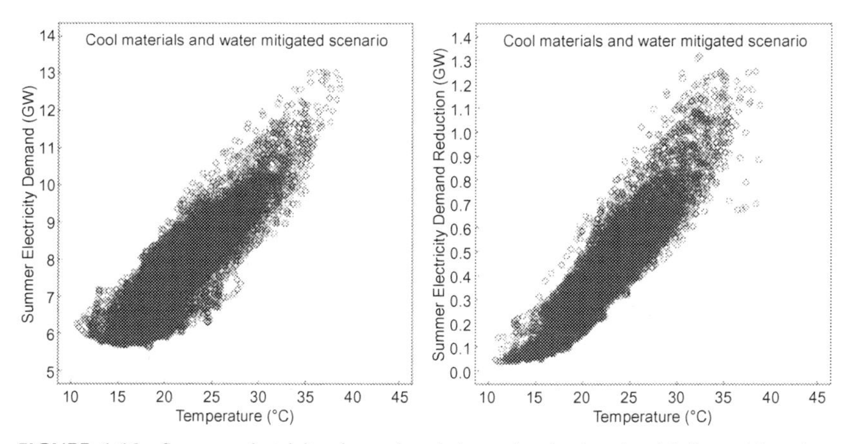

FIGURE 1.16 Summer electricity demand and demand reduction (modeled unmitigated – modeled mitigated) for the mitigation scenario with cool materials and water combined, plotted vs. the population weighted average temperature.

the electricity demand in the city of Sydney as well as the calculated reduction in electricity demand when a combination of mitigation technologies is applied. As shown, a wide implementation of urban mitigation technologies in Sydney can reduce peak electricity demand by up to 9%. Such a reduction in demand corresponds to about three medium size power plants.

High urban temperatures have a serious impact on heat related mortality and morbidity. Studies have shown that the impact of mitigation technologies on health is highly significant. Figs. 1.17–1.18 show the calculated distribution of heat related mortality in Western Sydney, Australia, under the current conditions as well as when urban mitigation technologies are applied. As shown, heat related mortality may decrease by up to 40%.

The most effective mitigation strategy, namely the combined use of cool materials and water, can lower the cumulative heat-related extra-deaths even 55 km from the coast to the levels that are met in the unmitigated scenario 15 km from the coast, namely to approximately 7.5 deaths per 100,000 inhabitants. With cool materials and water, the cumulative heat-related excess deaths close to the coast are lowered to approximately 5 per 100,000 people, including during hot years such as the summer of 2016/2017.

In a similar way, the potential decrease in hospital admissions in Darwin due to the implementation of urban mitigation technologies is assessed. Fig. 1.19 shows the calculated decrease of heat-related hospital admissions for the various considered mitigation technologies. It is calculated that heat-related hospital admissions may decrease by up to 39% because of the use of proper mitigation technologies.

In parallel, the same study has revealed that anomalies in heat-related mortality in Darwin increased by 5% and 4% for every 1°C increase in daily

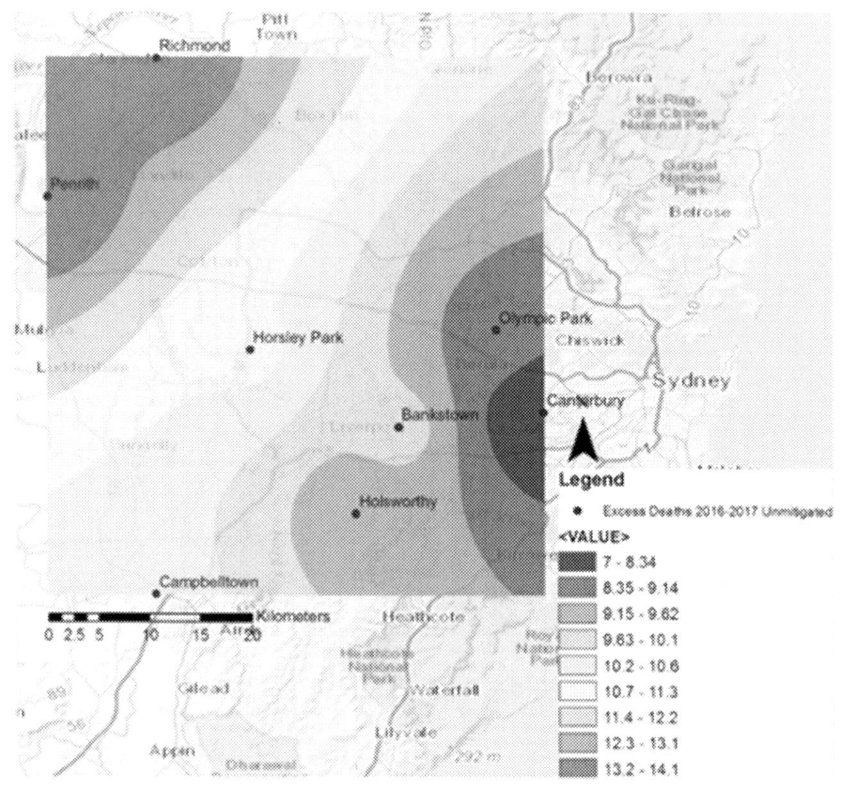

FIGURE 1.17 Excess deaths per 100,000 inhabitants during 2016/2017 in the unmitigated scenario.

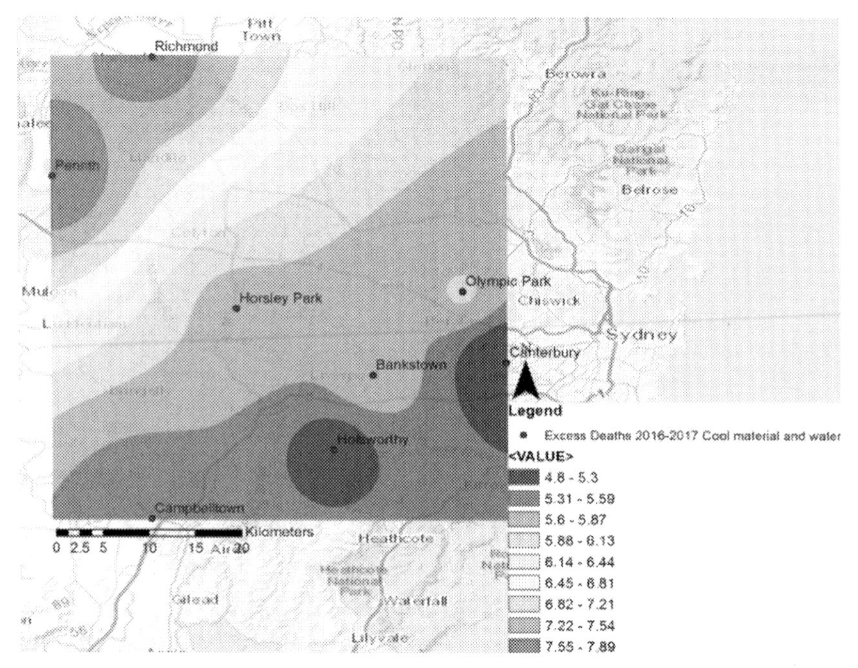

FIGURE 1.18 Excess-deaths per 100,000 inhabitants during 2016/2017 in the mitigation scenario with cool materials and water combined.

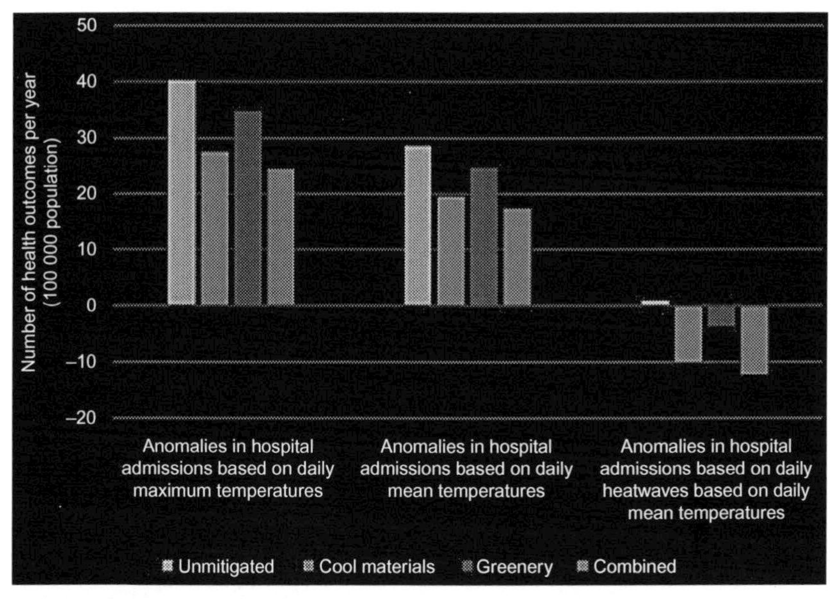

FIGURE 1.19 Expected decrease in heat-related hospital admissions when urban mitigation technologies are implemented.

maximum temperature, with the result that the combined scenario can save 9.66 excess deaths per year per 100,000 inhabitants within the Darwin Urban Health District (Fig. 1.20). Additionally, it has been calculated that the implementation of the combined mitigation scenario could result in the avoidance of four deaths annually in a population of 100,000 people aged above 65.

Energy Poverty: Status and Prospects

Energy poverty refers to the condition where households are not able to cover their basic energy needs. Although there are many definitions and thresholds to describe and measure energy poverty, there is not a universal and agreed definition that applies everywhere in the world. This is quite evident as economic, technological and social conditions vary tremendously from country to country.

In the developing world, energy poverty is very much related to the lack of provision of clean and proper energy and fuels, due in part to the nonexistence of networks and infrastructures, but mainly because of the economic inability of the population to cover its energy needs. Hundreds of million of people are living without any supply of electricity, and rely on polluting and inappropriate fuel sources for cooking. The lower the income of households, the more polluting the energy source used (Fig. 1.21). As a result, the

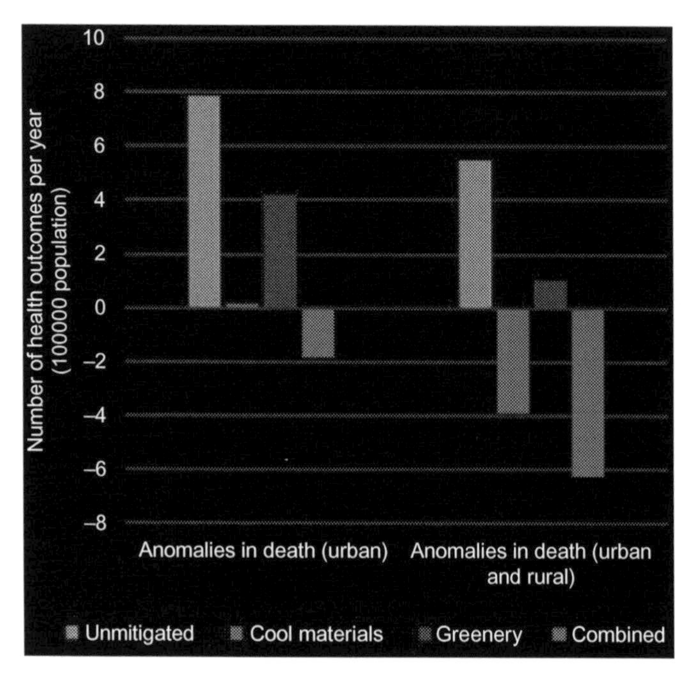

FIGURE 1.20 Calculated anomalies in death in the Darwin area for the different considered mitigation technologies.

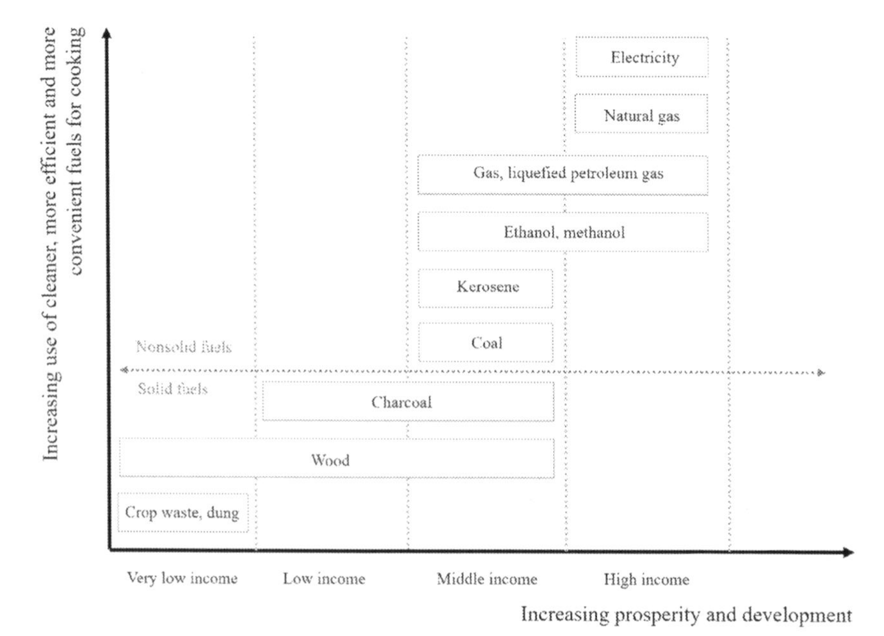

FIGURE 1.21 Types of fuel used as a function of the income. Source: *WHO (2006)*.

concentration of harmful indoor pollutants is extremely high and seriously affect the health of the population. According to the Institute for Health Metrics and Evaluation (2017), 2.6 million people died prematurely in 2016 from illness attributable to household air pollution. Fig. 1.22 reports the deaths per region attributed to the inappropriate indoor air quality (Roser and Richie, 2018). As shown, most of the deaths occur in Asia and Africa.

The very clear relation between income and deaths caused by inappropriate indoor air quality is demonstrated in Fig. 1.23 (Institute for Health Metrics and Evaluation [IHME] 2016). It is evident that death rates are highest at the lowest incomes, and decline as average incomes transition towards USD$20,000.

Energy poverty is a serious problem in the developed world, Given the nonexistence of a clear and universal definition of energy poverty, the exact number of energy poor people cannot be precisely defined. However, it is estimated that it exceeds 200 million people in all developed countries. Energy poverty in developed countries is mainly the result of economic and social drivers. Unemployment, low family income, high energy prices and especially the inappropriate quality of dwellings are the main parameters influencing energy poverty.

Energy poverty is very strongly related to energy conditions and characteristics and local climate change. Houses of low energy quality that are not well protected and present a high energy consumption are a serious burden for low income populations as the required budget to satisfy proper indoor temperature conditions is usually high and cannot be afforded given household income. In parallel, it is well known that low income populations mainly reside in relatively deprived urban areas where the phenomenon of the UHI tends to be quite strong. Higher summer outdoor temperatures increase the vulnerability of the low income population and put their health conditions under stress.

There are several ways to fight energy poverty. However, it is widely accepted that the provision of proper housing presenting a low energy consumption is the most efficient way to protect low income populations. Several major projects aiming to rehabilitate existing low income houses are carried out and it is concluded that it is possible to seriously decrease their energy consumption and improve indoor environmental conditions at a quite reasonable cost.

Energy consumption in the building sector, local climate change, urban overheating and energy poverty are among the major problems in the developed countries. Several sectoral policies have been defined and applied with quite limited (and in many cases unsuccessful) results. It is evident that a holistic approach aiming to tackle the three problems in an integrated way is the proper approach. Such an approach should be based on comprehensive and well-defined objectives and targets for the future, while a clear roadmap to achieve these objectives must be agreed upon.

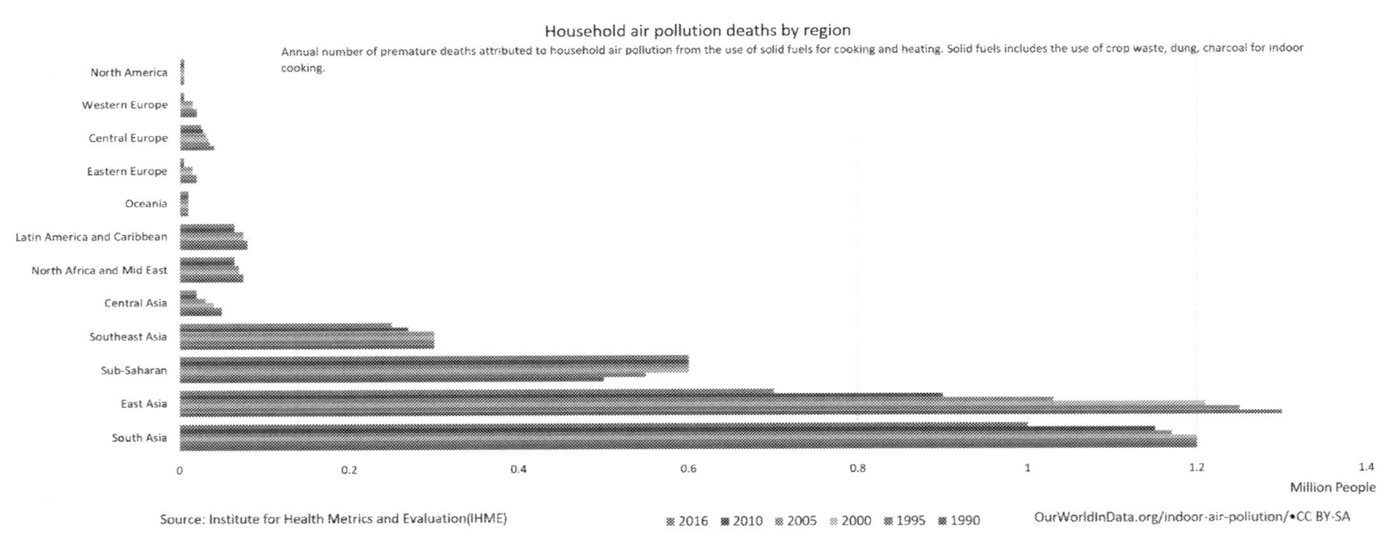

FIGURE 1.22 Household air pollution deaths by region. Source: *Roser and Richies (2018).*

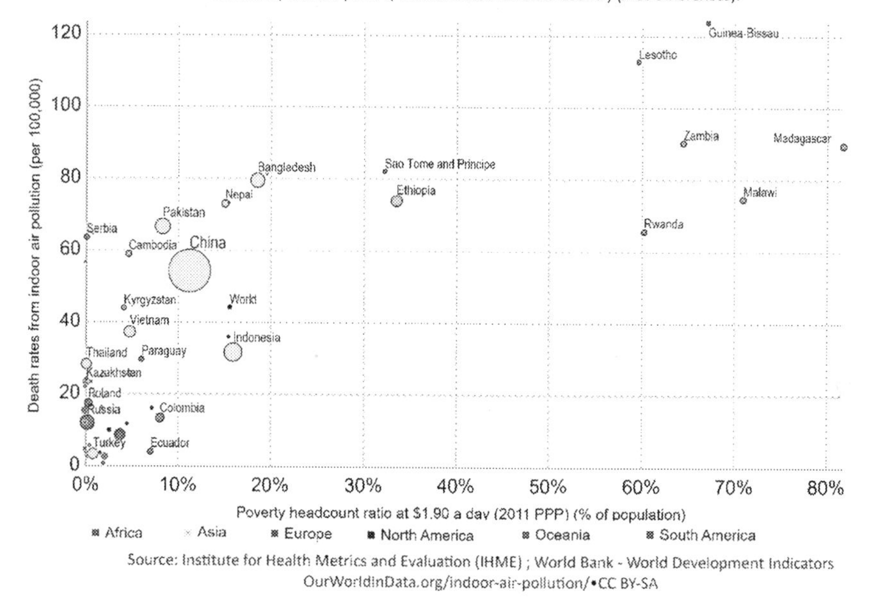

FIGURE 1.23 Death rates from indoor air pollution as a function of income. Source: *Roser and Richies (2018).*

The first part of this book explores the current state of the art regarding energy consumption in the built environment, local climate change and the problem of urban overheating, as well as problems related to energy poverty. In the second part of the book, future objectives and targets for the three specific problems are defined. How is it possible to minimize energy consumption, fully mitigate local climate change and completely eradicate energy poverty in the developed world ? The synergies between them are explored and analyzed to better define a roadmap for the future. The principal technological developments, policies, economic and social issues to be considered are explored and discussed, while the main benefits arising from such a perspective are investigated.

REFERENCES

Baccini, M., Biggeri, A., Accetta, G., Kosatsky, T., Katsouyanni, K., Analitis, A., et al., 2008. Heat effects on mortality in 15 European cities. Epidemiology 19 (5), 711–719.

BIMHow, 2015. Impact of the construction industry on the environment. http://www.bimhow.com/impact-of-the-construction-industry-onthe-environment/. Assessed January 2018.

BWI, Building and wood workers International, 2006a. Briefing on labour standards in construction contracts resented at the BWI World Council 7th December 2006.

BWI, Building and woodworkers International, 2006b. Defending workers rights in construction. http://www.bwint.org/pdfs/WCProcurementFiona.pdf. Assessed January 2018.

Emrath, P., 2015. Impact of home building and remodeling on the united states economy. Published by Housing Economics.com, http://www.nahb. org/fileUpload_details.aspx?contentTypeID = 3&contentID = 227858&subContentID = 577446&channelID = 311.

Global Construction Perspectives and Oxford Economics, 2013. Global construction 2025. http://www.globalconstruction2025.com/. Assessed January 2018.

Hultgren, A., 2011. SWCA environmental consultants, global buildings and construction: Energy and environmental impacts. http://www.kleiwerks.org/global-buildings-and-construction-energy-and-environmental-impacts-pdf/. Assessed January 2018.

IHS Economics, 2013. Global construction outlook: Executive outlook. https://www.ihs.com/pdf/IHS_Global_Construction_ExecSummary_Feb2014_140852110913052132.pdf. Assessed January 2018.

IIASA: Energy End-Use: Buildings, 2015. http://www.iiasa.ac.at/web/home/research/Flagship-Projects/Global-Energy-Assessment/GEA_CHapter10_buildings_lowres.pdf. Assessed January 2018.

Institute for Health Metrics and Evaluation (2017). Global burden of disease (GBDx Results Tool).

International Energy Agency, 2013. World energy outlook 2013. http://www.worldenergyoutlook.org/weo2013/. Assessed January 2018.

National Association of Home Builders, 2015. The economic impact of home building in a typical local area income, jobs, and taxes generated april 2015 housing policy department, 2015. https://www.nahb.org/~/media/Sites/NAHB/LMA/FileUploads/35601-1-REPORT-local-20150318115955.ashx?la=en. Assessed January 2018.

Pais, Jesim, 2002. Casualization of urban labor force: Analysis of recent trends in manufacturing. Economic and Political Weekly 46 (16), 631–652.

Papanikolas, C., Lange, M.A., Fylaktos, N., Montenon, Alaric, Kalouris, G., Fintikakis, N., et al., 2015. Design, construction and monitoring of a near zero energy laboratory building in Cyprus. Advances Building Energy Research 9 (1), 140–150.

Roser M. and H. Richie: Indoor Air pollution, available through: https://ourworldindata.org/indoor-air-pollution. Assessed January 2018.

Santamouris, M., 2014. On the energy impact of urban heat island and global warming on buildings. Energy and Buildings 82, 100–113. October 2014.

Santamouris, M., 2015. Regulating the damaged thermostat of the cities – Status impacts and mitigation strategies. Energy Build. 91, 43–56.

Santamouris, M., April 2016. Innovating to zero the building sector in Europe: Minimising the energy consumption, eradication of the energy poverty and mitigating the local climate change. Solar Energy 128, 61–94.

Santamouris M. and S. Garshasbi: Cooling the cities. Advances on urban mitigation technologies, Proc SET conference 2018.

Santamouris, M., Kolokotsa, D., 2015. On the impact of urban overheating and extreme climatic conditions on housing Eeergy comfort and environmental quality of vulnerable population in Europe. Energy and Buildings 98, 125–133. Available from: https://doi.org/10.1016/j.enbuild.2014.08.050.

Santamouris, M., Cartalis, C., Synnefa, A., Kolokotsa, D., July 2015. On the impact of urban heat island and global warming on the power demand and electricity consumption of buildings–A review. Energy and Buildings 98, 119–124. Available from: https://doi.org/10.1016/j.enbuild.2014.09.052.

Santamouris, M., Ding, L., Fiorito, F., Oldfield, P., Paul Osmond, Paolini, R., et al., September 2017. Passive and active cooling for the outdoor built environment – Analysis and assessment of the cooling potential of mitigation technologies using performance data from 220 large scale projects. Solar Energy 154, 14–33.

Santamouris, Mattheos, Haddad, Shamila, Saliari, Maria, Vasilakopoulou, Konstantina, Synnefa, Afroditi, Paolini, Riccardo, et al., 2018. On the energy impact of urban heat island in Sydney: Climate and energy potential of mitigation technologies. Energy & Buildings 166, 154–164.

UNEP, 2012. Building design and construction: Forging resource efficiency and sustainable development.

United Nations, 2009. Urban and rural areas, www.unpopulation.org. Assessed January 2018.

UN–Habitat, 2011. Cities and climate change: Global report on human settlements.

UN–Habitat, 2013. Streets as public spaces and drivers of urban prosperity, nairobi.

WHO (2006) – Fuel for life: Household energy and health.

WIEGO, 2015. Woman in informal economy: Globalizing and organising: Construction workers. http://wiego.org/informal-economy/occupational-groups/construction-workers. Assessed January 2018.

FURTHER READING

World Bank: Urban Development, 2015. <http://data.worldbank.org/topic/urban-development>. Assessed January 2018.

Chapter 2

Energy Consumption and Environmental Quality of the Building Sector

EVOLUTION OF ENERGY CONSUMPTION IN THE WORLD: TRENDS AND PROGRESS

The building sector is one of the major consumers of energy. According to the UNEP (2012), it counts for about 30%–40% of worldwide energy consumption, but this figure may increase up to 50% when the required energy consumption for construction and demolition is considered (WBCSD, 2009). According to the International Energy Agency (2013), the total amount of energy delivered to the building sector was close to 23.7 PWh in 2010, and it may rise to 38.4 PWh in 2040. Such a significant increase is mainly attributed to the dramatic increase of the energy consumption in non-OECD countries (IEA, 2013).

The building sector contributes greatly to improving the global economy and improving the quality of life of humans. The sector is associated with major benefits, as summarized in Fig. 2.1. Buildings represent almost 13% of the global Gross Domestic Product (GDP), while construction is a major employing sector. According to the available statistics, almost 180 million people are employed by the building sector.

However, the final contribution of the building sector to global employment is much higher. Buildings present a very high multiplier factor, and for each job generated in the construction sector many other jobs are generated in the global economy. Construction is a major economic sector, with a total budget in 2013 that was close to USD$8.3 trillion. Forecasts show that the total budget may increase to up to USD$15 trillion by 2025.

Although the building industry contributes highly to the global economy, they may cause substantial problems for human society and generate several negative impacts (Fig. 2.2). particularly in the areas of energy consumption, local and global climate change, poverty and human vulnerability.

In fact, apart from the very high energy consumption of buildings, the sector is responsible for serious global environmental damage. As mentioned

Minimizing Energy Consumption, Energy Poverty and Global and Local Climate Change in the Built Environment: Innovating to Zero. DOI: https://doi.org/10.1016/B978-0-12-811417-9.00002-7

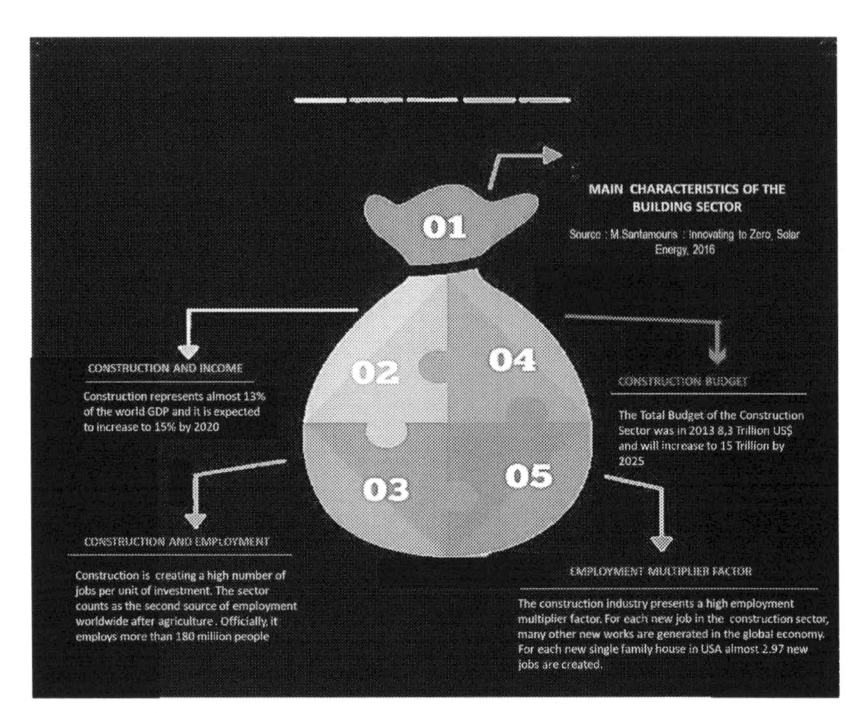

FIGURE 2.1 Main benefits of the building sector. Source: *Santamouris (2016a).*

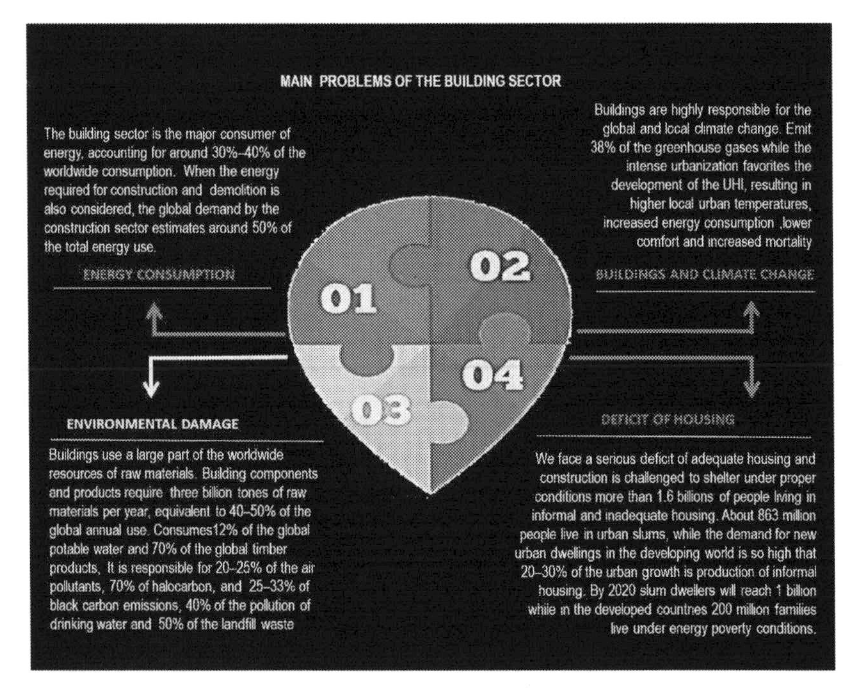

FIGURE 2.2 Main challenges and problems in the building sector. Source: *Santamouris (2017).*

in Fig. 2.2, buildings consume a very high amount of raw materials. Buildings in particular, and the construction sector in general, require about 3 billion tones of raw materials per year, which is equivalent to about 40%–50% of global annual consumption. In parallel, buildings consume almost 12% of the potable water in the world and about 70% of global timber products. Finally, buildings are responsible for about 20%–25% of air pollutants, 70% of halocarbon, 25%–33% of black carbon emissions, 40% of drinking water pollution and 50% of landfill waste (Hultgren 2011; BIMHow 2015; IIASA 2015).

Apart from the very high consumption of materials and resources, the building sector is significantly affecting local and global climate change. The sector is responsible for about 38% of greenhouse gases emissions, while intense urbanization favors the development of Urban Heat Islands (UHI), resulting in higher local urban temperatures, increased energy consumption, lower comfort and increased hospital admissions and mortality (UNEP 2012; Santamouris, 2015).

As shown in Fig. 2.2, humanity is challenged by a very serious deficit of proper housing, mainly in developing countries. The building sector faces the challenge of providing adequate housing conditions for about 1.6 billion people living in informal and inadequate housing. According to UN–Habitat (2013), there are almost 863 million people living in urban slums. In parallel, about 20%–30% of the new urban growth in developing countries can be attributed to the production of informal housing, mainly because of the very high demand for new urban dwellings. UN–Habitat (2013) estimate that by 2020 there will be more than 1 billion slum dwellers, while in the so-called developed countries there will be more than 200 million families living in energy poverty.

Developing countries are threatened by an exceptional level of urbanization, which challenges the capacity of the building and construction sector to comply with the additional need for housing and the related infrastructures for the new and existing urban shelters. According to the United Nations (2009), the urban population will reach almost 6.3 billion people (Fig. 2.3), which corresponds to an increase close to 84 compared to 2009 levels. As reported by UN–Habitat (2011), in 2020 there will be more than 527 cities in the world with over one million citizens.

ENERGY CONSUMPTION IN DEVELOPED AND DEVELOPING COUNTRIES

Energy Consumption in Developed Countries

Developed countries consume much more energy per person to satisfy the demand of buildings than countries under development. As shown in Table 2.1, while the average world energy consumption of the building sector is close to 4.57 MWH/c/year, the North American countries consume

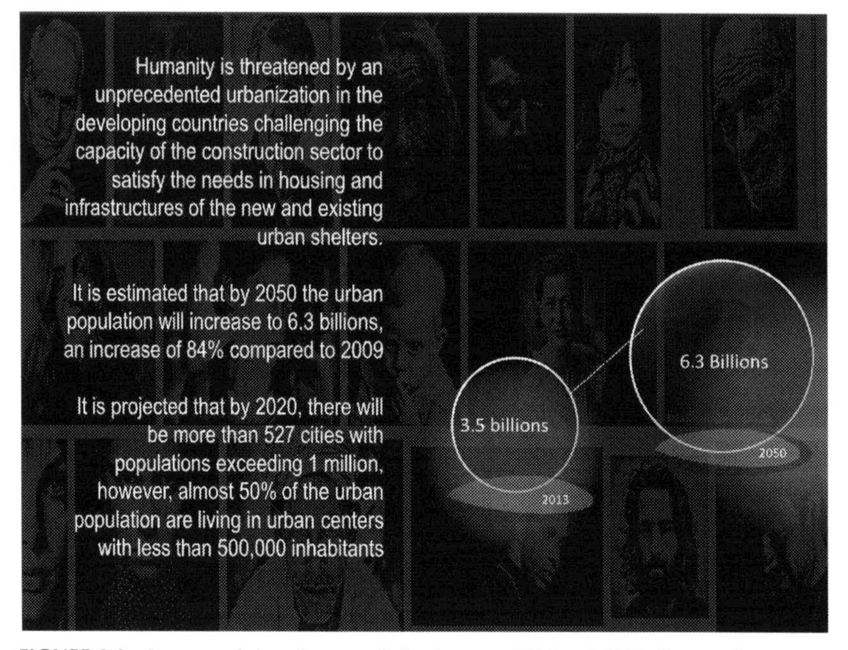

FIGURE 2.3 Increase of the urban population between 2013 and 2050. Source: *Santamouris (2016a).*

TABLE 2.1 Residential and Commercial Energy Demand per Capita in MWh/y/Capita

World Regions	Residential and Commercial Energy Demand per Capita, MWh/Capita-yr.
USA and Canada	18.6
Middle East	5.75
Latin America	2.32
Former Soviet Union	8.92
European Union-27	9.64
China	3.20
Asia excluding China	2.07
Africa	3.19
World	**4.57**

Source: IIASA (2015). Adapted from Energy End-Use: Buildings, 2015, Reproduced with permission from the International Institute for Applied Systems Analysis (IIASA).

close to 18.6 MWH/c/year, Europe 9.6 MWH/c/p, while Asia and Africa are around to 2.09 and 3.19 MWH/c/year respectively.

Energy Consumption in Europe

Buildings are the most important consumers of energy in Europe and are responsible for about 41% of global final energy consumption, while the total stock of buildings in Europe is close to 24 billion square meters. Almost 75% are residential buildings with an average floor area close to 87 m^2 per dwelling. Residential buildings are responsible for about 27% of the total energy consumption in Europe, while the commercial sector accounts for almost 14% of the total energy consumption. The mean energy consumption in Europe varies as a function of the local climatic conditions and the quality of the building stock (Fig. 2.4, European Environmental

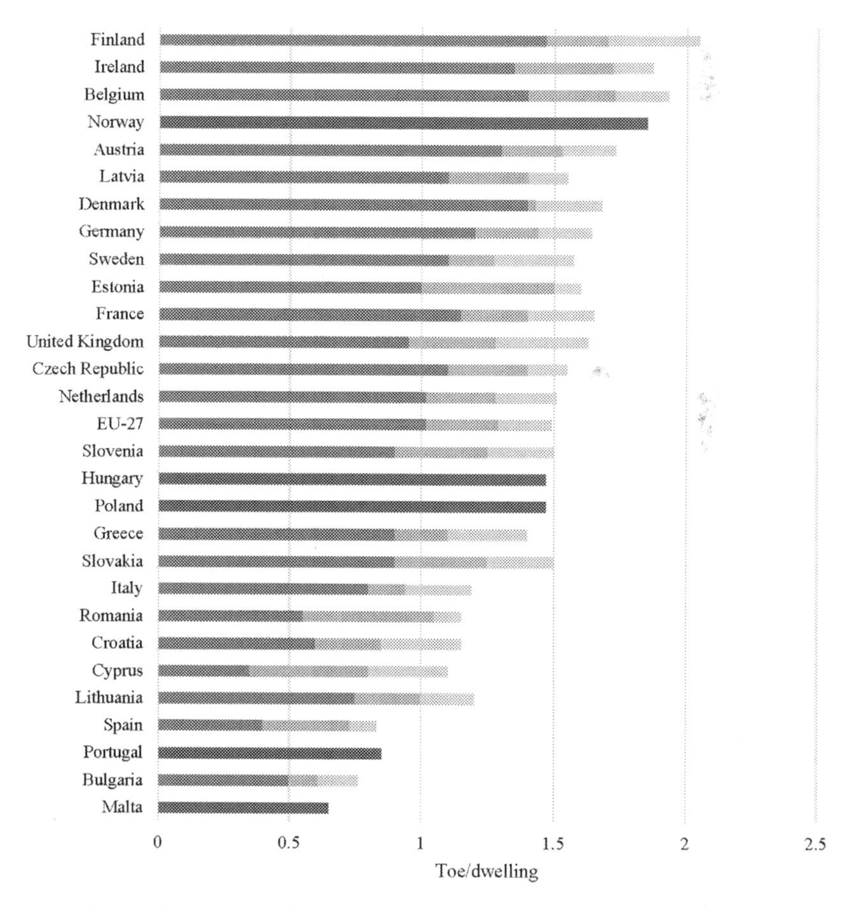

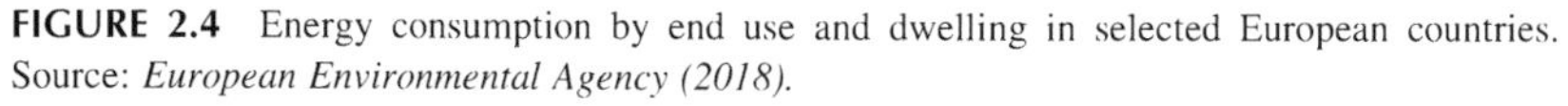

FIGURE 2.4 Energy consumption by end use and dwelling in selected European countries. Source: *European Environmental Agency (2018).*

Agency, 2018). The average consumption in 2009 was close to 220 kWh/m^2/year, and varied between 320 kWh/m^2/year in Finland and 150 kWh/m^2/year in Bulgaria (Odysee-Mure 2012). The corresponding average consumption of nonpresidential buildings in Europe is close to 295 kWh/m^2/year. As shown in Fig. 2.4, space heating represents almost 71% of the total residential energy consumption on the continent. Water heating, cooking, lighting and energy-consuming appliances represent almost 12%, 4% and 15% respectively (Odyssee-Mure, 2012). Energy consumption for cooling is important only in the Southern European countries, and represents about 10% of the total energy consumption per dwelling. The penetration of air conditioning is increasing rapidly in the south of Europe and it is expected to represent a much higher share of the energy consumption soon.

Energy consumption for heating purposes varies substantially between the various European member states. The lowest consumption is observed in Spain at 1.6 koe/m^2/dd (dd = Degree day), while the highest value is close to 4.1 koe/m^2/dd in the north of Europe. In parallel, the energy consumption for lighting and appliances varies between 1000 kWh/dwelling and year in Estonia and Romania, and 4000 kWh/m^2/y in Finland and Sweden.

Despite serious legislative measures adopted in Europe and the significant improvement of energy efficiency achieved during the recent years (almost 1.4% per year), residential energy consumption in Europe has increased by 14% between 1990 and 2014. During the same period, electricity consumption by the building sector increased by 60%. This is mainly attributed to the very rapid penetration of energy-consuming appliances, electronic equipment and air conditioning in Europe.

The observed increase in energy consumption in Europe is accredited to several technical, economic and social issues. It is well accepted that the increase in the total number of households in Europe has a significant impact on residential energy consumption. During the last 25 years, the number of households in Europe increased by about 1% per year, while the corresponding increase of the population was no higher than 0.3% per year (Odyssee-Mure, 2012). This is attributed to the decrease of the size of households observed during that period. In Finland, the percentage of households with just one person rose from 27% in 1981 to about 41% in 2008. In parallel, the size of households in Spain, decreased from 3.5 in 1980 to 2.7 in 2008 (House Statistics in European Union, 2010).

Another important issue is the increase of the total housing surface caused by the rise of GNP in Europe. From 1995–2009, household income in Europe increased by almost 0.9%. This as resulted in a significant increase of the total surface of residential stock in Europe. As reported by Odyssee-Mure (2012), the average size of dwellings increased by 3% over the period 1997–2009. According to the Housing Statistics of the European Union 2010, the mean size of dwellings in France increased from 88 m^2 in 1996 to 91 m^2 in 2000, in Germany from 86.7 m^2 in 1998 to

89.9 m^2 in 2006, and in Denmark from 108.9 m^2 in 2001 to 114.4 m^2 in 2009.

Between 1997 and 2009 there has been a decrease in heating demand of close to 15%. This decrease is the result of advanced research on topics related to energy conservation for heating purposes, the implementation of intensive legislative measures, the use of very efficient heating systems and fuels, and a significant rate of refurbishment New dwellings built after 1997 present a much lower energy consumption for heating at 30%–60% of the consumption levels of old housing stock, while efficient dwellings represent almost the 20% of the total residential stock (Odyssee-Mure, 2012). Dwellings built after 2009 in Germany, Sweden, Denmark, Slovakia and the Netherlands consumer about 50%–60% less energy consumption for heating than dwellings built in the 1990s (Odyssee-Mure, 2012). According to existing statistical data, energy consumption for heating in these countries is decreasing by 1.4%–2.2% per year, while the average rate of decrease in Europe is close to 1.4% per year.

The legislative measures that have been implemented over the last ten yearas to decrease energy consumption in Europe have decoupled the total energy consumption from the corresponding annual GDP (Fig. 2.5).

The decoupling of annual GDP and annual primary consumption was achieved through a serious decrease in energy intensity in Europe (Fig. 2.5). This has contributed to decline of energy consumption per person and primary energy consumption in the residential sector (Figs. 2.6–2.7)

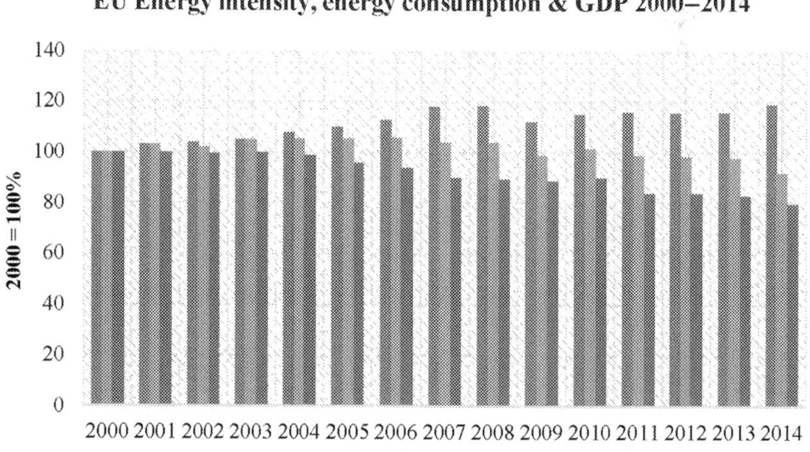

FIGURE 2.5 Evolution of energy consumption and GDP in the EU, 2005–2013. Source: *European Commission (2016)*.

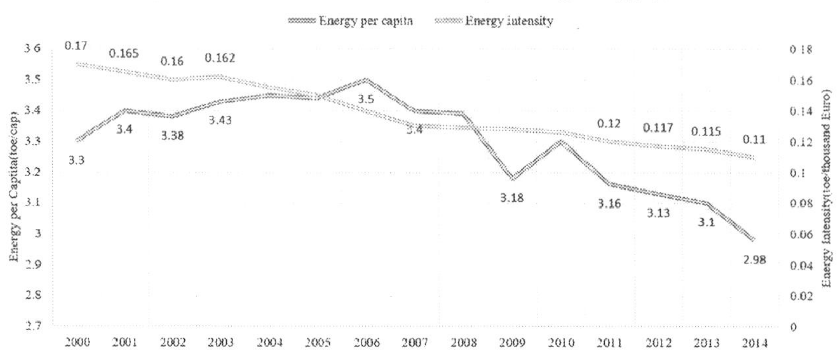

FIGURE 2.6 Energy indicators for total primary energy supply, EU-28. Source: *European Commission (2016).*

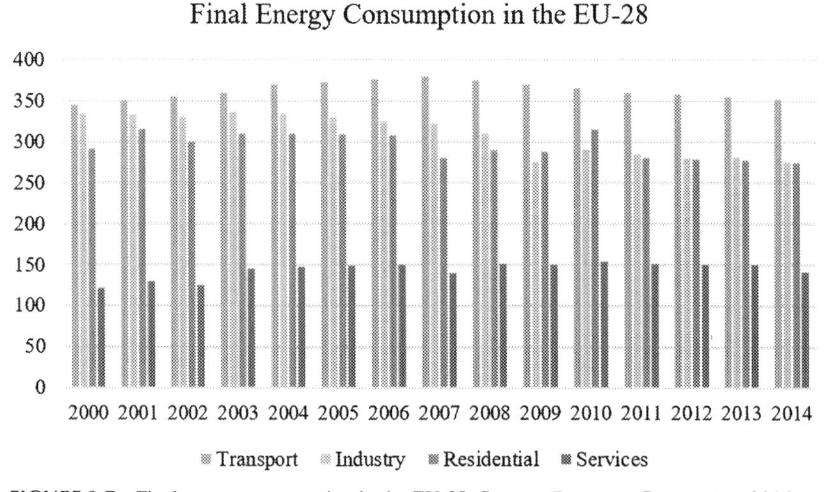

FIGURE 2.7 Final energy consumption in the EU-28. Source: *European Commission, 2016.*

During the period between 2000 to 2015, energy consumption in the residential sector fluctuated between positive and negative growth rates (Fig. 2.8) as a result of the specific economic and climatic conditions (Bertoldi et al., 2018). As mentioned, the residential energy consumption per capita in Europe during this period decreased by 57 koe, a drop of 9.5%. This important reduction in energy consumption by the residential sector during recent years was very much promoted by favorable weather conditions. Fig. 2.8 shows the relation between the annual heating degree days and the final energy consumption per dwelling in recent years (Bertoldi et al., 2018). The impact of weather conditions on the energy consumption of the dwellings is clear.

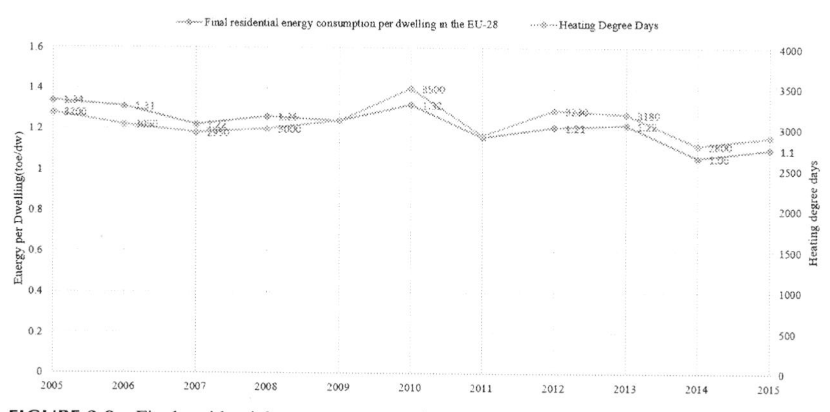

FIGURE 2.8 Final residential energy consumption per dwelling and heating degree days in the EU-28, 2005–2015. Source: *Bertoldi et al. (2018).*

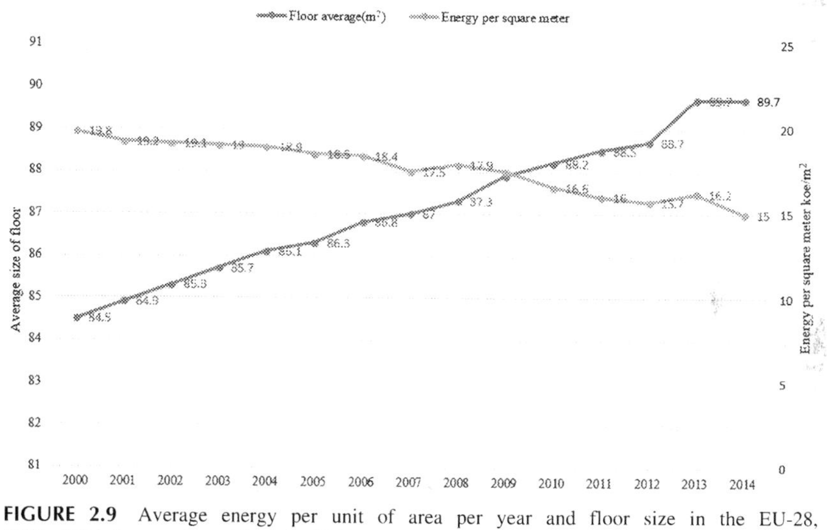

FIGURE 2.9 Average energy per unit of area per year and floor size in the EU-28, 2000–2014. Source: *Bertoldi et al. (2018).*

Although favorable weather conditions have a positive impact on residential energy consumption, a significant increase in the total surface of residential stock could substantially increase the final energy consumption. However, as shown in Fig. 2.9, this was not the case. Despite the increase in total residential area, primary energy consumption declined significantly (Bertoldi et al., 2018).

However, when primary energy consumption by the residential sector is both normalized regarding the weather conditions (heating degree days) and the total surface area (Fig. 2.10), it is evident that the normalized consumption has decreased substantially (Bertoldi et al., 2018).

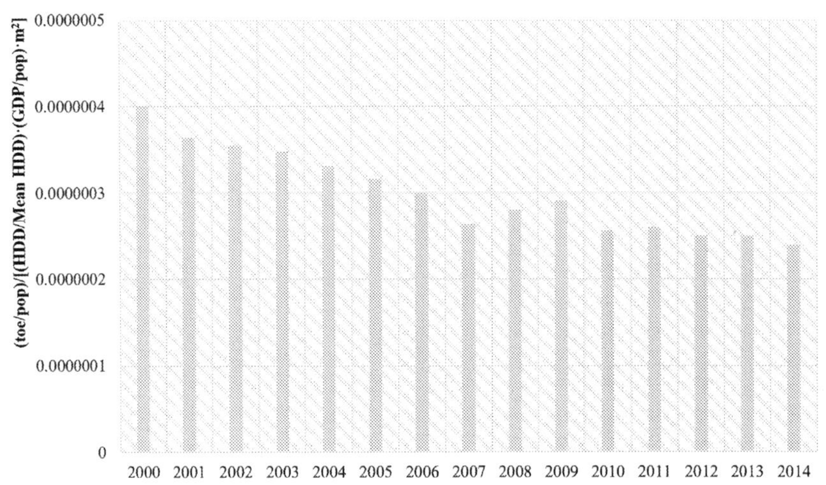

FIGURE 2.10 Residential energy per capita normalized by HDD, GDP per capita and average floor area in the EU-28, 2000–2014. Source: *Bertoldi et al. (2018).*

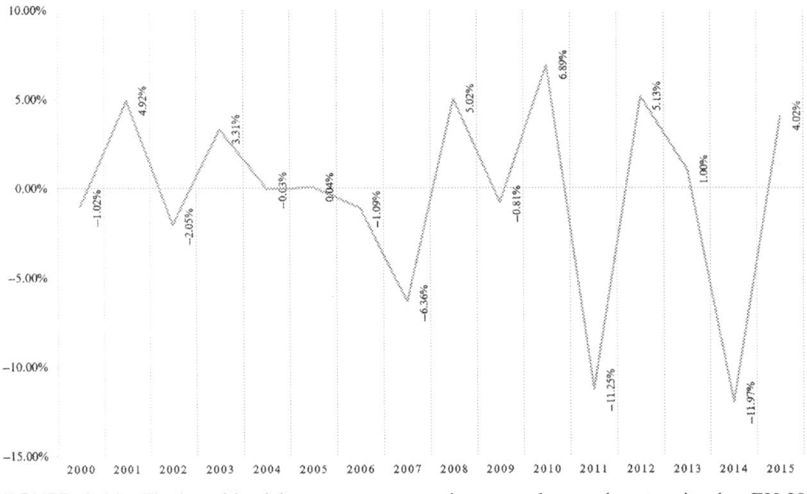

FIGURE 2.11 Final residential energy consumption annual growth rates in the EU-28, 2000–2015. Source: *Bertoldi et al. (2018).*

As already mentioned, residential energy consumption in the various European countries presents a very high variability (Figs. 2.11–2.12) (Bertoldi et al., 2018). As can be seen, the lowest energy consumption per person is present in Malta, while the maximum is in Finland. It is important to note that in almost all European countries the residential energy consumption in 2015 was substantially lower than that of 2010.

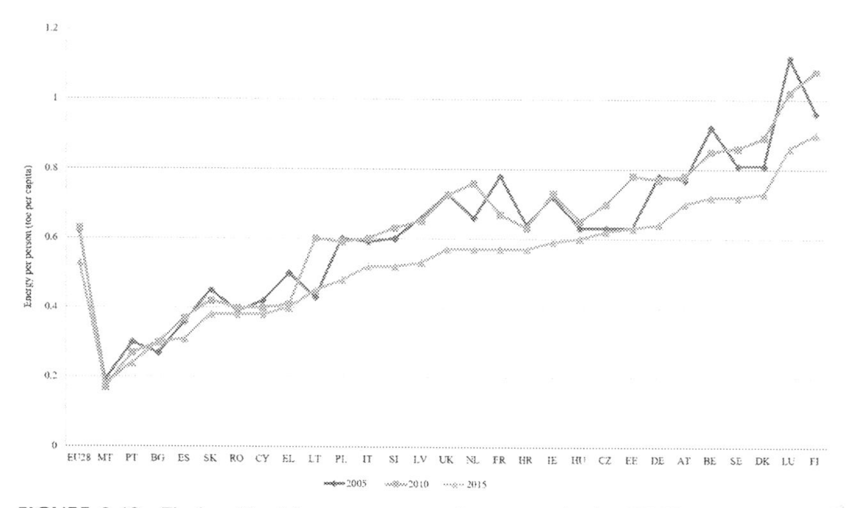

FIGURE 2.12 Final residential energy consumption per capita by EU-28 member state in 2005, 2010 and 2015. Source: *Bertoldi et al. (2018).*

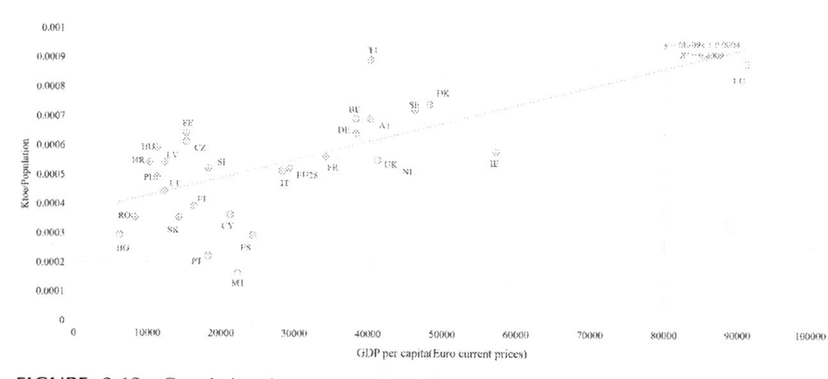

FIGURE 2.13 Correlation between residential energy consumption and GDP per capita (current prices) in the EU-28 member states, 2015. Source: *Bertoldi et al. (2018).*

Although primary energy consumption in Europe was decoupled from the corresponding GDP, a clear correlation between the two parameters exists for the individual member states (Fig. 2.13). However, when the normalized energy consumption of the residential sector concerning the weather conditions and dwelling floor area is plotted against the GDP the relation changes shape, and it can be concluded that the wealthier the country, the lower the normalized energy consumption of the residential sector (Fig. 2.14).

In 2009, the average energy consumption in Europe for lighting and appliances was close to 2550 kWh/dwelling. Consumption in the various countries varied between 1000 kWh/dwelling and 4200 kWh/dwelling. The lowest energy consumption was observed in Estonia and Romania, and the

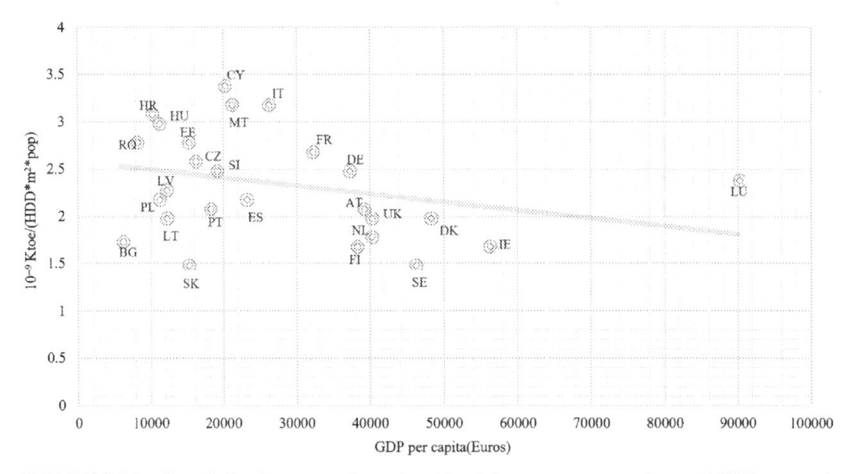

FIGURE 2.14 Correlation between adjusted residential energy consumption and GDP per capita in the EU-28 member states, 2015. Source: *Bertoldi et al. (2018).*

highest in Finland and Sweden. Despite the introduction of serious legislative measures aiming to decrease energy consumption in the residential sector, the corresponding consumption for lighting and appliances has increased since 1997, with a rate close to 1.7%, mainly because of the growing use of small appliances. Te growth rate of the energy consumption attributed to small appliances is close to 6.5% per year (Odyssee-Mure, 2012).

Total energy consumption for cooling purposes is quite low in Europe, however it increases rapidly in the south of Europe. Among the various European countries, Cyprus presents the highest energy consumption, with about 670 kWh per dwelling per year, followed by Malta with almost 540 kW per dwelling per year. The penetration of air conditioning in the southern European countries is very rapid. Between 2005 and 2009 energy consumption for cooling purposes increased in Bulgaria by 100%, while in Spain and Italy the corresponding increase reached 30%, (Odyssee-Mure, 2012).

Future energy consumption by the residential sector for heating and cooling purposes is forecasted by Isaac and Van Vuuren (2009). The estimated figures for the period up to 2050 and 2100 concerning heating consumption is given in Table 2.2, while the corresponding figures for cooling are given in Table 2.3. The modeled energy consumption for heating and cooling for the years,2000, 2050 and 2100 are given in Fig. 2.15 in comparison with the corresponding consumption by the rest of the world.

As shown, in western Europe, the expected energy consumption for heating may decrease by up to 25% by 2050 compared to the reference year of 2000, while a further reduction of up to 34% is expected by the end of the century. As concerns the rest of Europe, it is expected that by 2050, heating

TABLE 2.2 Change in Regional Drivers of Heating Energy Demand Between 2000–2050 and 2050–2100 (%)

	Western Europe	Russia	Rest of Europe
2000–2050			
Population	3	−22	−7
Floor space	11	93	83
HDD	−18	−11	−20
Intensity	−12	−25	−15
Efficiency	10	4	14
Energy use	−25	−3	1
2050–2100			
Population	−11	−21	−15
Floor space	4	26	32
HDD	−21	−12	−18
Intensity	−13	−33	−8
Efficiency	−5	4	4
Energy use	−34	−43	−18

Source: *Isaac and Van Vuuren (2009).*

energy consumption may present a very small increase of 1%, while the expected additional decrease by 2100 is close to 18%. The expected decrease of heating energy consumption is mainly attributed to the impact of the climate change, a decrease in the number of heating degree days, and an increase in energy efficiency.

As concerns future energy consumption for cooling, a tremendous increase is expected in western Europe. Forecasts indicate that consumption in this region will increase by 261% by 2050, while the forecasted additional increase up to the end of the century is close to 106%. In the rest of Europe, the calculated increase for 2050 and 2100 is close to 65% and 118%. This increase in cooling energy consumption is mainly driven by the tremendous penetration of air conditioning equipment, which is primarily due to an increase in family income and an increase in the ambient temperature caused by global and local climate change.

Similar results and conclusions have also been obtained by other studies aiming to forecast future energy consumption in Europe. Mirasdegis et al. (2007) predict that because of climate change, electricity demand in Greece

TABLE 2.3 Change in Regional Drivers of Cooling Energy Demand Between 2000–2050 and 2050–2100 (%)

	Western Europe	Russia	Rest of Europe
2000–2050			
Population	3	−22	−7
Average household size	−9	−11	−17
Climate max. saturation	86	84	148
Availability	29	1409	1190
UEC	134	227	293
Efficiency	46	46	46
Energy demand	261	4211	6537
2050–2100			
Population	−11	−21	−15
Average household size	−1	3	0
Climate max. saturation	46	32	31
Availability	0	4	16
UEC	100	95	112
Efficiency	25	25	25
Energy demand	106	74	118

Source: *Isaac and Van Vuuren (2009).*

will increase by 5.5%, while Christenson et al. (2006) estimated that the cooling energy consumption of office buildings in Switzerland may increase between 223% and 1050% by 2050. Ciscar et al. (2014) has also forecasted the future energy consumption of the building sector in Europe for the period between 2010 to 2050, using different climatic scenarios. They calculate that heating energy consumption may decrease by 9%−17%, while the corresponding cooling demand is expected to increase by up to 70%. It is evident that the expected decrease in heating demand will be more significant in the north of Europe, while the more serious increase in cooling demand is forecasted for the south of Europe.

Energy consumption by the tertiary sector in Europe has followed a continuously increasing trend over the last 30 years. As mentioned by the European Commission Services (EU-25 Energy and Transport, 2015), the annual increase in the rate of energy consumption by tertiary sector was

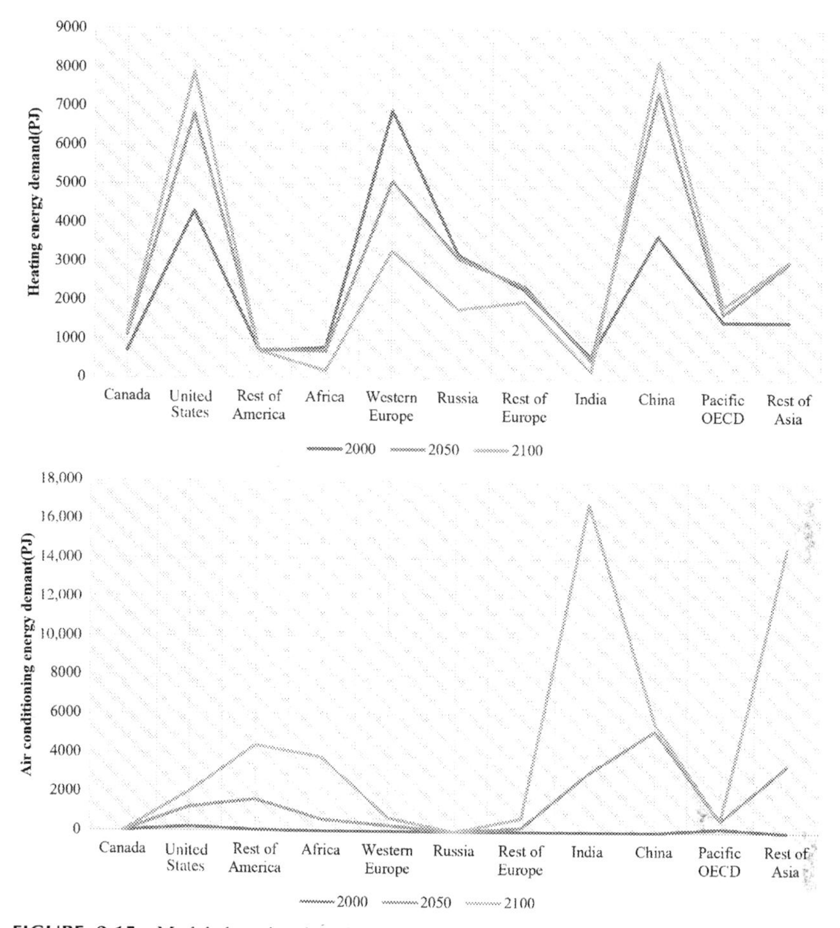

FIGURE 2.15 Modeled regional residential energy demand for heating (top) and for air conditioning (bottom) in the years 2000, 2050 and 2100 (reference scenario). Source: *Isaac and Van Vuuren (2009).*

close to 0.6%–1.2% for the period between 1990–2000. It increased to 1.2% between 2000–2010, while forecasts for 2010–2020 predict a rise of close to 1.1%. This considerable increase in energy consumption by the tertiary sector is mainly attributed to the very significant growth of the services sector. In fact, during the same period, services increased by 1.3% per year, while it is expected that services will drive future energy consumption by the tertiary sector in Europe, comprising almost 93% of the expected increase (EU-25 Energy and Transport, 2015). Out of the various types of tertiary buildings, office and trade buildings are the largest consumers of energy, comprising 26% of the total tertiary consumption. Space heating represents almost 70% of global energy consumption in Germany and almost 60% in

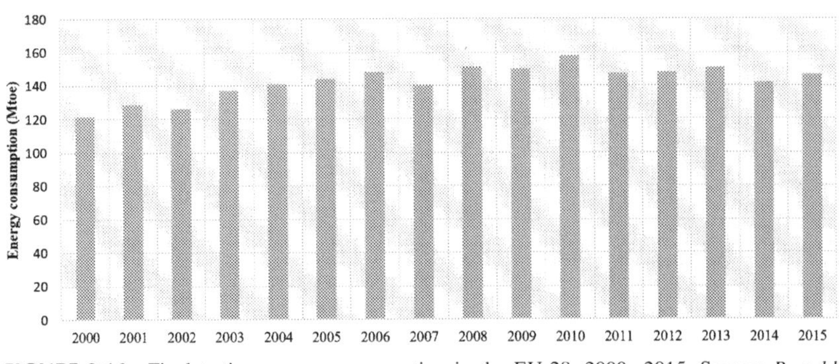

FIGURE 2.16 Final tertiary energy consumption in the EU-28, 2000–2015. Source: *Bertoldi et al. (2018).*

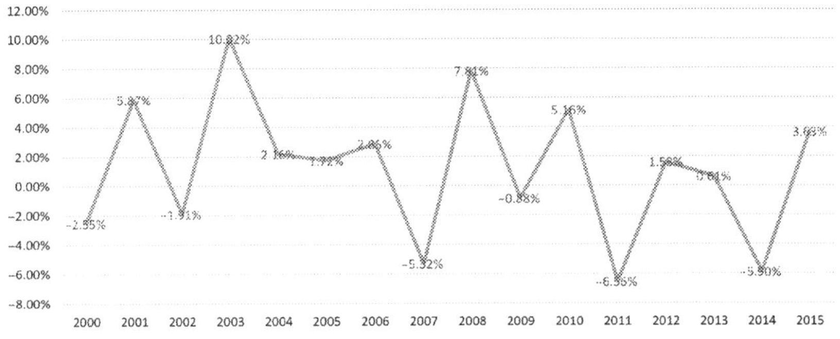

FIGURE 2.17 Annual growth rates in final tertiary energy consumption in the EU-28, 2000–2015. Source: *Bertoldi et al. (2018).*

France (Odyssee-Mure, 2012). Like the residential sector, energy consumption for heating has decreased over recent years, mainly because of the serious legislative measures that have been implemented and increased energy efficiency of heating-related systems and components.

Detailed figures for energy consumption by the tertiary sector in Europe for the years 2000–2015 are given in Fig. 2.16.

Similar to the residential sector, the annual variation in energy consumption by tertiary buildings was either negative or positive (Fig. 2.17). The variation was mainly attributed to the considerable variability of the climatic conditions and the specific conditions occurring in the tertiary sector. Malta, Romania and Cyprus were the member states that presented the highest energy consumption during the period.

The final increase in energy consumption by the tertiary sector during the specific period for all member states is given in Fig. 2.18. Most of the countries presented an increasing trend, with some exceptions coming mainly from eastern European countries. Energy spent per employee is an

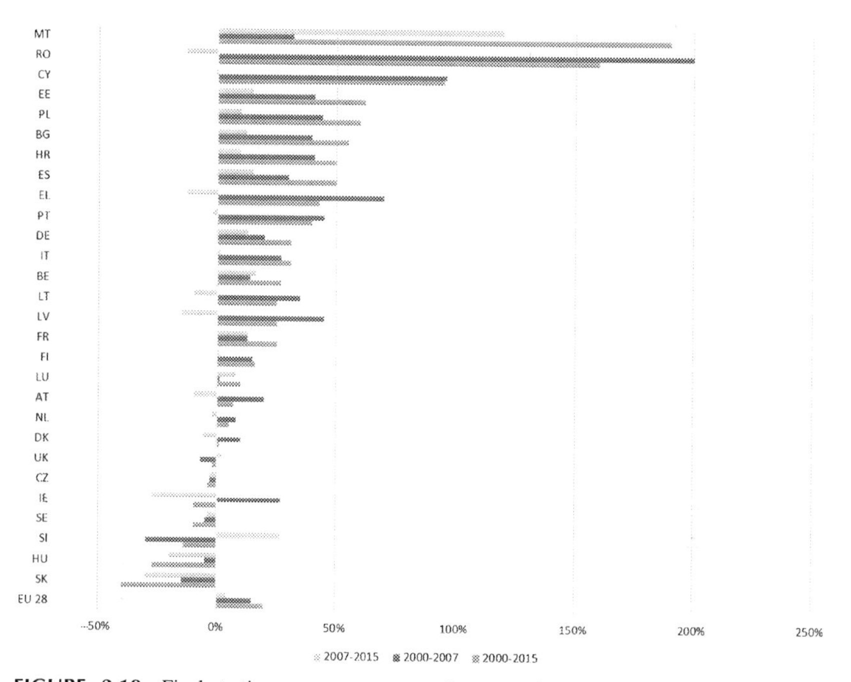

FIGURE 2.18 Final tertiary energy consumption growth rates by Ms in the EU-28: 2000–2015, 2000–2007 and 2007–2015. Source: *Bertoldi et al. (2018).*

interesting index to characterize energy consumption by the tertiary sector. Fig. 2.19 reports the final energy consumption per employee for all European member states. Significant differences occur between the various states. The highest energy consumption is observed in Finland and the lowest in Romania.

Energy Consumption in The United States of America

The evolution in energy consumption by residential and the commercial buildings in the United States is given in Fig. 2.20. As shown, the total energy consumption by the residential sector in 2017 was close to 1760 TWh per year, or 6 quadrillion British Thermal Units (BTUs), while the corresponding consumption by the commercial sector was close to 1465 TWh/year, or 5 quadrillion BTUs.

Future projections show a continuous slight decrease in energy consumption by the residential sector and a moderate increase in consumption by tertiary buildings (Fig. 2.20). Concerning the type of fuels used, electricity is foreseen to be the main source of energy followed by natural gas, while the use of petroleum and other liquids tends to decrease continuously. A similar picture is also observed in the commercial sector where the use

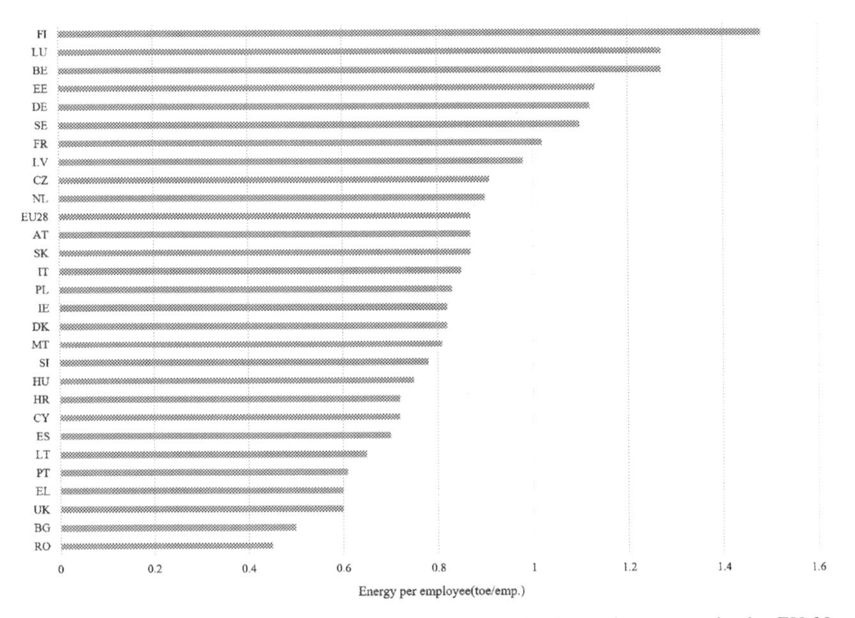

FIGURE 2.19 Final energy consumption per employee in the tertiary sector in the EU-28 member states, 2015. Source: *Bertoldi et al. (2018).*

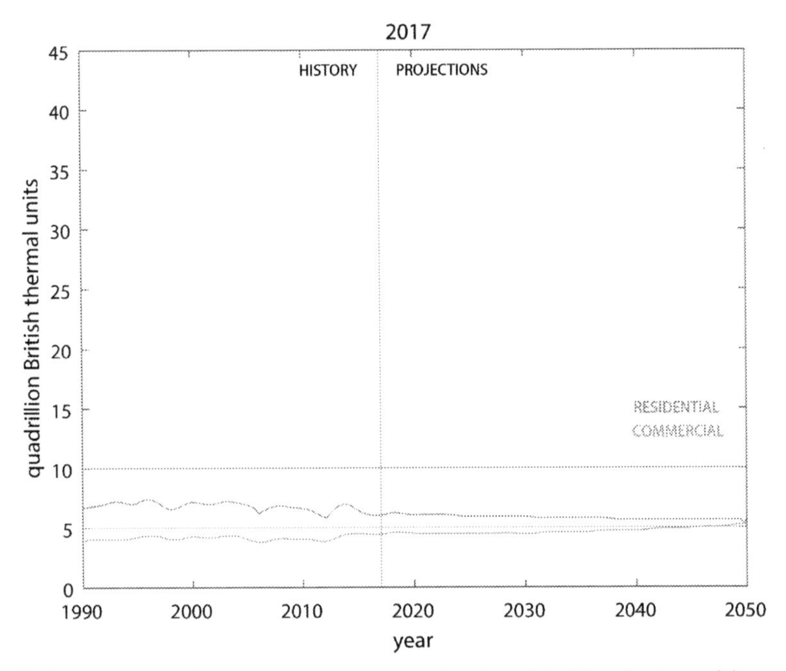

FIGURE 2.20 Past and future energy consumption by the residential and commercial sectors in the USA. Source: *Based on IEA data from World Energy Outlook 2013 © OECD/IEA 2015, http://www.worldenergyoutlook.org/weo2013/ Licence: www.iea.org/t&c.*

of electricity and natural gas seems likely to increase in the near future (Fig. 2.21).

Fig. 2.22 provides information on specific electricity uses by the residential sector in 2017 and predicted future consumption. As shown, other energy uses, mainly energy spent by appliances, present the highest energy consumption followed by cooling. While absolute electricity consumption in 2050 is expected to decrease, cooling is the only electricity use that will increase. In 2050, energy consumption for lighting is expected to decrease by more than 50%, mainly because of the extensive use of high-efficiency LED lighting systems.

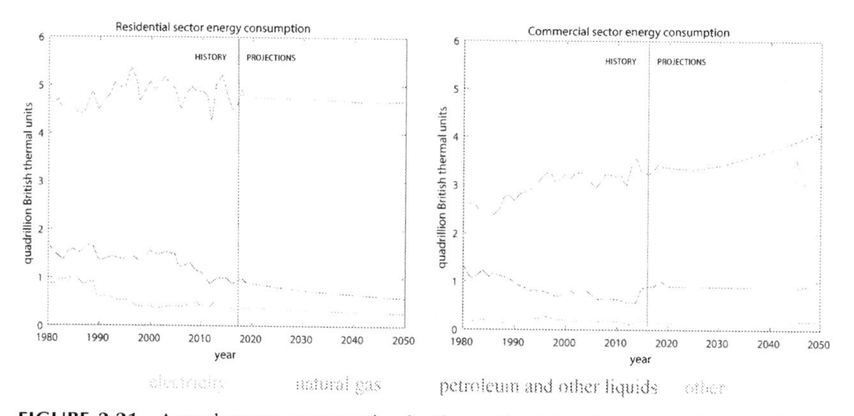

FIGURE 2.21 Annual energy consumption by the residential and commercial sectors for the different types of fuel. Source: *Based on IEA data from World Energy Outlook 2013 © OECD/IEA 2015, http://www.worldenergyoutlook.org/weo2013/ Licence: www.iea.org/t&c.*

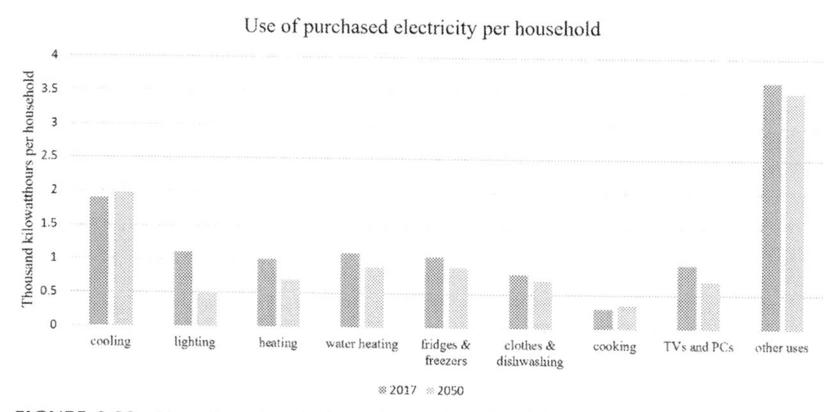

FIGURE 2.22 Use of purchased electricity per household in the United States of America. Source: *Based on IEA data from World Energy Outlook 2013 © OECD/IEA 2015, http://www.worldenergyoutlook.org/weo2013/ Licence: www.iea.org/t&c.*

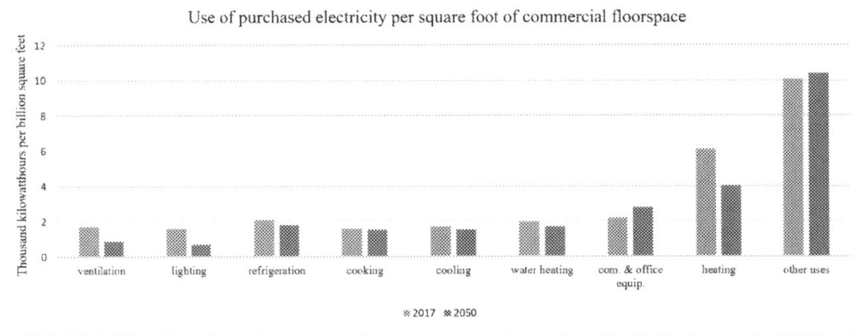

FIGURE 2.23 Use of purchased electricity per square foot of residential buildings in 2017 and forecasts for 2050. Source: *Based on IEA data from World Energy Outlook 2013 © OECD/IEA 2015, http://www.worldenergyoutlook.org/weo2013/ Licence: www.iea.org/t&c.*

The corresponding specific electricity consumption by the residential sector in 2017 and the expected consumption in 2050 are given in Fig. 2.23. The highest energy consumption is due to other energy uses, such as by appliances, followed by space heating. Because of the expected extensive future use of computer and office equipment, the energy spent for this purpose is expected to increase. All other types of energy use appear set to decrease substantially. Electricity use for heating, cooling and ventilation may drop by one third by 2050 due to serious energy conservation measures and the implementation of energy efficient technologies in commercial buildings. As in the case of residential buildings, energy consumption for lighting is expected to decrease by up to 56% by 2050, mainly because of the extensive use of high-efficiency lighting systems.

Solar electricity and photovoltaics may increase their share in electricity consumption in both the residential and commercial sectors. It is foreseen that the use of solar electricity will increase by 9% per year in the residential sector and by about 6% per year in the commercial sector (Fig. 2.24).

Energy Consumption in the Rest of the World

Energy consumption is influenced by many social and economic factors and drivers. Especially in the less developed countries, the tremendous increase of the population as well as the expected significant increase of domestic GDP may result in a significant increase in energy consumption. As reported by Exxon Mobil (2018), the world's population is expected to reach 9.2 billion people by 2040, compared to nearly 7.4 billion today. Most of the new population will be in Africa and Asia (Fig. 2.25).

In parallel, the GDP of the less developed countries is expected to increase significantly (Fig. 2.26). An increase in family income results in higher living standards and an increased energy consumption per person

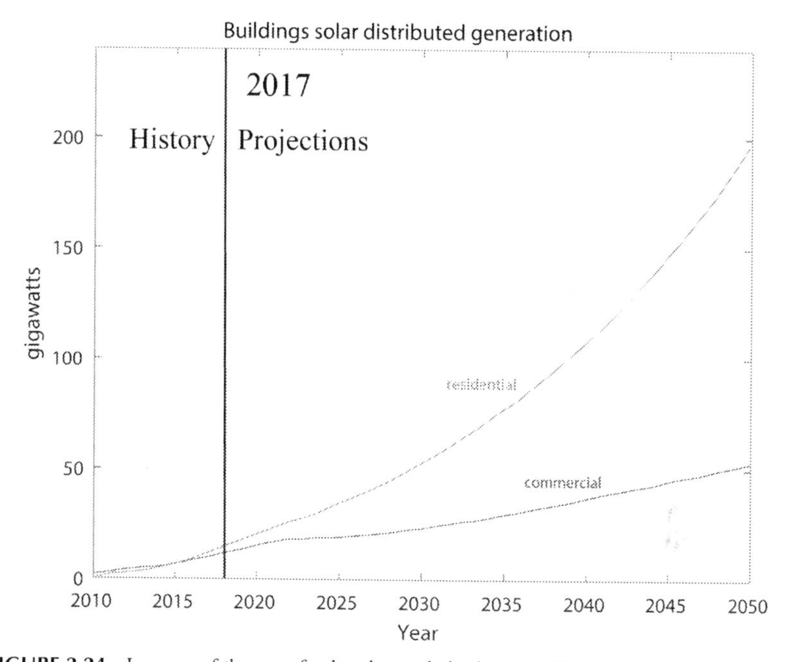

FIGURE 2.24 Increase of the use of solar photovoltaics between 2017 and 2050. Source: *Based on IEA data from World Energy Outlook 2013 © OECD/IEA 2015, http://www.worldenergyoutlook.org/weo2013/ Licence: www.iea.org/t&c.*

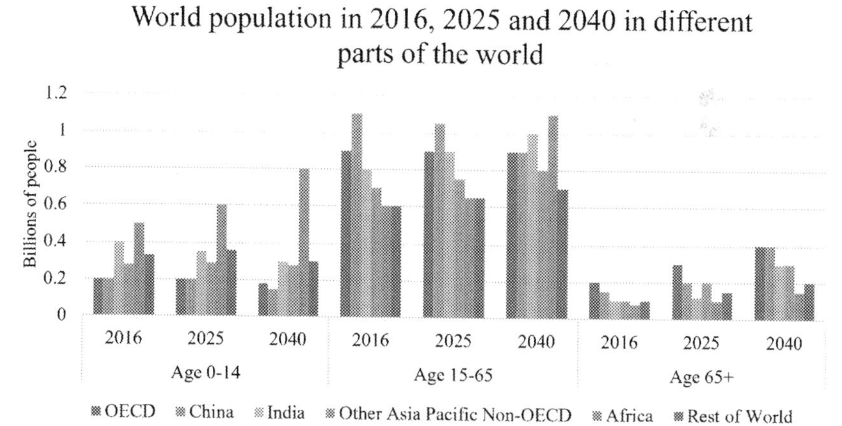

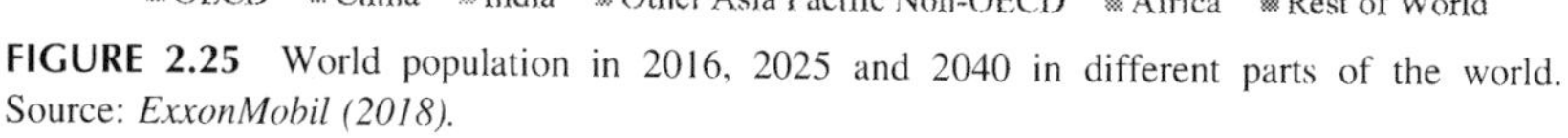

FIGURE 2.25 World population in 2016, 2025 and 2040 in different parts of the world. Source: *ExxonMobil (2018).*

(Fig. 2.27). It is estimated that by 2040, the global energy demand will increase by about 25% compared to actual consumption. As stated by ExxonMobil (2018), this "is roughly equivalent to adding another North America and Latin America to the world's current energy demand."

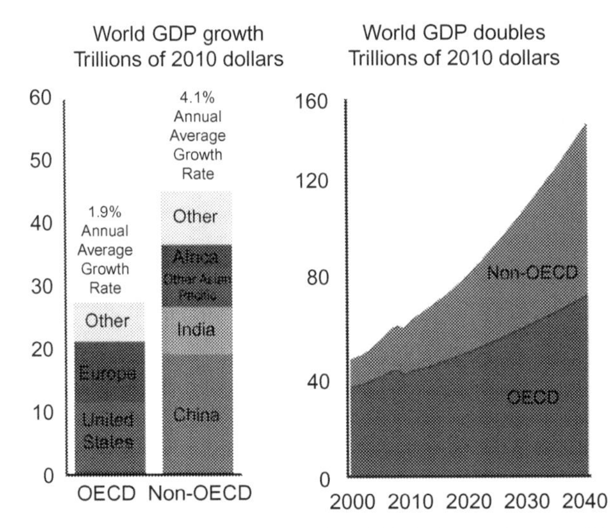

FIGURE 2.26 Increase of the GDP in the world until 2014. Source: *ExxonMobil (2018).*

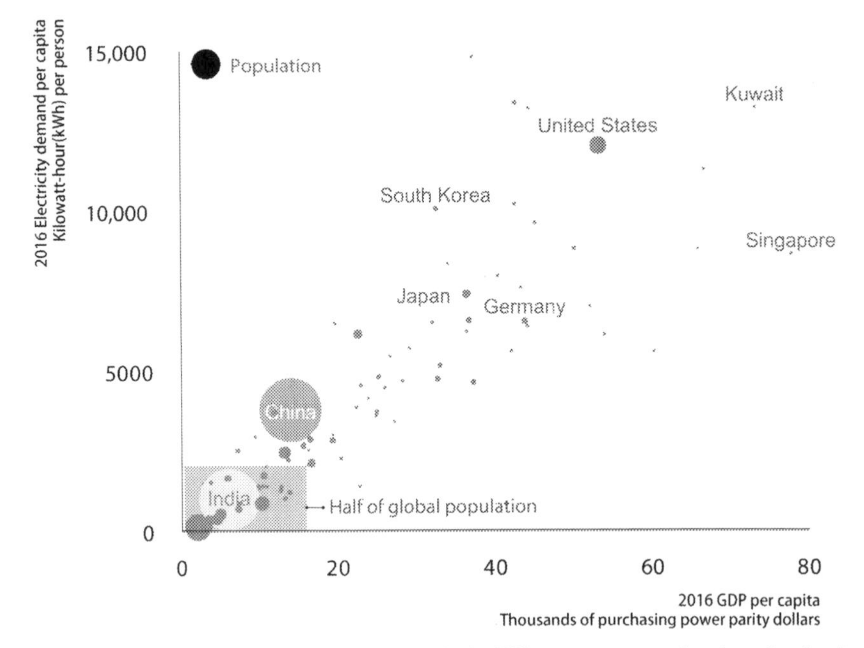

FIGURE 2.27 Electricity consumption per capita in kWh per person as a function of national GDP in 2016. Source: *ExxonMobil (2018).*

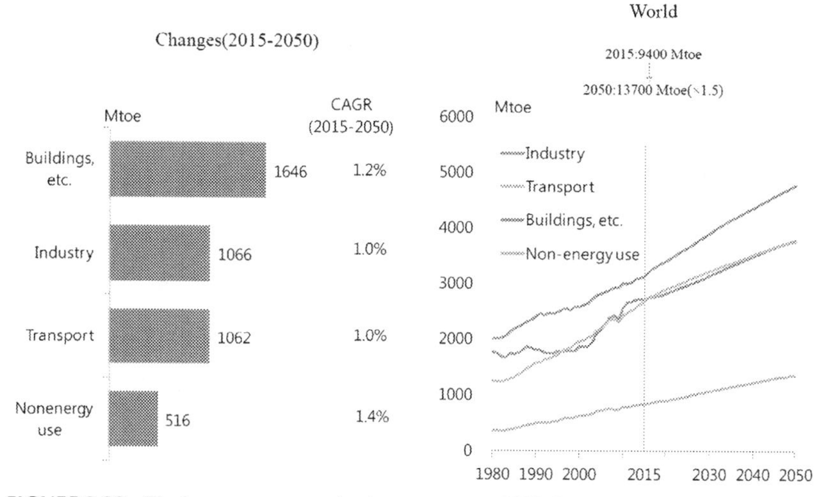

FIGURE 2.28 Final energy consumption by sector up to 2050. Source: *IEE Japan (2017).*

Global energy consumption by the residential and commercial sectors is expected to increase by more than 20% by 2040, while electricity will be the main energy provider in the building sector. About 90% of the growth will be met by electricity. Such an increase is significant, but not as high as the corresponding growth of consumptio by the transport and industrial sectors (Fig. 2.28). Electricity production is forecast to increase by almost 30%, largely due to the increase of the world population and the rise of GDP.

Several forecast have been developed to predict future energy consumption by the buildings sector. The Institute of Energy Economics in Japan (2017) published a reference and an advanced technology scenario regarding future energy consumption by the residential and commercial sectors by 2050. The specific energy consumption for the reference scenario is given in Fig. 2.28, while predictions for the advanced technological scenario are given in Fig. 2.29. The advanced technological scenario assumes that the energy efficiency of the residential sector will improve by almost 21% compared to the reference scenario, while the corresponding improvement considered for commercial buildings is 20%.

As shown, the reference scenario predicts an annual increase in energy consumption by the building sector of close to 1.8% per year. The advanced technological scenario predicts that by 2050, the reduction in energy consumption may exceed 25% compared to the reference scenario. This represents a serious reduction.

The expected growth in residential and commercial energy consumption up to 2040 is given in Fig. 2.29. While the expected increase in consumption by the residential and commercial sectors between 2016 and 2040 will be

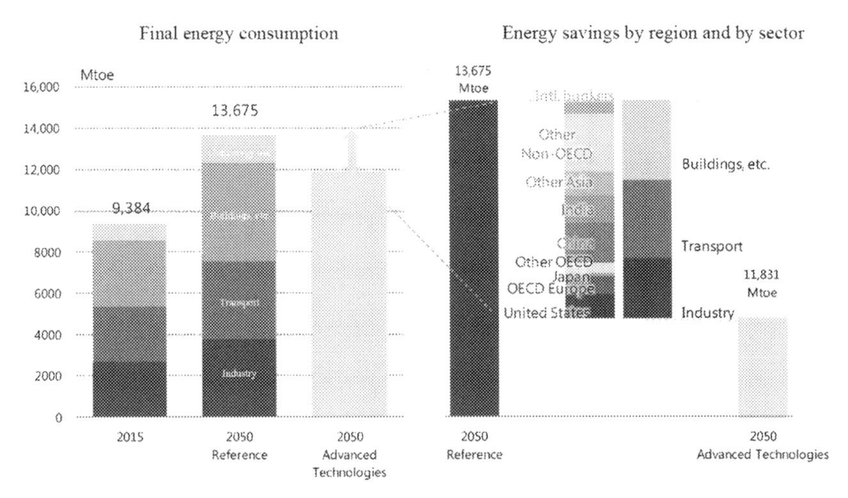

FIGURE 2.29 Final energy consumption of the various sectors by 2050 as predicted by the advanced technological scenario. Source: *IEE (2017).*

close to 20%, the corresponding growth for non-OECD countries will exceed 40%. As forecast by ExxonMobil (2018), Africa and Asia will drive the expected increase in energy consumption, each accounting for 30% of the expected rise.

Concerning the type of fuels to be used in the residential sector (Fig. 2.30), it is clear that a very significant rise in energy efficiency must be expected. However, most of the population in the less developed countries, especially in Africa, must rely on the use of biomass to satisfy their energy needs. As already mentioned, an increase in electricity consumption of close to 70% is expected by 2040 (Fig. 2.31). It is expected that electricity will cover almost 40% of the global residential and commercial demand by 2040. The use of natural gas is forecasted to grow by 20% by 2040, while it will cover almost 20% of the global residential and commercial energy demand. However, despite the significant increase in the number of people with access to electricity, a important percentage of the world's population will not be connected to the grid. In fact, there are around 2.5 billion people in the world without any access to clean energy, and about 1 billion people who are not connected to the electricity grid (ExxonMobil, 2018) (Fig. 2.32).

The expected increase in electricity consumption by the residential sector in the various parts of the planet is given for up to 2030 in Figs. 2.33 and 2.34. It is estimated that electricity consumption will increase by almost 75% by 2040. Most of the additional consumption will take place in less developed countries. For example, the expected growth in electricity demand in Africa and India will be close to 250%. In parallel, electricity consumption per household will increase by 30% as the number of households will grow tremendously until 2040.

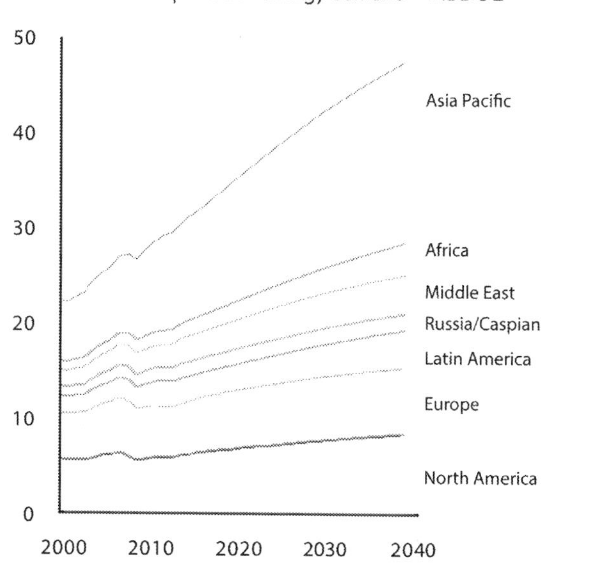

FIGURE 2.30 Global energy demand in 2016, 2025 and 2040 by energy sector. Source: *ExxonMobil (2018).*

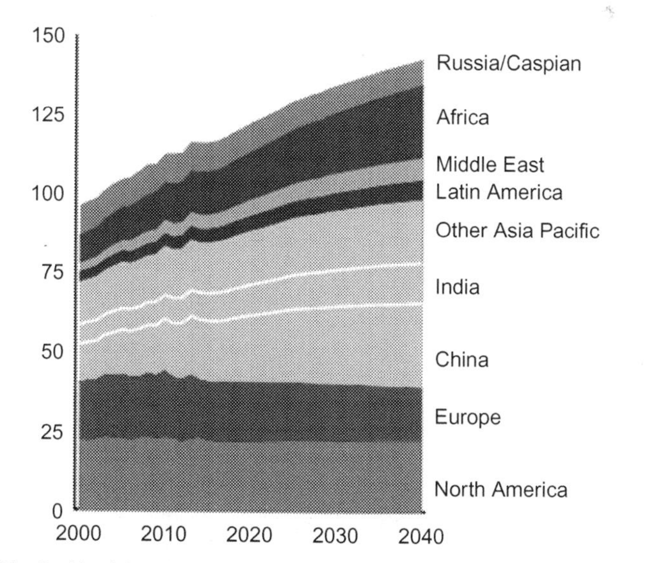

FIGURE 2.31 Residential and commercial demand shifts to non-OECD countries by 2040. Source: *ExxonMobil (2018).*

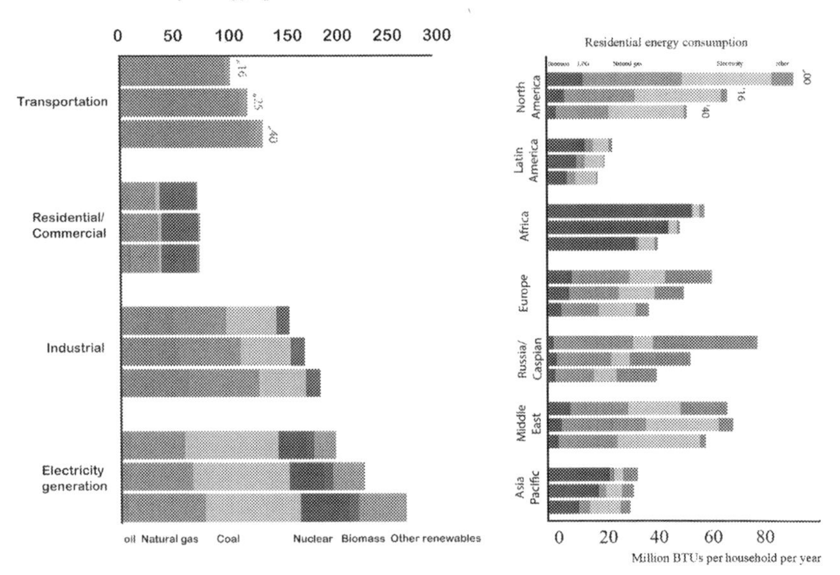

FIGURE 2.32 Residential energy consumption and type of fuel used in 2000, 2016 and 2040 in the various parts of the planet. Source: *ExxonMobile (2018)*.

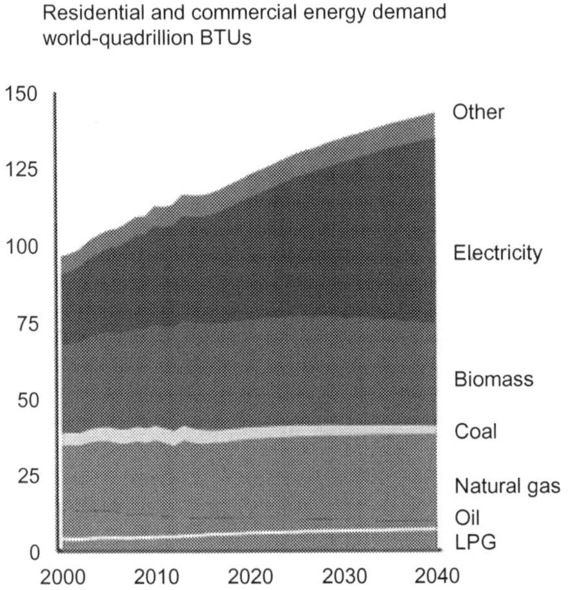

FIGURE 2.33 Increase in the consumption of the various energy fuels in the world by 2040. Source: *ExxonMobil (2018)*.

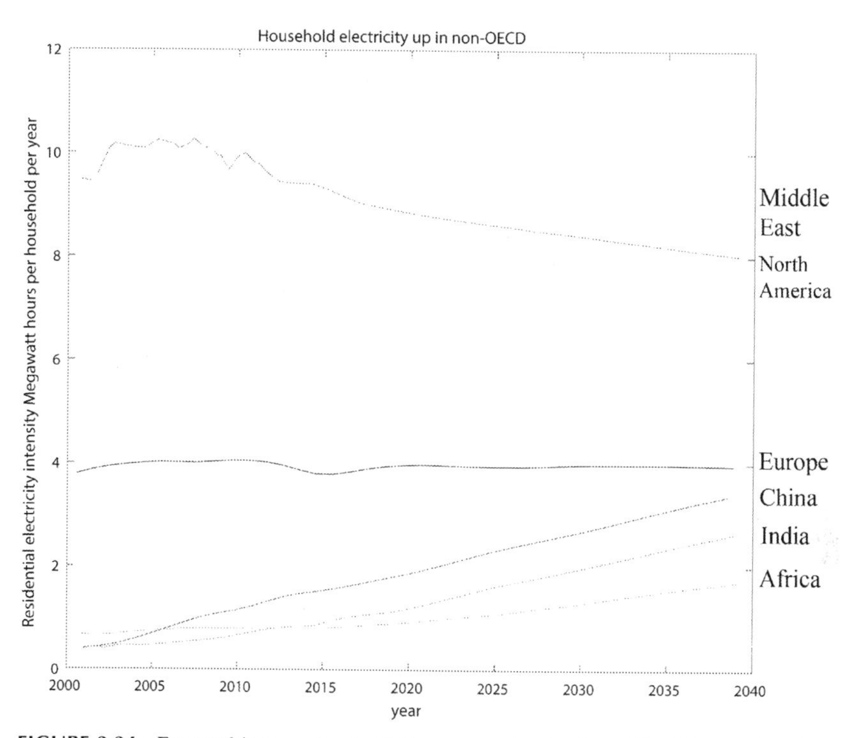

FIGURE 2.34 Expected increase in electricity consumption by the residential sector by 2040. Source: *ExxonMobil 2040.*

THE IMPACT OF GLOBAL ECONOMY ON THE ENERGY CONSUMPTION OF THE BUILDING SECTOR

The provision of energy is a must for our societies. Energy is a commodity that controls and determines the quality of life of people in the modern world. It is essentially an obligation for all countries and states to ensure that sufficient energy is available to all citizens to cover at least basic energy needs. The fulfillment of such a commitment requires that states design and implement proper policies that satisfy energy needs and protect citizens from all possible external economic, social and technological and environmental risks associated with the supply and use of energy, while assuring a reasonable energy consumption to satisfy the specific energy needs.

As the present book has mentioned several times, energy consumption by the building sector is influenced by many economic, financial, environmental, technological and social factors. Their specific perturbation greatly affects energy consumption by residential and tertiary buildings. Santamouris (2016b) has collected information on the elasticity of the main parameters and drivers that affect the cooling energy consumption of buildings,

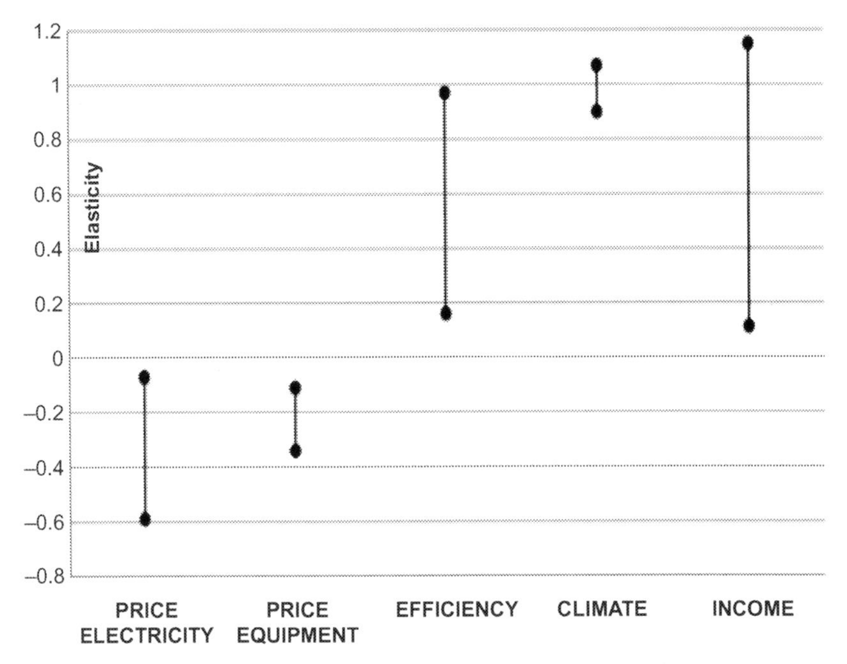

FIGURE 2.35 Calculated upper and lower values of elasticity of air conditioners with respect to family income, price of electricity and equipment, equipment efficiency and climate. Data on income elasticity are from Ormrod (1990), Rapson (2011); Davis and Gertler (2015), Golder and Tellis (1998), Auffhammer (2014); on electricity prices from Biddle (2008), Rapson (2011), Rapson (2013); on equipment price from Rapson (2011), Rapson (2013); on climate from Auffhammer and Mansur (2014), and on efficiency from Phadke et al. (2014). Source: *Santamouris (2016b).*

including the price of electricity and cooling equipment, the energy efficiency of the equipment, climate and income. The results are given in Fig. 2.35.

As shown, climatic and financial drivers highly affect the energy consumption of buildings, perhaps more so than the energy efficiency or the price of electricity and equipment. It is well known that the higher the household income, the higher the penetration of air conditioning (Fig. 2.36). In parallel, the income of households is strongly correlated to the size of dwellings. As shown in Fig. 2.37, an increase in income per capita results in a greater size of dwellings (Isaac and van Vuuren, 2008).

A sensitivity analysis aiming to evaluate the potential impact of economic, financial and the other drivers influencing the cooling energy consumption of buildings has been performed by Santamouris (2016b). The expected cooling energy consumption of the residential sector for 2050 is shown in Fig. 2.38. Three scenarios are examined, and the sensitivity

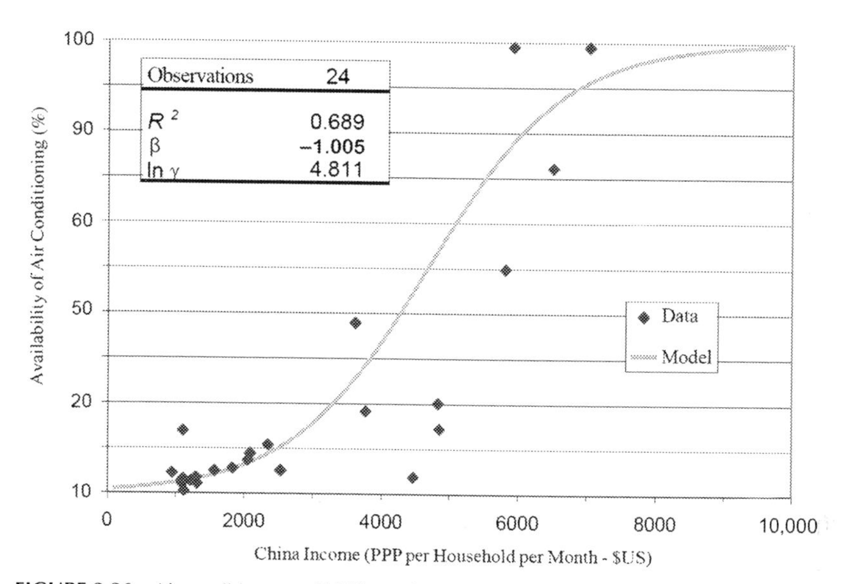

FIGURE 2.36 Air conditioner availability vs income. Source: *McNeil and Letschert (2008).*

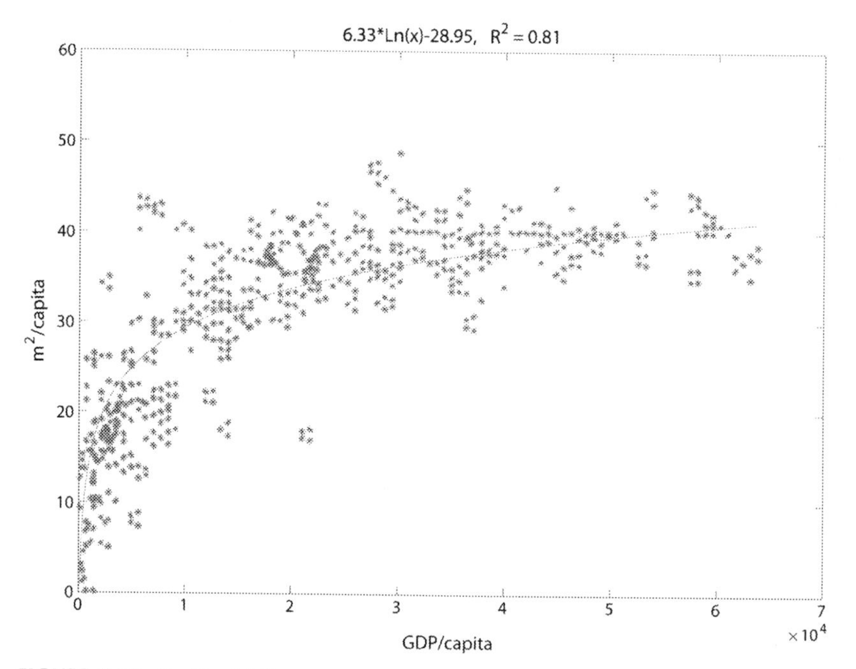

FIGURE 2.37 Residential floor area per capita vs. GDP per capita. Source: *Isaac and van Vuuren (2008).*

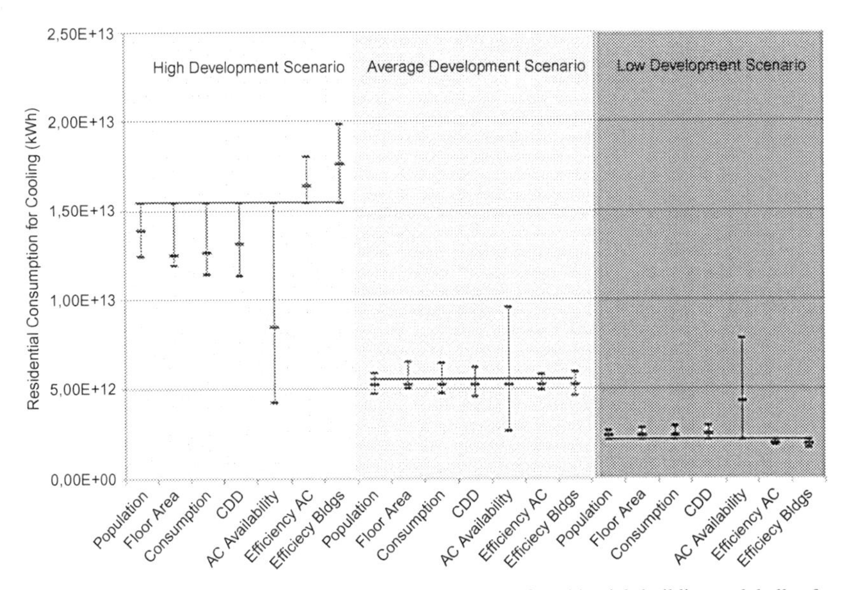

FIGURE 2.38 Predicted cooling energy consumption of residential buildings globally for 2050. Results of low, average and high development scenarios and variability as a function of the main parameters and drivers. Source: *Santamouris (2016b).*

regarding each of the main drivers affecting energy consumption is shown in the vertical bars. It is evident that economic factors affecting the penetration and the future availability of air conditioning in the residential sector have a significant impact on final energy consumption for cooling.

In parallel, significant economic factors affect the final energy consumption of the low income population and determine the degree to which the energy requirements are met. Analysis of energy consumption by the building sector in Europe during the period 2007–2012 reveals that the serious economic crisis in that period decreased energy consumption by the residential sector of 4% on average. However, countries experiencing serious economic crisis, such as Portugal, Slovakia and Ireland, presented a decrease between 16% and 22%. In Greece, a country suffering from a serious economic crisis and strict austerity measures, energy consumption for heating purposes was reduced by 68,7% in just one year (2012–2013), while energy consumption between 2010 and 2013 was close to 20%. Santamouris et al. (2013) analyzed energy consumption for heating purposes in many low income households for the winter periods of 2010–2011 and 2011–2012. As reported, due to serious financial problems, energy consumption for heating decreased by 15% in one year although the winter of 2011–2012 was much colder than the previous one. The corresponding decrease in heating consumption by low income households was much higher, at close to 50%.

STANDARDS AND REGULATIONS ON THE ENERGY CONSUMPTION OF THE BUILDING SECTOR

Standards and regulations applied in the building and construction sectors, have a serious impact on the economic and social life of our societies. The construction sector is responsible for about 9% of the GDP of the European Union countries and it provides more than 18 million jobs (European Commission, 2016). In parallel, most of the firms involved in the construction sector are small to medium enterprises (SMEs), which have a serious impact on the economic life of their home countries. A proper legislative framework can promote business in the construction sector and enhance economic growth. Legislative measures aimed at rehabilitating the existing building stock could significantly decrease the energy consumption of buildings, decrease the emission of greenhouse gases, create new jobs and enhance global economic activity. The European Commission (2016) estimates the energy saving potential of the rehabilitation of existing buildings ito be between 33–80.5 MToe up to 2030.

To achieve such a target, the European Commission is implementing a detailed legislative frame. The whole program, known as the Europe 2020 strategy, began in 2010. It aims to achieve three specific targets by 2020:

1. Reduce the emissions of greenhouse gases by 20% compared to 1990 levels,
2. Increase energy efficiency by 20%, and
3. Increase the contribution of renewable energy by 20% in the final energy consumption.

Additionally, the European strategy for 2050 aims to reduce greenhouse gas emissions by 40% in 2030, 60% by 2040 and 80% by 2050, compared to 1990 levels. It is assumed that the building sector will account for a substantial portion of the expected energy reductions. The legislation already implemented is part of the Energy Performance Directive (EPBD) of the European Commission (European Concerted Action, 2015).

According to the Directive, all new private buildings and all new public buildings should be near zero energy by 2020 and 2018 respectively. Additionally, the member states should define and implement minimum energy performance requirements for the building sector, including the energy systems of the buildings. As an example, and in the frame of the Energy Performance Directive, France has defined specific energy consumption targets for energy consumption by buildings (Fig. 2.39). Another positive example is reported by Gaglia et al. (2017) concerning the increase inthe energy consumption of residential buildings in Greece (Fig. 2.40). Table 2.4 reports the energy consumption of the buildings before the introduction of any energy performance legislation in the country, and also during the previous and actual energy legislation. As shown, because of the

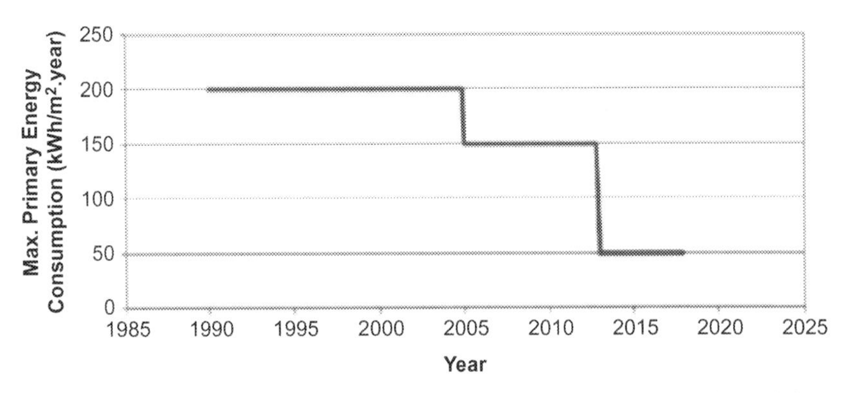

FIGURE 2.39 Maximum primary energy consumption foreseen due to French regulations. Source: *European Commission, 2016.*

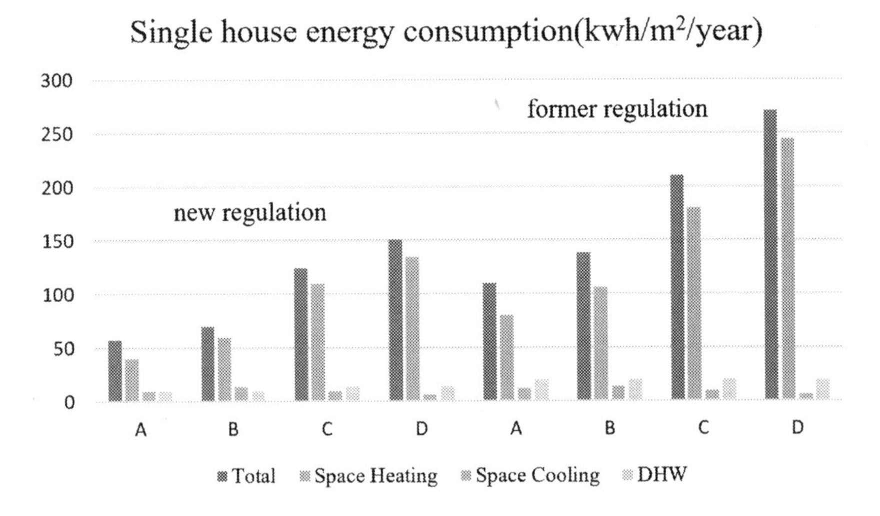

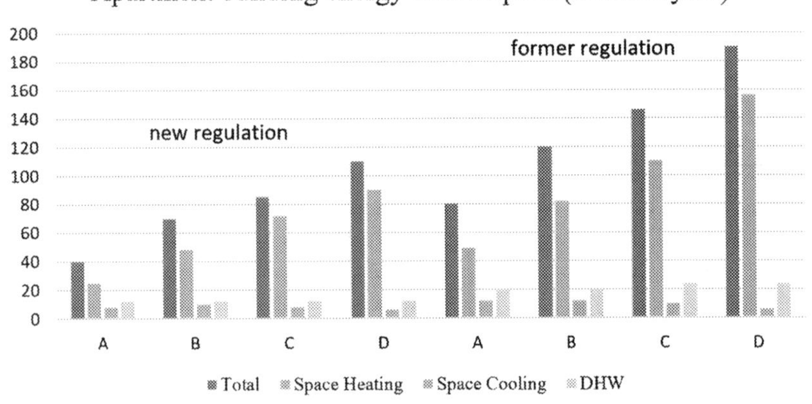

FIGURE 2.40 Energy consumption by a single house and an apartment per regulation and climate zone in Greece. Source: *Gagglia et al. (2017).*

TABLE 2.4 Percentage (%) of Dwelling Certifications per Energy Class with a Construction or Renovation Period up to 1980 (Before the TIR), 1980–2010 (According to the TIR), and After 2010 (According to the REPB)

Regulation/Construction Period	Energy Consumption[a]	Percentage (%) of Dwelling per Energy Class								
	(kWh m^{-2} yr^{-1})	A+	A	B+	B	C	D	E	F	G
No regulation (before 1980)	175–525	0.0	0.0	0.1	0.5	5.0	11.5	15.6	21.9	45.4
TIR regulation (1980–2010)	120–350	0.0	0.1	0.4	4.0	27.4	32.8	18.0	7.8	9.6
REPB regulation (after 2010)	50–270	0.1	0.7	7.5	34.5	37.6	14.6	3.4	0.9	0.8

[a] *Space heating, cooling and DHW.*
Source: *Gaglia et al. (2017).*

introduction of the Building Performance Directive, energy consumption has been reduced substantially, and most of the buildings are classified in higher energy classes than previously. This is clear when comparing the energy consumption of a typical house and a typical apartment designed within the scope of the previous legislation and the European Directive (Fig. 2.40). It is evident that residential buildings designed according to the European Directive present almost 40% lower energy consumption (Fig. 2.40).

Both the Energy Performance Directive and the Energy Efficiency Directive have been implemented. The Energy Efficiency Directive primarily defines the policies needed to renovate the national stock of public and private buildings in the member states. The Energy Efficiency Directive requires the Member States to renovate 3% of their governmental buildings per year to minimize their energy consumption, while measures must be defined to achieve annual energy savings close to 1.5%.

Additionally, to the two above Directives, a third one, the Directive on Renewable Energy, aims to define specific targets regarding the use of renewable energies in buildings. The Directive sets specific targets for new and renovated buildings, and promotes the use of renewable energies to cover the specific energy needs of buildings. A fourth Directive, the so called Ecodesign, provides standards regarding the minimum performance of the energy systems and products used in buildings, boilers, ventilators, pumps, and so on. Finally, the Energy Labelling Directive aims to define the minimum requirements regarding the implementation of labelling schemes for building-related energy technologies.

The whole legislative frame, as described above, is completed by a set of actions and legislative measures aiming to improve social and territorial cohesion in Europe regarding access to appropriate heating and cooling energy for all the population groups in Europe, especially the low income groups. The impact of the energy legislation has already been quite significant, with the consumption of residential buildings decreasing by up to 1.4% per year.

REFERENCES

Auffhammer, Maximilian, 2014. Cooling China: The weather dependence of air conditioner adoption. Front. Econ. China 9 (1), 70–84.

Maximilian Auffhammer, Erin T. Mansur, Measuring climatic impacts on energy consumption: A review of the empirical literature, April 25, 2014.

Bertoldi P., Diluiso F., Castellazzi L., Labanca N., Serrenho T. 2018: Energy consumption and energy efficiency trends in the EU-28, 2000–2015 efficiency trends of energy related products and energy consumption in the EU-28, JRC Ispra, 2018.

Biddle, J., 2008. Explaining the spread of residential air conditioning 1955–1980. Explor. Econ. Hist. 45 (4), 402–423.

BIMHow, 2015. Impact of the construction industry on the environment. http://www.bimhow.com/impact-of-the-construction-industry-onthe-environment/. Assessed in February 2018.

Christenson, M., Manz, H., Gyalistras, D., 2006. Climate warming impact on degree-days and bulding energy demand in Switzerland. Energy Convers. Manage. 47 (6), 671–686.

Ciscar, J.C., Feyen, L., Soria, A., Lavalle, C., Raes, F., Perry, M., et al., 2014. Climate impacts in Europe. The JRC PESETA II Project. JRC scientific and policy reports, EUR 26586EN.

Davis, Lucas W., Gertler, Paul J., 2015. Contribution of air conditioning adoption to future energy use under global warming. PNAS 112 (May (19)), 5962–5967.

EU-25 Energy and Transport, 2015. Outlook to 2030. https://ec.europa.eu/energy/sites/ener/files/documents/trends_to_2030_update_2009.pdf. February 2018.

Energy Information Administration of USA: Annual Energy Outlook 2018, with projections to 2050. February 6, 2018.

European Commission, 2016: Commission staff working document, good practice in energy efficiency, accompanying the document: Proposal for a directive of the European parliament and of the council amending directive 2012/27/EU on energy efficiency {COM(2016) 761 final}.

European Concerted Action, 2015. Energy performance of buildings. http://www.epbd-ca.eu/. Assessed in February 2018.

European Environmental Agency: Energy consumption per use and dwelling. Assessed in 2018.

ExxonMobil: 2018 Outlook for energy: A View to 2040.

Gaglia, Athina G., Tsikaloudaki, Aikaterini G., Laskos, Costantinos M., Dialynas, Evangelos N., Argiriou, Athanassios A., 2017. The impact of the energy performance regulations' updated on the construction technology, economics and energy aspects of new residential buildings: The case of Greece. Energy and Buildings 155, 225–237.

Golder, P., Tellis, G., 1998. Beyond diffusion: An affordability model of the growth of new consumer durables. J. Forecast. 17, 259–280.

Housing Statistics in the European Union, 2010. Edited by Kees Dol and Marietta Haffner OTB research institute for the built environment, delft university of technology.

Hultgren, A., 2011. SWCA environmental consultants, global buildings and construction: Energy and environmental impacts. http://www.kleiwerks.org/wp-content/uploads/2012/11/Buildings-Energy-and-Carbon.pdf. February 2018.

IIASA: Energy End-Use: Buildings, 2015. http://www.iiasa.ac.at/web/home/research/Flagship-Projects/Global-Energy-Assessment/GEA_CHapter10_buildings_lowres.pdf. Assessed in February 2018.

International Energy Agency, 2013. World energy outlook 2013. http://www.worldenergyoutlook.org/weo2013/. Assessed in February 2018.

Isaac, M., Van Vuuren, D.P., 2009. Modeling global residential sector energy demand for heating and air conditioning in the context of climate change. Energy Policy 37 (2), 507–521.

Mirasdegis, S., Sarafidis, Y., Georgopoulou, E., Kotroni, V., Lagourvardos, K., Lalas, C.P., 2007. Modeling framework for estimating impacts of climate change on electricity demand at regional level: Case of Greece. Energy Convers. Manage 48 (5), 1737–1750.

Odyssee-Mure, 2012. Energy efficiency trends in buildings in the EU lessons from the ODYSSEE MURE project. http://www.odyssee-mure.eu/publications/br/energy-efficiency-trends-policies-buildings.pdf. Assessed in February 2018.

Ormrod, R.K., 1990. Local context and innovation diffusion in a well-connected world. Econ. Geogr 66 (February (2)), 109–122.

Phadke, Amol, Abhyankar, Dr. Nikit, Shah, Dr. Nihar, 2014. Avoiding 100 new power plants by increasing efficiency of room air conditioners in India: Opportunities and challenges. Lawrence Berkeley National Laboratory, Berkeley, USA.

David Rapson, Durable goods and long-run electricity demand: A case study of air conditioner purchase behavior. Working Paper, UC Davis, 2011.

David Rapson, 2014. Durable goods and long-run electricity demand: Evidence from air conditioner purchase behavior. J. Environ. Econ. Manage. 68 (1), 141–160.

Santamouris, M., 2015. Regulating the damaged thermostat of the cities – Status impacts and mitigation strategies. Energy Build. 91, 43–56.

Santamouris, M., 2016a. Innovating to zero the building sector in Europe: Minimising the energy consumption, eradication of the energy poverty and mitigating the local climate change. Solar Energy 128, 61–94.

Santamouris, M., 2016b. Cooling the buildings. Past, Present and Future, Energy and Buildings 128, 617–638.

Santamouris, M., Paravantis, J.A., Founda, D., Kolokotsa, D., Michalakakou, P., Papadopoulos, A.M., et al., 2013. Financial crisis and energy consumption: A household survey in Greece. Energy Build. 65, 477–487.

UN–Habitat, 2011. Cities and climate change: Global report on human settlements.

UN–Habitat, 2013. Streets as public spaces and drivers of urban prosperity, Nairobi.

UNEP, 2012. Building design and construction: Forging resource efficiency and sustainable development.

United Nations, 2009. Urban and rural areas, http://www.unpopulation.org. Assessed in February 2018.

WBCSD, 2009. Transforming the market: Energy efficiency in buildings. world business council on sustainable development. August 2009.

FURTHER READING

Institute of Energy Economics of Japan: Energy Outlook 2018. Prospects and challenges until 2050. Japan, 2017.

McNeil, M.A., Letschert, V.E., del Rue du Can, S., Ke, J., 2013. Bottom-up energy analysis system (BUENAS)—an international appliance efficiency policy tool. Energy Effic. 6 (2), 191–217.

Chapter 3

Urban Heat Island and Local Climate Change

DEFINITION OF LOCAL CLIMATIC CHANGE AND URBAN OVERHEATING

The urban heat island phenomenon is one of the more documented issues relating to climate change. It was reported at the beginning of the 19th century in London (Howard, 1833). The phenomenon is related to the development of higher ambient temperatures in dense central urban zones compared to surrounding suburban and rural areas (Fig. 3.1).

The Urban Heat Island (UHI) results from a positive energy balance in urban areas, which increases ambient urban temperature. The magnitude of the phenomenon is influenced and determined by several parameters, which have been summarized by Oke et al. (1991):

a. The thermal properties of the materials in the urban environment. Most of the materials used in buildings and open spaces present a high absorptivity of solar radiation. This results in higher surface temperatures and heat storage in cities. The stored heat is then released into the atmosphere through convection and radiation processes. In parallel, the replacement of the vegetation and of the natural soil decreases the evaporation processes and results in higher ambient temperatures.

b. The anthropogenic heat released by heating systems, air conditioning, cars, industry and other combustion systems.

c. The radiative geometry of canyons. Because of the specific geometry of urban canyons, the infrared radiation emitted by buildings, streets and pavements cannot escape into space. It is reflected and absorbed by the canyon's surfaces and contributes to a more positive thermal balance in the canyons. In parallel, due to multiple reflections in the canyon, a high part of the initially reflected solar radiation is absorbed, increasing the surface temperature of the vertical walls and of the other canyon surfaces.

d. The urban greenhouse effect. Because of the high concentration of atmospheric pollutants, part of the infrared radiation emitted from ground surfaces is reflected back to the earth's surface by pollutants and the

Minimizing Energy Consumption, Energy Poverty and Global and Local Climate Change in the Built Environment: Innovating to Zero. DOI: https://doi.org/10.1016/B978-0-12-811417-9.00003-9

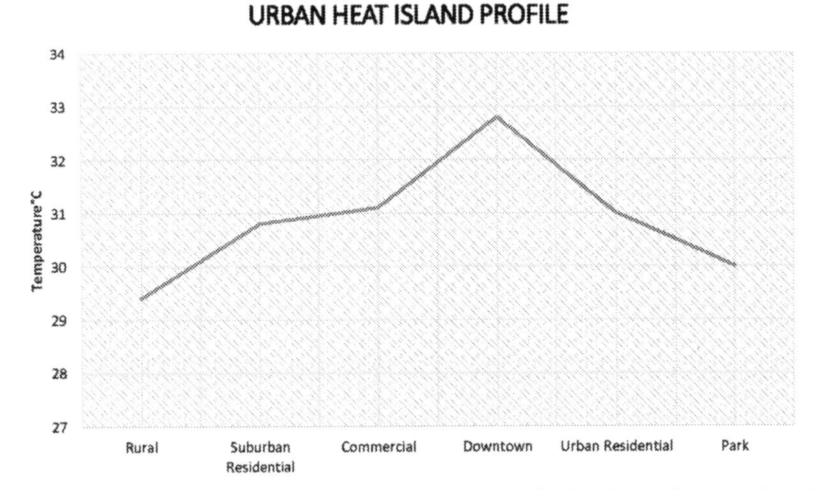

FIGURE 3.1 Urban heat island: Ambient temperature profile in urban, suburban and rural areas. Source: *LBNL Web site, USA.*

clouds, while the infrared radiation emitted by pollutants is also added to the thermal balance of the urban environment.

e. Reduced turbulent transfer of heat within streets.

f. The reduction of evaporating surfaces in the urban environment, which decreases latent heat and increases sensible ones.

Higher urban temperatures increase energy consumption for cooling, deteriorate outdoor and indoor thermal comfort conditions, raise the concentration of harmful pollutants such as tropospheric ozone, have a serious impact on human health and well-being, and increase the ecological footprint of cities (Santamouris, 2016a). To document the amplitude and the characteristics of the urban heat island, measurements have been taken in hundreds of cities, and the phenomenon is fully documented in more than 400 major cities around the world.

Local climate change is a more complex phenomenon than the urban heat island. It is related to the increase of the ambient temperature, but it is also linked to the important change of the frequency of heat waves and the prolongation of hot spells.

The present chapter aims to provide all the necessary and available information about the characteristics, magnitude and impact of urban heat islands and local climate change.

SYNERGIES BETWEEN LOCAL AND GLOBAL CLIMATE CHANGE

Local and global climate change must be seen as separate phenomena. Both contribute to increasing the ambient temperature, but through different mechanisms. The possible interactions between local and global climate is still an open scientific question. Very few studies have attempted to identify and explain the possible synergies and the corresponding mechanisms that drive them.

According to Li and Abu-Zeid (2013), three are the mechanisms that determine the interrelationships between the two phenomena. Two of the mechanisms intensify temperature differences between urban and rural areas, while the third has a negative feedback on the magnitude of the temperature difference.

Two key mechanisms contributing to increasing the magnitude of temperature differences between the cities and the rural areas: (1) Heat waves are associated with low wind speeds. It is well known that under specific synoptic conditions, the intensity of urban heat islands increase, and (2) During heat waves, surface temperatures increase, resulting into higher thermal storage and evapotranspiration from the surfaces. Given that soil in the rural areas has a higher moisture availability, the temperature difference between the city and rural areas tends to increase.

Other mechanisms decrease the amplitude of the temperature difference between cities and rural areas. It is well known that heat waves strengthen possible secondary circulations. Under specific conditions, warm air in the city moves upwards, while cool air enters the city from the neighboring rural areas, decreasing the urban temperature.

Just a few studies have attempted to analyze the synergies between the urban heat island phenomenon and global climate change. Oke (1997) predicts that even intensive conditions of global climate change may not modify the amplitude of the urban heat island. In parallel, Brázdil and Budíková (1999) predicts that under intensified global climate change conditions, the amplitude of the urban heat island will decrease due to increased vertical instability and heat dissipation in the urban environment. Similar conclusions are drawn by McCarthy et al. (2010). They predict that intensified global climate change may reduce the magnitude of the urban heat island by 6%. However, it is also predicted that in cities presenting an intensive urbanization, the amplitude of the urban heat island may increase by up to 30%.

Contrary to the above conclusions, several other studies suggest potential synergistic reactions between global climate change and the urban heat island phenomenon (Li and Abu-Zeid 2013; Founda et al. 2015; Li et al. 2015). A detailed analysis of urban heat island intensity under heat wave conditions in a coastal city, Athens, shows that the magnitude of the phenomenon is increasing both during the day and night (Founda and Santamouris, 2017). It is found that urban heat island intensity is increasing alongside increasing maximum ambient temperatures. The corresponding magnitude of the temperature rise is found to depend highly on wind speed and direction. When a sea breeze develops, the corresponding cooling mechanisms increase the UHI intensity considerably. The intensity of the

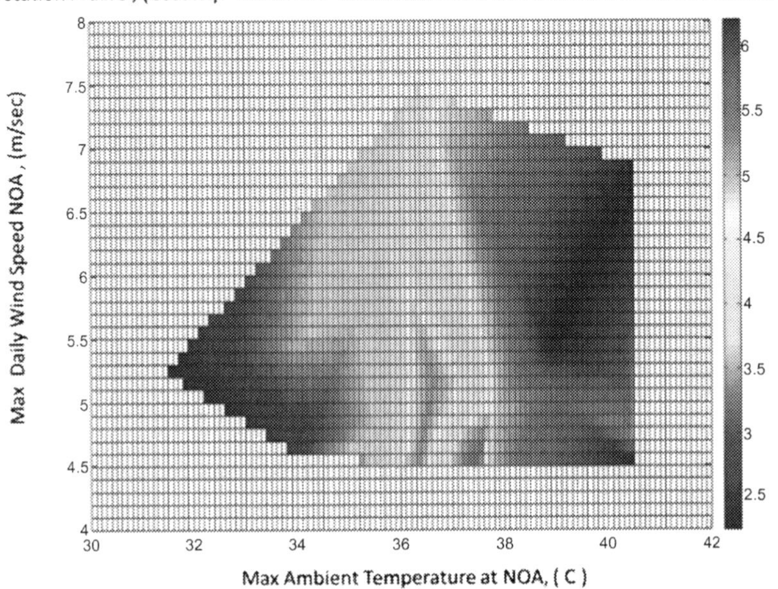

FIGURE 3.2 UHI intensity in relation to the maximum wind speed and ambient temperature at the center of the city (NOA), for winds from the sea in both urban and coastal stations. Source: *Founda and Santamouris (2017).*

UHI phenomenon can be up to 4 K higher during a period of heat waves (Fig. 3.2).

In conclusion, most of the experimental studies carried out show an important intensification of the UHI phenomenon under synoptic heat wave conditions. It is evident that synergistic effects between the urban heat island phenomenon and global climatic change is still an open scientific topic and further studies are necessary.

ON THE MAGNITUDE OF THE URBAN HEAT ISLAND IN THE WORLD

The characteristics and amplitude of UHIs have been determined experimentally in more than 400 cities around the world. The experimental methodology followed to identify the magnitude and the characteristics of UHIs is not the same in every location. There are three main experimental protocols used to measure a UHI's intensity (Santamouris, 2015a):

a. Monitoring studies using standard experimental stations and equipment,
b. Monitoring studies using non-standard experimental equipment,
c. Studies based on mobile traverses around the concerned area.

In parallel, given that the duration of the experimental period differs highly from city to city, and the available data may cover a limited period of time, studies based on mobile traverses and non-standard meteorological stations usually report the measured maximum temperature difference during the experimental period. Studies based on multi-year measurements using standard equipment, usually report either the average annual UHI intensity, the annual average maximum UHI intensity, or the absolute maximum intensity of the phenomenon. The diversity of the experimental methods used and the variety of reported UHI intensities makes inter-study comparisons difficult. It points to a problem with the authenticity of existing measurements, the accuracy and representativeness of results, and the overall validity of the scientific conclusions given.

Despite the above problems, most of the existing data have been analyzed, classified and reported. Santamouris (2015a) analyzed and classified the experimental data collected from 101 Asian and Australian cities. For all cities where UHI data were collected using mobile traverses (Fig. 3.3), the intensity of the phenomenon varies between 0.4°C to 11 K, with an average close to 4.1°C and a standard deviation of around 2.3°C. It is characteristic that for about 27% of the reported data, the UHI intensity exceeds 5°C.

When non-standard measuring equipment is used, the magnitude of the urban heat island varies between 1.5°C to 10.7°C, with an average value close to 5°C (Fig. 3.4). The corresponding standard variation is 2.9°C.

Finally, Figs. 3.5–3.7, report the annual average, average maximum and absolute maximum UHI intensities when standard monitoring equipment is

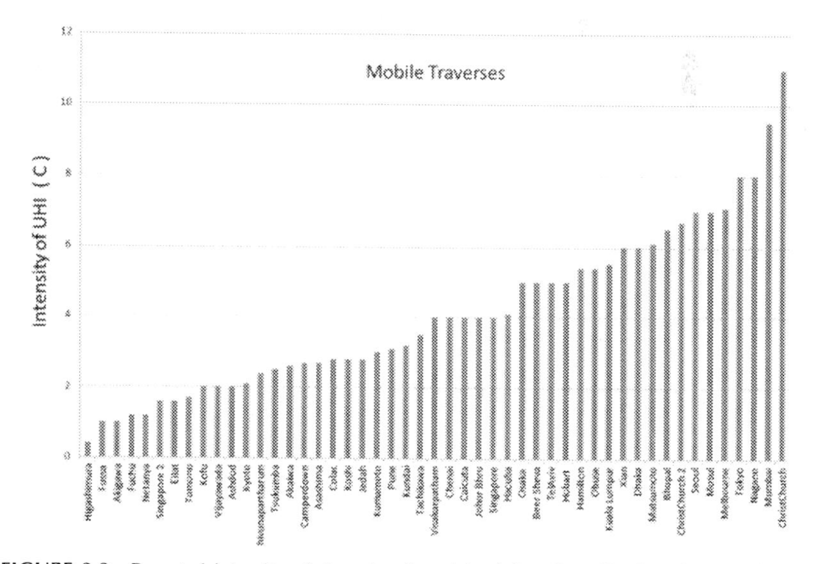

FIGURE 3.3 Reported intensity of the urban heat island for all studies based on mobile traverses. Source: *Santamouris (2015a)*.

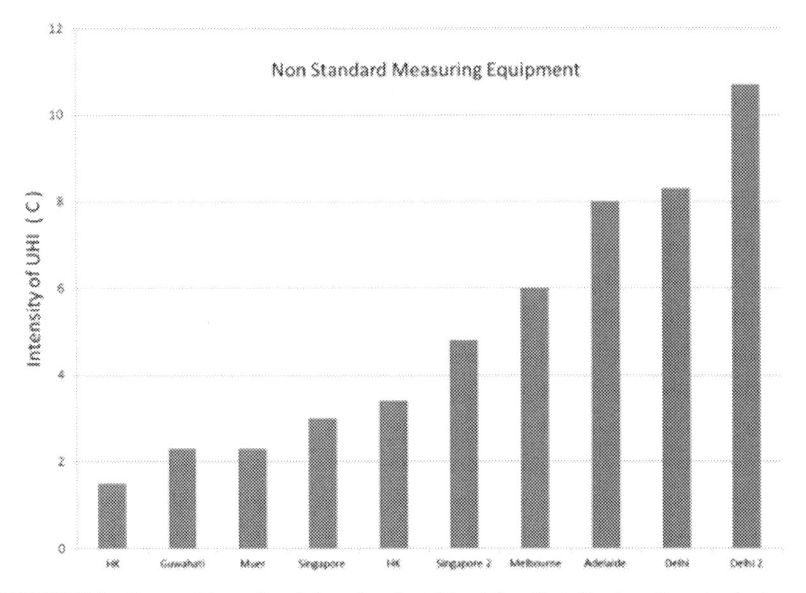

FIGURE 3.4 Reported intensity of the urban heat island for all studies based on standard measuring equipment. Source*: Santamouris (2015a).*

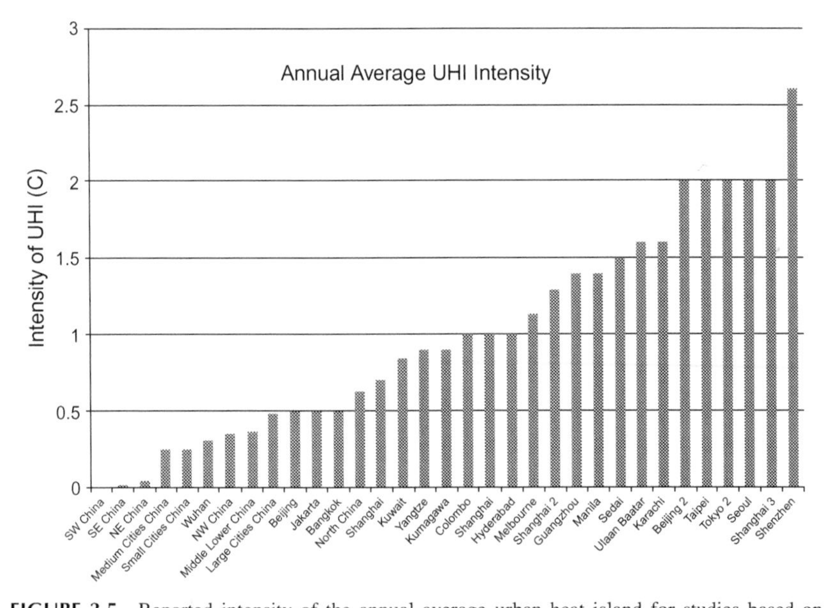

FIGURE 3.5 Reported intensity of the annual average urban heat island for studies based on standard measuring equipment. Source*: Santamouris (2015a).*

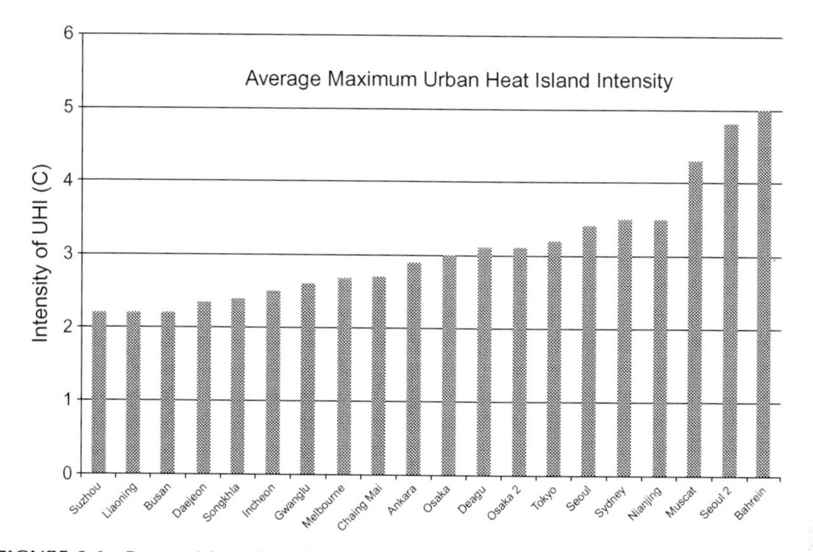

FIGURE 3.6 Reported intensity of the average maximum urban heat island for studies based on standard measuring equipment. Source: *Santamouris (2015a)*.

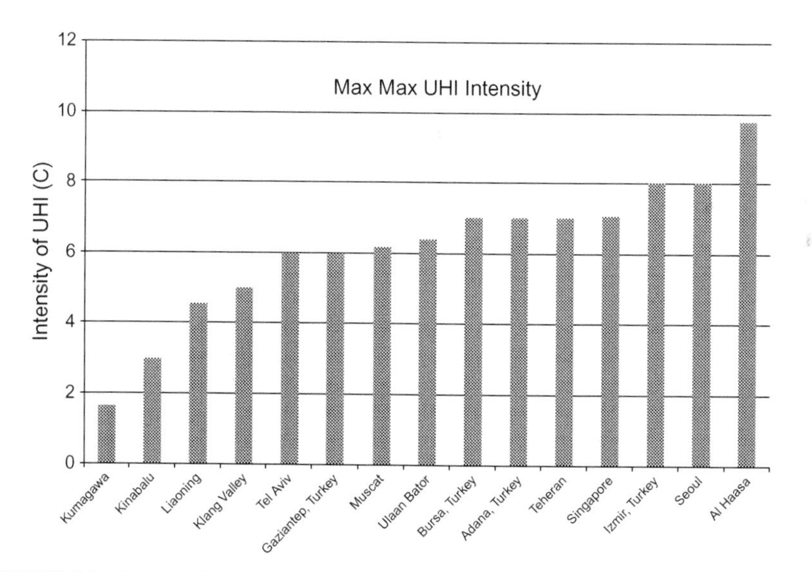

FIGURE 3.7 Reported intensity of the max–max urban heat island for studies based on standard measuring equipment. Source: *Santamouris (2015a)*.

employed. The mean magnitude of the average annual UHI intensity was found close to 1°C, while the corresponding values for the average and absolute maximum UHI intensity were close to 3.1°C and 6.2°C. For about 20% of the cities, the absolute maximum UHI intensity exceeds 8°C.

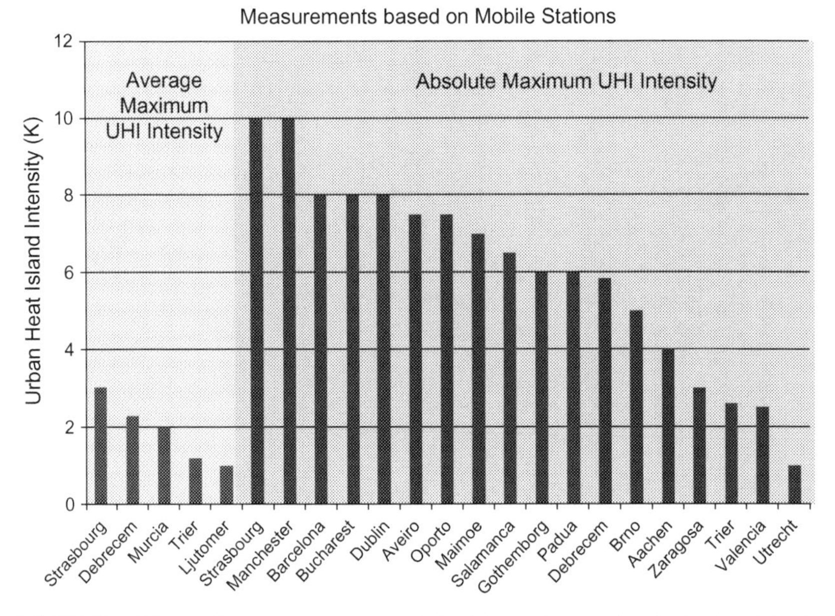

FIGURE 3.8 Measured average maximum and absolute maximum urban heat island intensity for all experiments based on the use of mobile traverses. Source: *Santamouris (2016a).*

In parallel, Santamouris (2016a) has analyzed and classified existing UHI data collected in 110 European cities. The reported UHI intensity when mobile traverses are used is given in Fig. 3.8. The absolute maximum UHI intensity varies between 1°C to 10°C, with an average close to 6°C. The corresponding figures when non-standard meteorological stations are used are reported in Fig. 3.9. As shown, the average maximum UHI intensity varies between 2°C and 5°C with an average value close to 3.2°C The absolute maximum varies between 0.8°C and 10°C with a mean value equal to 5.2°C. Finally, Fig. 3.10 reports the mean annual, mean maximum and absolute maximum UHI intensities when standard monitoring equipment is used. It is found that the mean annual UHI intensity varies between 0.1°C and 2°C with an average value equal to 1.08°C. The mean maximum UHI intensity varies between 0.3°C and 6.8°C with an average equal to 2.6°C, and finally the absolute maximum is between 2.8°C and 12°C with an average equal to 6.2°C.

Analysis of the previously reported data shows clearly that when mobile and non-standard stations are used, the measured UHI intensity is considerably higher than when standard equipment is used. This is because mobile stations and non-standard equipment are mainly used in dense urban areas where the intensity of the UHI is considerably higher than in low density urban areas where standard meteorological equipment is used.

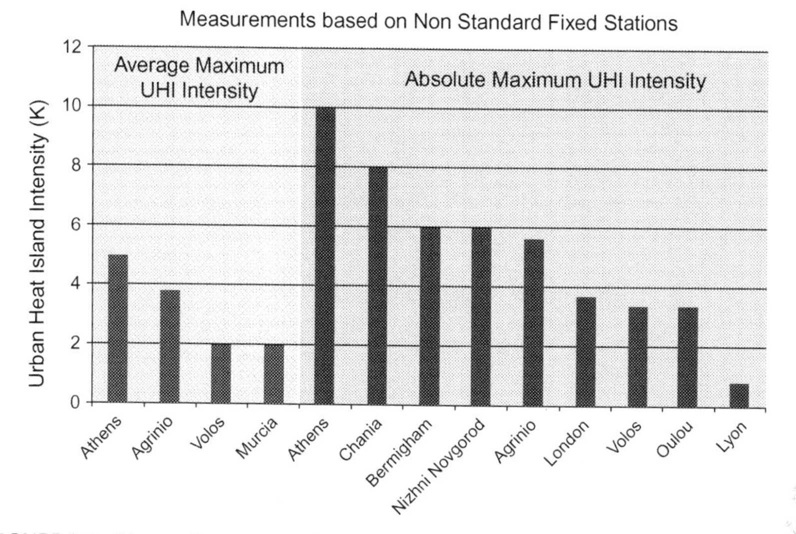

FIGURE 3.9 Measured average maximum and absolute maximum urban heat island intensity for all experiments based on the use of non-standard fixed meteorological stations. Source: *Santamouris (2016a).*

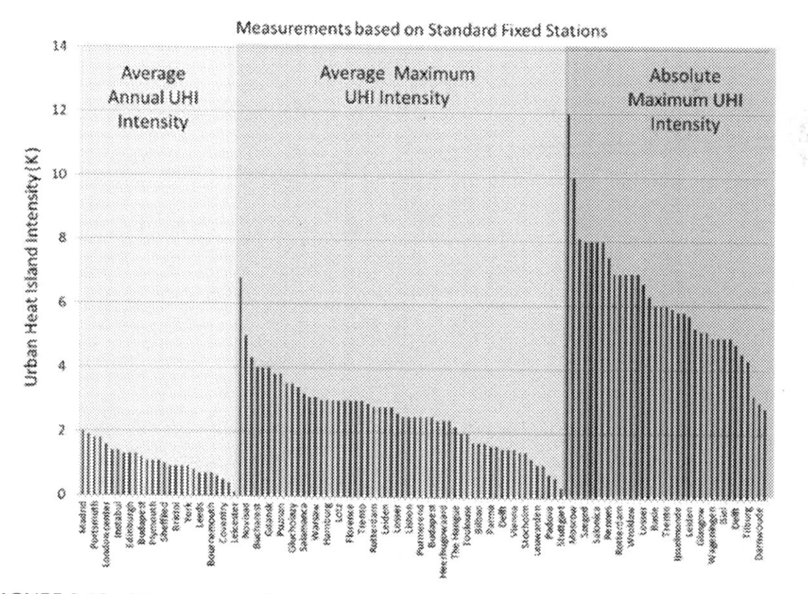

FIGURE 3.10 Measured annual average, average maximum and absolute maximum urban heat island intensity for all experiments, based on the use of standard fixed meteorological stations. Source: *Santamouris (2016a).*

ANALYSIS OF THE MAIN PARAMETERS DEFINING THE MAGNITUDE AND THE CHARACTERISTICS OF LOCAL CLIMATE CHANGE

The magnitude and the characteristics of the urban heat island in a specific place depend on many parameters. The most important of them are synoptic climatic conditions, humidity levels, precipitation, wind speed and cloud cover. In coastal cities, sea breeze highly influences the characteristics of the UHI.

It is well known that the UHI is more likely developed under anticyclonic synoptic climatic conditions. Analysis of data from a relatively high number of cities shows that under cyclonic synoptic conditions, the magnitude of the UHI is quite low (Unwin, 1980; Morris and Simmonds, 2000; Mihalakakou et al., 2002).

Previous research on the impact of humidity on the characteristics and the amplitude of the UHI has shown that ambient relative humidity is negatively correlated with the amplitude of the urban heat island. In general, higher values of relative humidity correspond to lower UHI amplitudes. The intensity of the UHI tends to decrease as high ambient relative humidity levels are associated with higher evaporation in the urban environment. Because of the evaporative cooling, this results in reduced surface and air temperatures (Kim and Baik, 2002). In a similar way, the magnitude of the UHI is reduced during rainy days. Precipitation decreases cooling rates in rural areas, resulting in a reduction of the intensity of urban heat islands (Chow and Roth, 2006; Kubota and Ossen, 2014).

It is well-accepted that high UHI intensities correspond to low wind speeds. High wind speeds modify cooling rates in both urban and rural areas, and reduce the amplitude of the UHI (Erell and Williamson, 2007; Papanikolaou et al., 2008). The critical wind speed over which the UHI intensity is insignificant differs from place to place and may vary from 1–5 m/sec (Park, 1986; Morris et al., 2001).

UHIs become more intense under clear sky conditions. Under cloudy conditions, terrestrial infrared radiation cannot escape as it absorbed and re-emitted back by the clouds, affecting the thermal balance at the earth's surface (Mohan et al., 2012). However, important intensities of the UHI are also reported under cloudy sky conditions, in particular when wind speed is quite low (Erell and Williamson, 2007).

Sea breeze in coastal areas, when present, transfers cooler air into the mainland and decreases the intensity of UHIs (Khan and Simpson, 2001; Freitas et al., 2007). Several theoretical and experimental studies have concluded that there is a strong relationship between urban heat and the sea breeze. It is concluded that UHIs can accelerate the speed of the sea breeze towards the city (Freitas et al., 2007), but the front of the sea breeze stalls in the center of the city and continues to move when the heat island disappears (Freitas et al., 2007).

IMPACT OF LOCAL CLIMATE CHANGE ON ELECTRICITY POWER DEMAND

Higher ambient temperatures have a serious impact on the electricity consumption of buildings, and in particular on peak electricity demand. The sensitivity of electrical networks to the additional cooling demand induced by the urban overheating depends highly on the thermal quality of the building stock in the considered area, the degree of penetration of the air conditioning, the used thermostat temperature for cooling, and the various specificities of the particular electricity network.

It is well-known that the relation between the ambient temperature and electricity consumption follows an assymetrical U-shape curve. The curve presents its maximum during the winter period in heating-dominated climates, and in the summer in cooling-dominated areas. In both zones, the minimum corresponds to the period where both heating and cooling needs are not significant (Santamouris et al., 2016). The inflection point of the response curve is defined as the threshold temperature over which the use of electricity starts to increase. In general, it is considered that the deflection point is close to 18.3°C. However, recent research carried out in 15 European countries shows that in heating dominated climates it is close to 15°C, whereas in cooling dominated zones it is around 22.4°C (Bessec and Fouquau, 2008).

Existing data on the relation between ambient temperature and peak electricity demand report either the hourly, daily, monthly or even annual correlation between the two parameters. Table 3.1 (Santamouris et al., 2016) summarizes the data from most of the existing studies around the world.

The data presented in Table 3.1 are highlighted in Fig. 3.11, where the inflection point of the response factors is shown together with the base electrical load and the increase rate of electricity demand per degree of ambient temperature increase.

Analysis of the above data shows that the increase in the peak electrical load per degree of ambient temperature varies between 0.45% and 4.6%. The mean increase rate is close to 2.65% and the average rise of the electric load is 226 MW per degree of temperature increase. As calculated by Santamouris et al. (2016), the mean peak electricity penalty per person is 21 ($\pm$ 10.4) W per degree of temperature increase.

Increased ambient temperatures considerably increase the total electricity consumption. Specific data on the relation between the ambient temperature and total electricity consumption are available from 15 cities and states (summarized in Fig. 3.12). As shown, the hourly, daily or monthly electricity penalty varies between 0.5% and 8.5%. The mean value is close to 4.6%. The threshold temperatures over which electricity demand starts to increase varies in average between 11.7°C and 22°C.

TABLE 3.1 Existing Studies on the Impact of Ambient Temperature on Peak Electricity Demand

No.	City/Country	Reference Year	Additional Load per K	Percentage Increase in the Base Electricity Load per Degree of Temperature Increase	Threshold Inflection Temperature (°C)	Reference
Increase of the Peak Electricity Power Demand – Impact of 1K Increase of the Ambient Temperature						
1	Tokyo Japan	2004	Peak additional demand of 180 MW/K	0.45%	22°C	Yabe (2005)
2	Thailand	2004	A 1 K temperature increase raises the peak demand by 810 MW and the average demand by 577 MW (2004 levels)	4.6% of the peak and 3.8% of the average demand	Not reported	Parkpoom and Harrison (2008)
3	Ontario East Canada	1991–1995	Above 23°C, the peak daily electricity demand increases by 233 MW/K	1.5%	23°C	Colombo et al. (1999)
4	Los Angeles, USA	1986	Increase in daily peak Electricity demand by 545 MW/K	3.3%	183°C	Akbari et al. (1992)
5	Washington DC, USA	1986	Increase in daily peak electricity demand by 181 MW/K or 3.6% of the basic peak load	3.6%		Akbari et al. (1992)

6	Dallas TX, Fort Worth, USA	1986	Increase in daily peak electricity demand by 454 MW/K	3.1%	13°C	Akbari et al. (1992)
7	Colorado Springs, CO, USA	1986	Increase in daily peak electricity demand by 7.3 MW/K	1.8%	13°C	Akbari et al. (1992)
8	Phoenix, AZ, USA	1986	Increase in daily peak electricity demand by 101 MW/K	3.6%	24°C	Akbari et al. (1992)
9	Tuscon AZ, USA	1986	Increase in daily peak electricity demand by 22 MW/K or 1.8% of the basic peak load	1.8%	21°C	Akbari et al. (1992)
10	Israel	1987–1988	Increase in daily peak electricity demand by 90 MW/K	2.9%–3.1%	Not reported	Segal et al. (1992)
11	Part of Carolina USA	1985–1991	Not reported	3.5%–4%	18°C	Franco and Sanstad (2007)
Increase in Electricity Consumption – Impact of 1 K Increase of the Ambient Temperature						
1	Spain	1998	Daily additional electricity consumption of 8 GWh/K	1.6%	18°C	Pardo et al. (2002)
2	Athens, Greece	1993–2001	Increase in daily energy consumption of 1300 MWh/K,	4.1%	22°C	Giannakopoulos and Psiloglou (2006)

(*Continued*)

TABLE 3.1 (Continued)

No.	City/Country	Reference Year	Additional Load per K	Percentage Increase in the Base Electricity Load per Degree of Temperature Increase	Threshold Inflection Temperature (°C)	Reference
3	New Orleans USA	1995	Increase in daily average electrical load by 15 MWh/K	3%	22°C	Rosenfeld et al. (1995)
4	Hong Kong	2002	Monthly increase 111 GWh/K	4%	18°C	Fung et al. (2006)
5	Ohio, USA	1984–1993	Increase in monthly consumption by 30 kWh/person/K	7.5%	16°C	Wong et al. (2010)
6	California San Hose, Sacramento Pomona and Fresno	2004–2005	Increase in daily consumption by 18500 MWh/K	2.9%	15°C	Franco and Sanstad (2007)
7	Greece	1993–2002	Increase in daily electricity consumption between 1.7–3 GWH/K	1.1%–1.9%	18.5°C	Sailor and Munoz (1997)
8	Chicago, USA	1993–2004	Increase in hourly load by 200 MWh/K	Not reported	15°C–17°C	Mirasgedis et al. (2006)

9	California, USA	1984–1993	Increase in monthly consumption by 27 kWh/K	7.7%	17°C	Wong et al. (2010)
10	Louisiana	1984–1993	Increase oin monthly consumption by 40 kWh/person/K	8.5%.	20°C	Wong et al. (2010)
11	Maryland, USA	1989–2001	Increase inmonthly electricity consumption by 22 kWh/p/K for the residential sector	8.5%	15.6°C for residential and 11.7°C for commercial	Hayhoe et al. (2010)
12	Massachusetts, USA	1977–2001	For residential buildings, increase in monthly electricity demand equal to 9 kWh/person/K; For commercial buildings, increase of 12.7 kWh/person/K	Around 6.5% for the residential sector and 3.0% for the commercial sector	15.5°C for the residential sector and 12.8°C for the commercial sector	Ruth and Lin (2006)
13	Bangkok, Thailand	1986–2006	Increase in average monthly electricity consumption by 56.7 GWh/K	7.49%	Not defined	Wangpattarapong et al. (2008)
14	Singapore	2003–2012	Data on hourly electricity demand	1%–2.5%	Varies during the day	Amato et al. (2005)
15	Netherlands	1970–2007	Data on daily electricity consumption	0.5%	Variable and below 18°C	Doshi et al. (2014)

(Continued)

TABLE 3.1 (Continued)

No.	City/Country	Reference Year	Additional Load per K	Percentage Increase in the Base Electricity Load per Degree of Temperature Increase	Threshold Inflection Temperature (°C)	Reference
Increase in Electricity Consumption – Impact of 1% Increase of the Ambient Temperature						
1	Mild countries: Austria, Belgium, Denmark, France, Germany, Ireland, Luxembourg, Netherlands, New Zealand, Switzerland, Greece, Hungary, Italy, Japan, Korea, Portugal, South Africa, Spain, Turkey, United Kingdom, United States	1978–2000	Not reported	0.54%	Not reported	De Cian et al. (2007)
2	Hot countries: Australia, India, Indonesia, Mexico, Thailand, Venezuela.	1978–2000	Increase of the annual electricity demand by 1.659%	1.7%	Not reported	De Cian et al. (2007)
3	Cold Countries Canada, Finland, Norway, Sweden	1978–2000	Decrease of the annual demand by 0.508%	0.51%	Not reported	De Cian et al. (2007)

Source: *Santamouris (2015).*

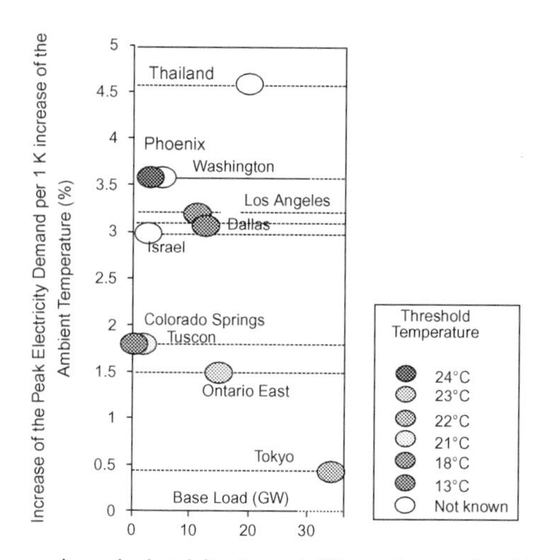

FIGURE 3.11 Increase in peak electricity demand (%) per degree of ambient temperature rise. Source: *Santamouris et al. (2016).*

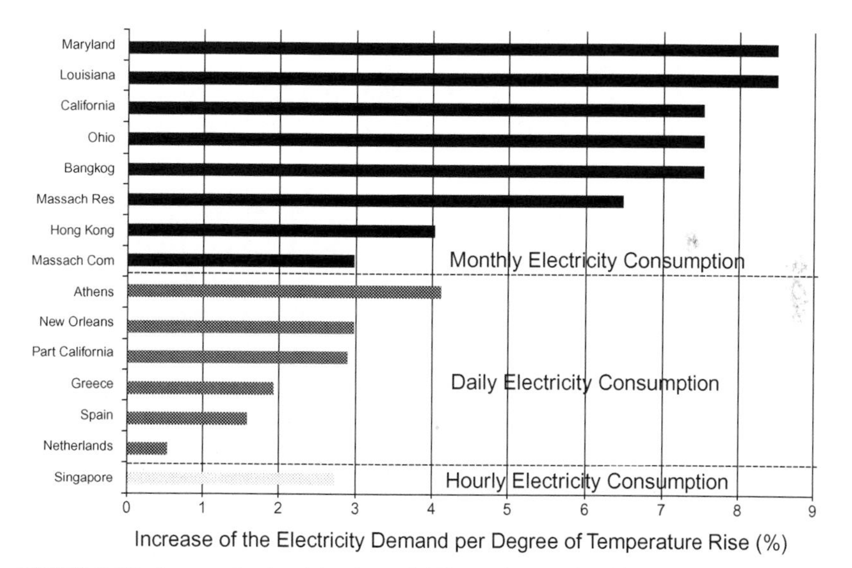

FIGURE 3.12 Increase in electricity demand (%) per degree of ambient temperature rise for various countries. Source: *Santamouris et al. (2016).*

Local and global climate change has a serious impact on total electricity load, as well as on global electricity energy consumption. The estimated increase of the electricity demand per degree of temperature rise varies between 0.5% and 8.5%. Additionally, peak electricity demand increases from 0.45% to 4.6% per degree. It is estimated that on average it corresponds to a penalty of close to 21 (±10.4) W per degree of temperature rise and per person.

IMPACT OF LOCAL AND GLOBAL CLIMATE CHANGE ON THE ENERGY CONSUMPTION OF BUILDINGS

Increased ambient temperatures have a serious impact on the heating and cooling energy demand of buildings. Several studies have evaluated the energy impact of global and local climate change on the energy consumption of buildings. These studies may be classified into two major groups:

a. Studies dealing with the evaluation of the energy impact of local climate change and in particular of the urban heat island. These studies are mainly based on the use of simultaneous urban and rural climatic data to estimate in a comparative way the specific energy needs of several reference buildings.
b. Studies dealing with the evaluation of temporal change in the energy consumption of buildings induced by global climate change. These studies usually employ long time series of climatic data collected in reference meteorological stations. For both types of studies, two kind of methodologies are applied. The first methodology focuses on the evaluation of the specific energy consumption of a reference building, while the second one focuses on the estimation of the global energy impact on the whole building stock in the targeted geographical zone.

Based on the above, Santamouris (2014) classifies the existing studies into three main categories:

a. Studies aiming to analyze the energy impact of the heat island on various types of buildings, using simultaneous urban and rural climatic data.
b. Studies investigating the energy impact of the UHI and global climate change on the global energy consumption of the total building stock in a city.
c. Studies analyzing the multi-annual impact of urban warming on the specific energy consumption of various types of buildings.

The main results and conclusions are summarized and discussed below.

Studies Aiming to Analyze the Energy Impact of the Heat Island on Various Types of Buildings, Using Urban and Rural Climatic Data

According to Santamouris (2014) there are thirteen published studies analyzing the impact of the climate change on the energy consumption of defined reference buildings. These studies concern five cities in Europe (Athens, London, Munich, Rome, Volos), four United States cities and states

(Boston, New York, California, Texas), and two other cities: Melbourne in Australia, and Bahrain in the Middle East. Details of all studies are given in Santamouris (2014).

a. In cooling-dominated areas, UHIs considerably increase the total energy load of buildings. In this case, the increase in cooling demand is much higher than the corresponding decrease in heating load. On the contrary, in heating-dominated zones, total energy consumption decreases due to the significant reduction of heating demand. Fig. 3.13 shows the existing relation for all studies between the cooling energy demand of individual buildings as calculated for the reference and the urban climatic stations.

The average increase of the cooling load induced by the UHI is close to 13.1%. This figure refers to the corresponding cooling load calculated using the meteorological data of the rural stations (reference cooling demand).

b. The calculated annual cooling penalty per degree of average UHI intensity is strongly correlated against the logarithm of the corresponding reference cooling demand (Fig. 3.14). It is clear that the cooling penalty follows the reference cooling demand, and the higher the reference cooling, the higher the absolute value of the cooling penalty per degree of UHI intensity.

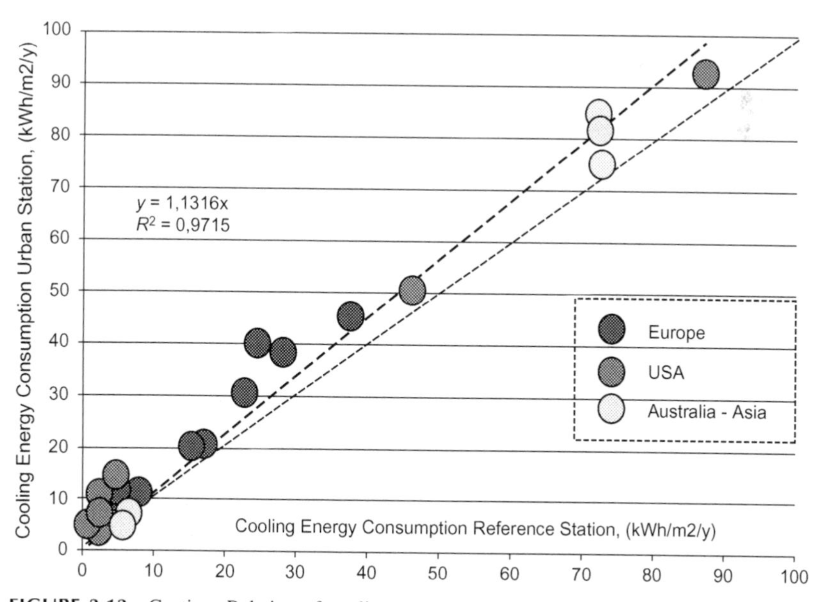

FIGURE 3.13 Caption: Relation of cooling energy demand of individual buildings as calculated for the reference and the urban stations. Source: *Santamouris (2014)*.

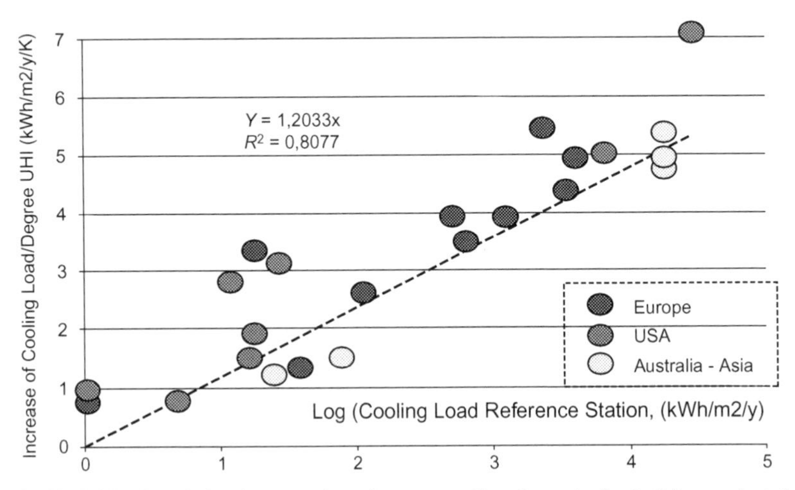

FIGURE 3.14 Correlation between the references cooling demand of a building against the corresponding cooling penalty per degree of UHI intensity. Source: *Santamouris (2014).*

Studies Investigating the Energy Impact of UHIs and Global Climate Change on the Total Energy Consumption of the Building Stock in a City

Although it is a reasonably simple procedure to calculate the impact of local and global climate change on the energy consumption of a reference building, it is quite complicated to calculate the global impact of ambient overheating on the whole building stock of a city. Several methodologies have been developed and implemented, and their main characteristics and results are summarized by Santamouris (2014).

Four large scale studies are available in the international literature. Two of them concerns specific areas of the cities of Athens, Greece; Tokyo, Japan; and Beijing, China. To calculate the global energy penalty of the UHI of a city, Santamouris (2014) proposes using four main homogenized indicators:

a. The global energy penalty per unit of city surface (kWh/m^2),
b. The global energy penalty per unit of city surface and per degree of the UHI intensity (kWh/m^2/K),
c. The global energy penalty per person (kWh/p), and
d. The global energy penalty per person and per degree of the UHI intensity (kWh/p/K).

The results of the whole analysis and the variability of the four indices are given in Fig. 3.15. Based on the results of the available studies, it can be concluded that the UHI causes an average energy penalty per unit of city surface (GEPS) that is close to 2.4 ($\pm$1.5) kWh/m^2, a global energy penalty per

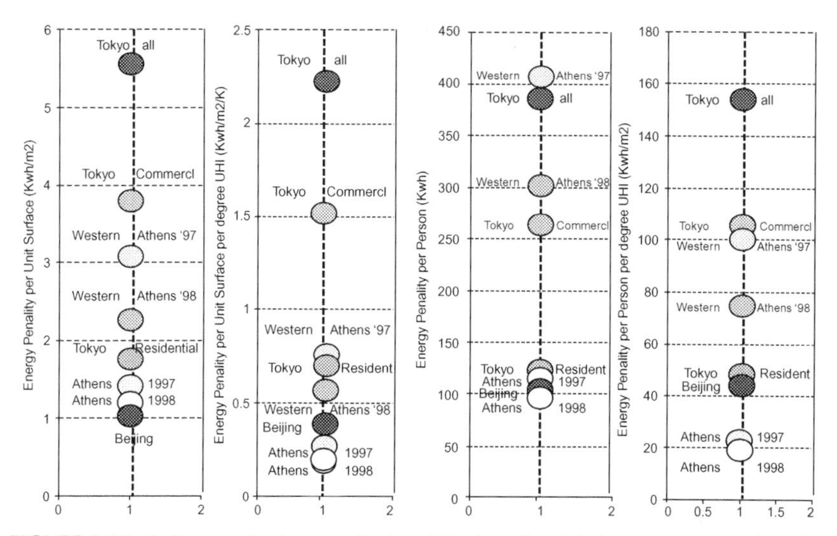

FIGURE 3.15 Indices on the impact of a heat island on the global energy consumption of a city. Source: *Santamouris (2014).*

unit of city surface and per degree of the UHI intensity (GEPSI) of around 0.74 (±0.67) kWh/m^2/K, a global energy penalty per person (GEPP) of around 237 (±130) kWh/p, and a global energy penalty per person and per degree of UHI intensity (GEPPI) that is close to 70 (±45) kWh/p/K.

Studies Analyzing the Multi-Annual Impact of the Urban Warming on the Energy Consumption of Various Types of Buildings

The impact of global climate change on the energy consumption of buildings is usually estimated using multi-annual time series of the ambient temperature. The variability of the annual energy consumption of one or more reference buildings is calculated. Santamouris (2014) analyzed the results of 18 relevant studies and calculated the average energy penalty induced by global climate change over recent years. The studies were carried out for the cities of Athens, Larisa, Corfu and Heraklion in Greece; Nicosia, Paphos, Limassol, Larnaka, Famagusta and Kerynia in Cyprus; Zurich, Geneva, Lugano and Davos in Switzerland; Phoenix, Washington DC, Puerto Rico in the United States; Hong Kong in China, and Resolute in Canada. All studies considered small and large offices as well as residential buildings. Santamouris (2014) provides detailed information on all of these studies. The analysis calculated, homogenized and compared the cooling energy demand for the period 1970–2010 and for all the existing studies. Fig. 3.16

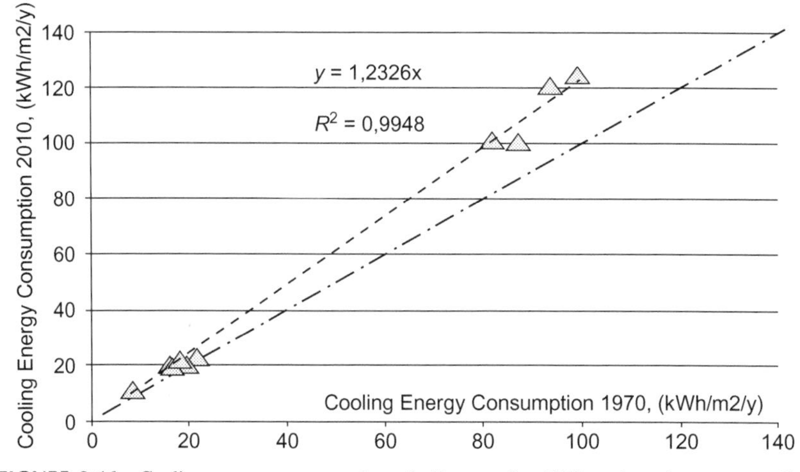

FIGURE 3.16 Cooling energy consumption of all cases for 1970 against the corresponding consumption for 2010. Source: *Santamouris (2014)*.

shows the relation between the calculated cooling demand for the years 1970 and 2010. There is a highly significant correlation between the two sets of data. The average cooling load of the typical buildings considered increased by almost 23% during the period between 1970 and 2010.

Expected Future Increase in the Cooling Energy Demand of Buildings Caused by Global Climate Change

A large number of studies have attempted to estimate the potential future cooling energy consumption of buildings as caused by global climate change. Santamouris (2016b) analyzed 144 relevant studies, calculating and reporting on the characteristics and the magnitude of the expected increase of the cooling demand. The studies refer to 40 different cities in Europe, America, Asia and Australia. Four main parameters were considered and analyed:

a. The undisturbed actual cooling energy consumption of each building studied, Q, which is the reference building cooling at time zero, and was not affected by global climate change,

b. The expected increase of the ambient temperature caused by global climate change during the reference period,

c. The reference cooling degree days, DD, at each place as calculated at time zero, at a temperature base equal to the considered set point temperature, and

d. The increase of the cooling load estimated for the end of the considered period and induced by climate change, Q.

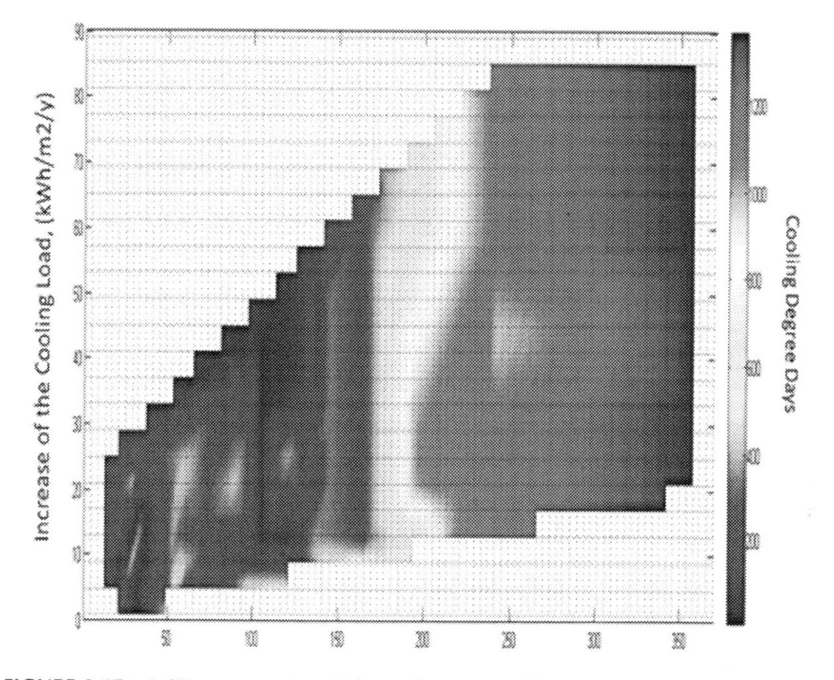

FIGURE 3.17 A 3D representation of the used current cooling energy consumption of office buildings (kWh/m^2/year), the expected increase of cooling energy consumption (kWh/m^2/year), and the current cooling degree days for each of the considered locations. Source: *Santamouris (2016b).*

Figs. 3.17 and 3.18, report the relation and variability between the four parameters considered. As observed, the lower the cooling degree days, the lower the reference cooling demand and the lower the calculated increase of cooling consumption. In parallel, high cooling degree days correspond to high reference cooling loads. For a similar reference consumption and degree days, the potential increase of the cooling demand varies considerably as a function of the considered climate change scenario. The impact of the global overheating becomes more important when the reference cooling demand of the buildings is higher.

A strong nonlinear correlation exists between the reference cooling demand, Q, and the calculated absolute increase in cooling consumption per degree of temperature rise, (Fig. 3.19). The size of the signs in Fig. 3.19 show the percentage increase of the reference cooling demand. For reference cooling demands around 50 kWh/m^2/year, the average increase of the cooling load per degree of temperature rise Q/T is around 6 kWh/m^2/y. It increases to about 12 kWh/m^2/year for reference cooling loads close to 150 kWh/m^2/year. When the reference cooling demand is up to 300 kWh/m^2/year, the expected increase of the Q/T parameter approaches 17 kWh/m^2/year/K.

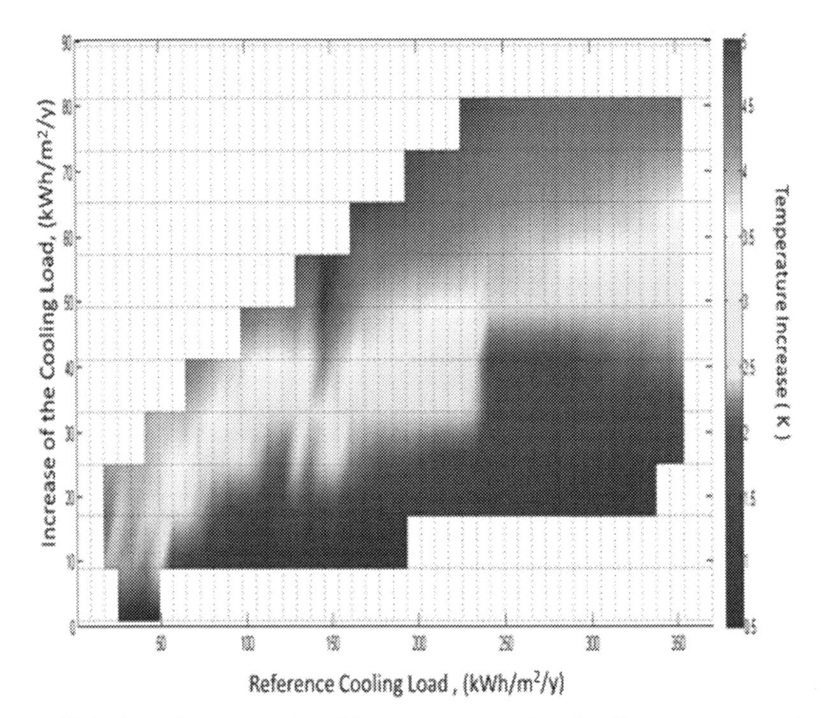

FIGURE 3.18 A 3D representation of the current consumption of cooling energy consumption of office buildings (kWh/m^2/year), the expected increase in cooling energy consumption (kWh/m^2/year), and the corresponding increase in the ambient temperature for each of the considered locations. Source*: Santamouris (2016b).*

The above conclusions are more apparent when the reference cooling demand, Q, is reported against the relative increase in the cooling load per degree of temperature rise, $\Delta Q/Q/\Delta T$ (Fig. 3.20). It is evident that the higher the reference cooling demand, the lower the relative increase of the cooling load per degree of temperature rise.

IMPACT OF LOCAL CLIMATE CHANGE ON THE ENVIRONMENTAL QUALITY OF CITIES

Increased urban temperatures affect the environmental quality of cities in three different ways (Santamouris, 2015b):

- High ambient temperatures act as a catalyst and hasten the photochemical reactions that generate tropospheric ozone,
- Urban heat islands affect the turbulent exchanges and the air flow in cities, causing an increase of the harmful atmospheric pollutants,

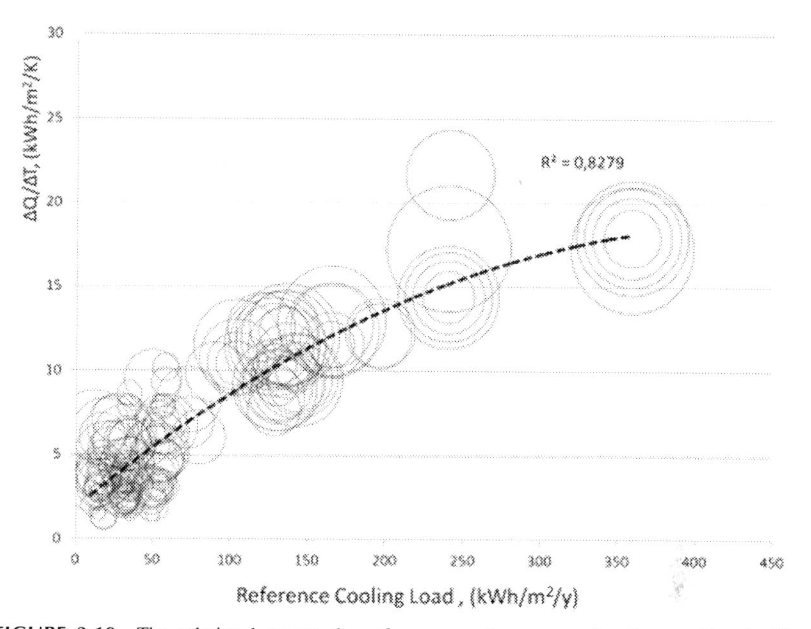

FIGURE 3.19 The relation between the reference cooling energy for the considered office buildings and of the increase of the cooling energy demand per degree of ambient temperature (Q/T). The size of the circles illustrates the percentage increase of the reference cooling load. Source*: Santamouris (2014b).*

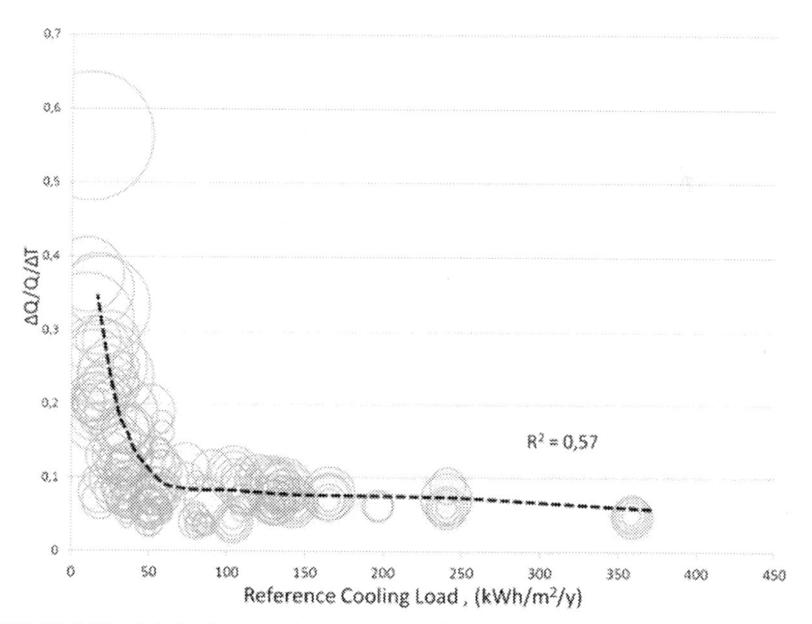

FIGURE 3.20 Relation between the reference cooling energy for the considered office buildings and the ratio of the cooling energy increase and the reference cooling demand per degree of ambient temperature (Q/Q/T). The size of the circles illustrates the percentage increase of the reference cooling load. Source*: Santamouris (2016b).*

– Higher ambient temperatures during the summer period increase the energy consumption required for cooling and the corresponding emissions from power plants.

As a result of high ambient temperatures and corresponding environmental degradation, the ecological footprint of cities is increasing considerably. Studies performed for the city of Athens, Greece (Santamouris et al., 2007), have estimated the level of the additional ecological footprint of the city necessary to compensate the environmental and energy impact of the UHI. It is estimated that this is almost equal to 1.5–2 times the city's political area and may exceed up to 110,000 hectares.

The impact of urban overheating on the concentration of tropospheric ozone in cities is well documented. As reported by Akbari et al. (1992), when the ambient temperature increases to 1°F, the number of days exceeding the threshold of the ozone concentration may increase by 10%. In parallel, Sathopoulou et al. (2008) found that in Athens, Greece, there is a strong correlation between ambient temperature and ozone concentration (Fig. 3.21). The authors found that the number of days exceeding the ozone threshold increase by 18% because of the urban heat island. Similar results and conclusions are also reported in Taiwan and Paris, France, under strong heat island conditions and similar synoptic conditions (Sarrat et al., 2006; Li-Wei Li and Wan-Li Cheng, 2009).

Apart from tropospheric ozone, the UHI affects the concentration of other atmospheric pollutants. As reported by Yoshikado and Tsuchida (2009), the UHI effect delays the penetration of the sea breeze in the Tokyo bay area and favors the blocking of atmospheric pollutants in the city. Higher concentrations of NO_2, NO, PM 2.5, PM10, CO, CO_2 and SO_2 are also measured in Taiwan under strong heat island conditions (Li-Wei Li, Wan-Li Cheng, 2009), while in Paris the higher concentration of NOx is reported under high UHI intensities (Sarrat et al., 2006).

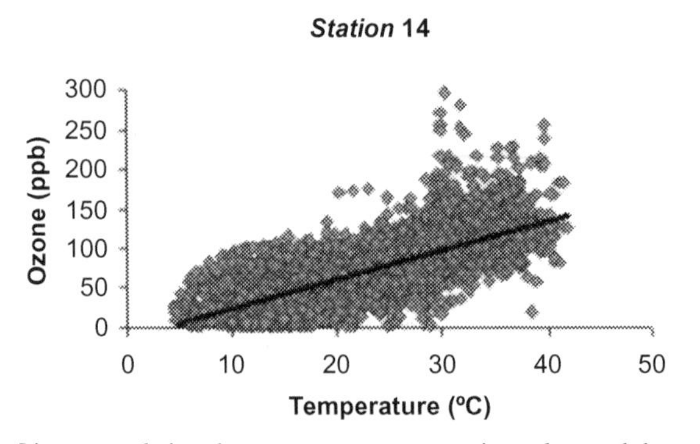

FIGURE 3.21 Linear correlations between ozone concentration values and the ambient air temperature ones in Athens, Greece. Source*: Stathopoulou et al. (2008).*

IMPACT OF LOCAL CLIMATE CHANGE ON HEALTH, MORTALITY AND MORBIDITY

It is widely accepted that exposure to high ambient temperatures may have a significant impact on human health and may increase mortality and morbidity (World Health Organization, 2007; DETR, 2000). Recent research has shown that high ambient temperatures may cause serious cardiovascular and respiratory problems, cerebrovascular disorders, problems of thermoreguala-tion and impaired kidney function, and also decrease the viscosity of blood and increase the risk of thrombosis (Flynn et al., 2005; Linares and Dıaz, 2007).

UHIs intensify health problems in cities. During the 2003 heat wave in France, the excess mortality in Paris surpassed 140%, while in smaller cities it was much lower at close to 40% (Dousset et al., 2010).

Most existing studies agree that hospital admissions related to heat exhaustion and heat stroke increase considerably during heat wave periods and during periods of high urban overheating (Kovats et al., 2004; Pirard et al., 2005). In Melbourne, Australia, hospital admissions of people suffering from acute myocardial infractions increased by 10.8% during days when the ambient temperature exceeded 30°C (Loughnan et al., 2010). When the average temperature of two consecutive days exceeded 27°C, then the corresponding increase of hospital admissions was close to 37.7% (Loughnan et al., 2010).

Higher ambient temperatures have a strong impact on human mortality. The relation between the ambient temperature and mortality rates follows a U shape. Mortality increases considerably above a threshold ambient temperature, which varies from between 22°C to 32°C (Wilkinson et al., 2001). The threshold temperature varies as a function of climate in various geographical areas, specific local parameters, and the physiological characteristics and age of the population (Conti et al., 2005). For example, a recent study carried out in 15 European cities found that the average threshold temperature in southern Europe was 29.4°C, while in northern Europe it was 23.3°C (Baccini et al., 2008). The specific relation between the ambient maximum apparent temperature and mortality in the 15 cities is given in Fig. 3.22.

As reported, the increase of the ambient temperature by 1°C above the critical temperature levels, increased mortality by 3.12% in southern Europe and 1.84% in northern Europe. Other, similar studies have reported a much higher mortality rate above the critical temperature. Hajat et al. (2006) found that the additional heat-related mortality above the threshold temperature was 15.2% in Milan, Italy; 5.5% in London, UK; and 9.3% in Budapest, Hungary. In parallel, Diaz et al. (2002) found that in Spain, the additional heat-related mortality increased by 28.4% above 36.5 K, while Goggins et al. (2002) reported that in Montreal, Canada, the excess mortality above 26°C was 28% in the heat island affected zones of the city, compared to 13% in

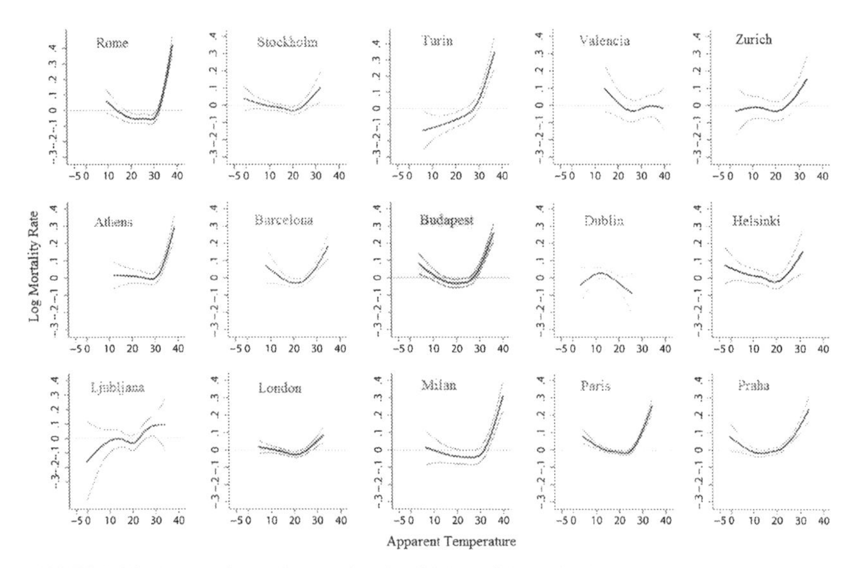

FIGURE 3.22 Regression splines (pointwise 95% confidence bands) describing, on a log scale, maximum apparent temperature (lag 0–3) and natural mortality in 15 European cities. Source: *Baccini et al. (2008).*

the rest of the city. Finally, other studies carried out in several Asian cities, such as Bangkok, Thailand; Delhi, India, and urban areas of Bangladesh and Hong Kong, found that above 29°C the additional mortality increase ranges between 4.1% and 7.5% per degree of temperature rise (Goggings et al., 2012; Chan et al., 2012)

Finally, it is important to mention that urban overheating seriously affects human mental health. Hansen et al. (2008) report that in Australia, hospital admissions for mental and behavior disorders during heat waves increase for ambient temperatures above 27°C.

IMPACT OF LOCAL CLIMATE CHANGE ON THE GLOBAL ECONOMY

It is evident that UHIs have an important impact on the economic life of cities. However, existing studies are limited and the available knowledge and information is incomplete. All studies agree that local climate change has a serious negative impact on a country's Gross National Product (GNP). The magnitude of the damage varies considerably as a function of local conditions. In particular, the Garnaut Review on Climate Change carried out in Australia [Garnaut, 2011] concluded that a 5°C temperature rise by 2030 may reduce the Australian GNP by 1.3%. Another study (Dell et al., 2012) analyzing the impact of climate change on the GNP of developing countries during the latter years of the last century concluded that urban warming and extreme climatic events reduce the GNP of developing countries by 1.3%.

Urban overheating has a serious impact on urban transport systems. AECOM (2012) found that during heat waves the transport system of Melbourne, Australia, faces significant disruptions because of the failure of air conditioning systems. Finally, it is reported that extreme ambient temperatures in cities have a significant impact on social behavior (Cohn, 1990). It is reported that above 32.2°C, the correlation between the ambient temperature and the probability of collective assault and domestic crimes is positive and linear.

IMPACT OF LOCAL CLIMATE CHANGE ON INDOOR AND OUTDOOR COMFORT CONDITIONS

Indoor and outdoor thermal comfort is highly influenced by the ambient temperature. Several international standards define the impact of ambient temperature on thermal comfort (EN, 2007; ANSI/ASHRAE, 2013). Levels of ambient temperatures above a threshold value cause discomfort and heavily impact the physiological conditions of humans. The specific impact of urban warming on indoor thermal comfort has been studied quite extensively (Wright et al., 2005; Summerfield et al., 2007; Sakka et al., 2012; Lomas and Kane, 2013). Most of these studies concentrate on the impact of urban overheating on the thermal conditions occurring in low income houses where the energy required to achieve thermal comfort is not always available. Most of the studies concluded that during heat waves, indoor temperatures considerably exceed temperature thresholds set for health considerations. In particular, measurements performed during a heat wave in nine low income houses in London and Manchester, UK, found that indoor temperatures exceeded 37°C, while high temperatures occurred during the whole summer period (Wright et al., 2005). In a similar study, 268 homes were monitored in Leicester, UK, during the summer of 2009 (Lomas and Kane, 2013). The authors observed that about 88% of the bedrooms and 28% of the living rooms were severely overheated according to national thresholds.

Other measurements performed in fifty low income houses in Athens during a heat wave (Sakka et al., 2012) found that when the ambient temperature was close to 35°C, indoor maximum temperatures reached almost 45°C. In parallel, the average indoor temperature was almost 4.2 K higher than the temperature recorded during the rest of the summer. Spells of more than 216 hours above 30°C were recorded in many of the houses (Fig. 3.23).

The impact of urban warming on outdoor thermal comfort levels is a well-investigated subject. Four main clusters of relevant studies have been identified. The first group of studies aims to identify differences in the spatial distribution of thermal comfort in a city caused by the local uUHI (Krüger et al., 2012; Hedquist and Braze, 2014). The second cluster involves

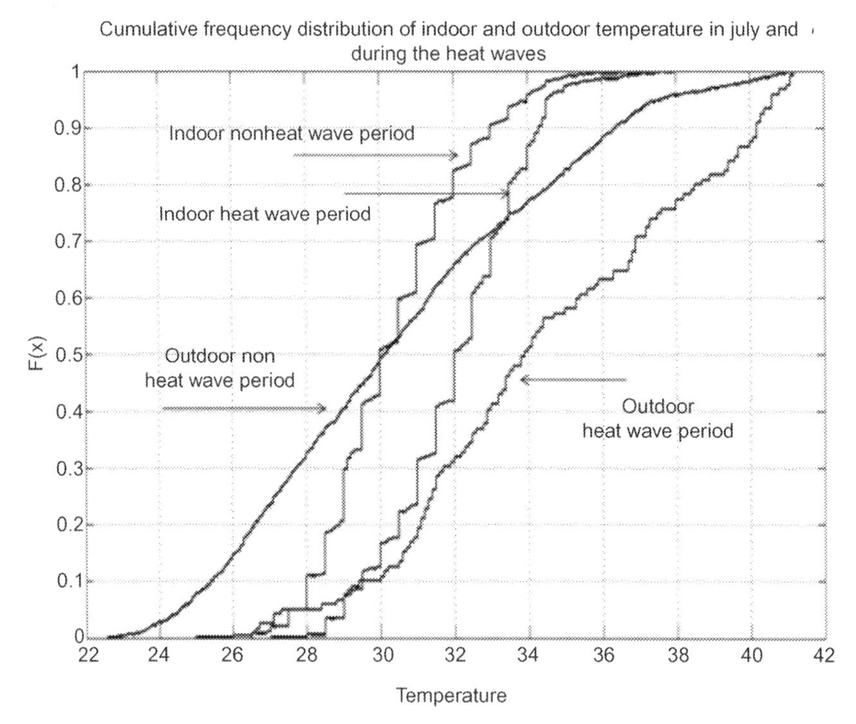

FIGURE 3.23 Cumulative frequency distribution of indoor and outdoor temperatures during the heat wave period and the rest of July. Source: *Sakka et al. (2012).*

studies aiming to investigate the temporal evolution of outdoor thermal comfort in a specific place (Bartzokas et al., 2013), while the third group includes studies analyzing the impact of heat waves on thermal comfort (Papanastasiou et al., 2015). Finally, the fourth cluster includes studies evaluating the potential change of thermal comfort levels because of global climate change.

Almost all of the existing studies assessing the spatial variability of outdoor thermal comfort conditions conclude that the UHI phenomenon significantly affects outdoor thermal comfort levels. Detailed measurements performed in Pheonix, Arizona (USA) show that during periods of high ambient temperatures, the levels of outdoor thermal comfort in the areas where the UHI was strong were quite reduced (Hedquist and Braze, 2014). Experiments carried out in Glasgow, UK (Krüger et al., 2012), found that the UHI was responsible for low thermal comfort levels during the summer period. Similar results are also reported by two experimental studies in Athens, Greece. Giannopoulou et al. (2013) have calculated the humidex thermal comfort index using data from 26 urban and suburban stations. As reported, urban zones suffering from a strong magnitude of UHI presented much lower comfort levels. In Western Athens, where higher ambient temperatures were measured, dangerous discomfort conditions occurred for 27.9% and 19.4% of the total time in July and August respectively. Katavoutas et al. (2015) also reported quite similar conclusions for the city of Athens using the PET thermal comfort index.

The temporal variability of thermal comfort conditions in Athens, Greece, is reported for the period 1954–2012 by Bartzokas et al. (2013). They found that after 1980, the magnitude and frequency of thermal discomfort conditions in the city both increased significantly. Fig. 3.24 shows the temporal variation of the number of summer days presenting a PMV >2 (very warm conditions) under calm and windy conditions. As shown, the frequency of high discomfort days doubled during the period under study. Thermal discomfort conditions occurred between mid-June to mid-September, while before the 1980s discomfort was mainly observed in July and August.

The specific thermal comfort conditions that occur during a period of heat waves are reported for the city of Athens, Greece by Papanastasiou et al. (2015). It is found that conditions of discomfort occurred during the whole period of heat waves, while severe heat stress conditions occurred for almost half of this period. Additionally, for about the 13% of the heat wave period medical emergency conditions were established in the city.

The potential impact of the global climate change on future thermal comfort conditions has been studied by Orosa et al. (2014), Thorsson et al. (2011) and Cheung and Hart (2014). In the city of Gotherborg, Sweden, the expected increase of the ambient temperature by the end of the century is 3.2°C, which may triple the hours in which there are extreme heat stress in the city (Thorsson et al., 2011). Orosa et al. (2014) estimated that the expected ambient temperature increase over the next 20 years may increase the humidex thermal comfort index in Galicia, Spain from 18 to 21. In parallel, Cheung and Hart (2014) report that in the city of Hong Kong, the expected ambient temperature increase caused by global climate change may shift thermal conditions from "no thermal stress" to "moderate heat stress" by 2065.

TARGETS FOR URBAN CLIMATIC RESILIENCE

Climate change has increasing importance for urban governance issues. The endless flux of evidence on the magnitude and characteristics of the urban climate change endorsed existing concerns and knowledge, and deepen our understanding of the colossal impact of local and global climate change on city life. Since the signs of urban overheating have become apparent, several local authorities have mobilized and implemented local mitigation programs aiming to decrease the magnitude of urban overheating. Several years later, the results of the monitoring and global evaluation of these programs have

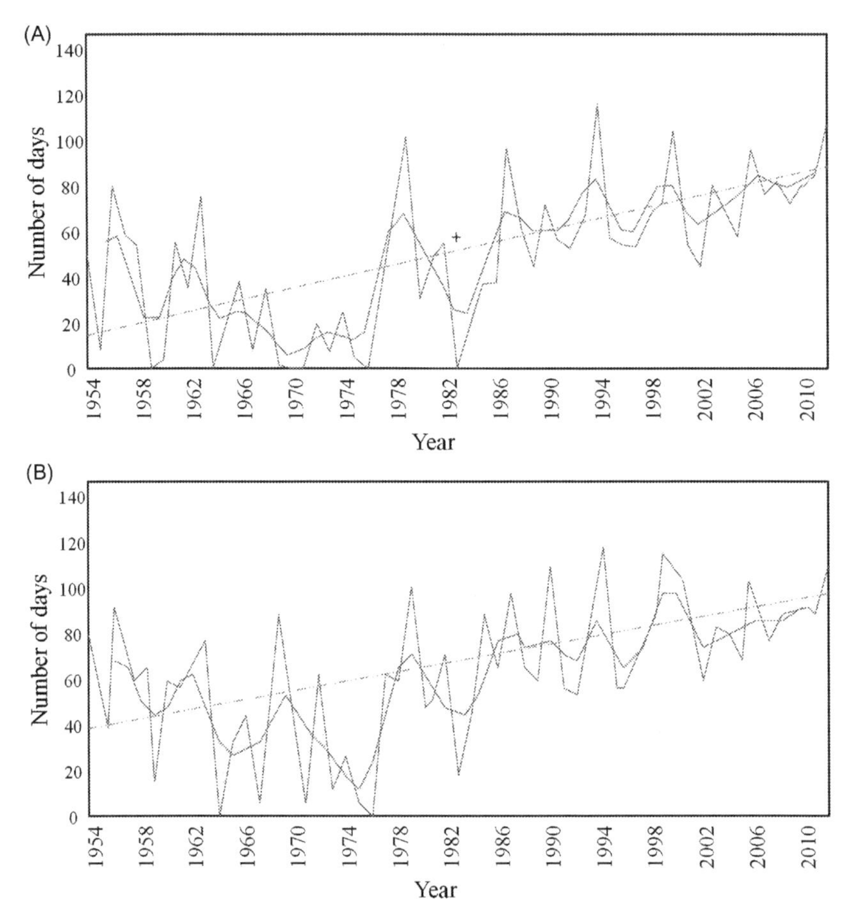

FIGURE 3.24 Inter-annual variation of duration of PMV ≥ 2 summer discomfort in Athens at 14:00 for (A) calm and (B) light wind conditions for the period 1954–2012. Source: *Bartzokas et al. (2013).*

validated the potential of existing technological knowledge and confirmed our capacity to decrease the maximum ambient temperature by up to 2.5°C (Santamouris et al., 2015).

The expected significant increase of the global urban population, combined with the accelerated spread of poverty and urban vulnerability in developing countries, has served to bring the issue of the local climate change to the forefront of the urban agenda. In fact, the world's urban population may increase by 6.4 billion people by 2050 (United Nations, 2014). Most of this increase is expected to happen in Asia and Africa, where cities are not prepared to receive such population growth. As a result of economic incapacity, lack of infrastructure and massive unemployment levels, most of the newcomers will settle in inadequate shelters lacking basic sanitary infrastructures. UN-Habitat, estimates that by 2050 about 2 billion people will live in slums (UN-Habitat, 2014), a tremendous figure that corresponds to about 31.6% of the global population on earth (UN-Habitat, 2014).

Climate change and other environmental disasters may put about 3.1 billion people in conditions of absolute poverty (United Nations Human Development Report, 2013). Even if predictions are eventually proved to be excessive, it is without doubt that over the next few decades some billion people will live under conditions of excessive vulnerability in developing countries.

Although it is reasonable to place emphasis on the future of cities in developing countries, poverty and energy poverty are expected to rise in developed countries as well. Actually, some hundred million of people live under energy poverty conditions in the developed world, possessing a very limited "if any" capacity to cover their basic energy needs (Santamouris, 2016a). Climatic change and urban overheating are expected to place low income populations in these countries under extreme stress and increase their vulnerability.

Actual developments and future projections call for an emergent strategic plan for future cities. Although some of the causes of global climate change lie far beyond urban boundaries, city governance networks and urban stakeholders must forestall global strategic rescue plans and actions to protect the urban population from extreme climatic events and decrease their levels of vulnerability.

REFERENCES

AECOM, Economic assessment of the urban heat island effect, AECOM, prepared for city of melbourne, Australia, 2012.

ANSI/ASHRAE Standard 55–2013, thermal environmental conditions for human occupancy.

Akbari, H., Davis, S., Dorsano, S., Huang, J., Winnett, S., 1992. Cooling our communities: A guidebook on ree planting and light-colored surfacing. Cooling Our Communities: A Guidebook on Tree Planting and Light-Colored Surfacing, Environmental Protection Agency. EPA, USA.

Amato, A.D., Ruth, M., Kirshen, P., Horwitz, J., 2005. Regional energy demand responses to climate change: Methodology and application to the commonwealth of massachusetts. Climatic Change 71, 175–201.

Baccini, M., Biggeri, A., Accetta, G., Kosatsky, T., Katsouyanni, K., Analitis, A., et al., 2008. Heat effects on mortality in 15 European cities. Epidemiology 19 (5), 711–719.

Bartzokas, A., Lolis, C.J., Kassomenos, P.A., McGregor, G.R., 2013. Climatic characteristics of summer human thermal discomfort in Athens and its connection to atmospheric circulation. Nat. Hazard. Earth Syst. Sci. 13, 3271–3279.

Bessec, M., Fouquau, J., 2008. The non-linear link between electricity consumption and temperature in Europe: a threshold panel approach. Energy Economics 30, 2705–2721.

Brázdil, R., Budíková, M., 1999. An urban bias in air temperature fluctuations at the lementinum, Prague, the Czech Republic. Atmospheric Environment 33 (24–25), 4211–4217.

Chan, E.Y.Y., Goggins, W.B., Kim, J.J., Griffiths, S.M., 2012. A study of intracity variation of temperature-related mortality and socioeconomic status among the Chinese population in Hong Kong. J. Epidemiol. Community Health 66 (4), 322–327.

Cheung, C.S.C., Hart, M.A., 2014. Climate change and thermal comfort in Hong Kong. Int. J. Biometeorol. 58 (2), 137–148.

Chow, W.T.L., Roth, M., 2006. Temporal dynamics of the urban heat island of Singapore. Int. J. Climatol. 26, 2243–2260.

Cohn, E.G., 1990. Weather and crime. Br. J. Criminol. 30 (1), 51–63.

Colombo, A.F., Etkin, D., Karney, B.W., 1999. Climate variability and the frequency of extreme temperature events for nine sites across Canada: Implications for power usage. Journal of Climate 12, 2490–2502.

Conti, S., Meli, P., Minelli, G., Solimini, R., Toccaceli, V., Vichi, M., et al., 2005. Epidemiologic study of mortality during the summer 2003 heat wave in Italy. Environ. Res. 98 (3), 390–399.

DETR., English house condition survey 1996: Energy report, department of the environment, transport and the regions, London, 2000.

De Cian, E., Lanzi, E., Roson, R., 2007. The impact of temperature change on energy demand: A dynamic panel analysis, No. I. The Fondazione Eni Enrico Mattei Note di Lavoro series index.

Dell, Melissa, Jones, Benjamin F., Olken, Benjamin A., 2012. Temperature shocks and economic growth: Evidence from the last half century. Am. Econ. J.: Macroecon. 4 (3), 66–95.

Diaz, R., Garcia, F., Velazquez de Castro, E., Hernandez, C., Lopez, A. Otero, 2002. Effects of extremely hot days on people older than 65 years in Seville (Spain) from 1986 to 1997. Int. J. Biometeorol. 46 (3), 145–149.

Doshi, T.K., Fellow, P., Rohatgi, A., Bin Zahur, N., Analyst, E., Hung, Y.K., et al., 2014. Impact of climate change on electricity demand of Singapore 1–15. Retrieved from https://editorialexpress.com/cgi-bin/conference/download.cgi?dbname = SERC2013&paper id = 405.

Dousset, B., Gourmelon, F., Laaidi, K., Zeghnoun, A., Giraudet, E., Bretin, P., et al., 2010. Satellite monitoring of summer heat waves in theParis metropolitan area. Int. J. Climatol. 31 (2), 313–323.

EN 15251 Standard 2007, Indoor environmental input parameters for design and assessment of energy performance of buildings addressing indoor air quality, thermal environment, lighting and acoustics.

Erell, E., Williamson, T., 2007. Intra-urban differences in canopy layer air temperature at a mid-latitude city. Int. J. Climatol. 27, 1243–1255.

Flynn, A., McGreevy, C., Mulkerrin, E.C., 2005. Why do older patients die in a heatwave? [Commentary]. QJM 98, 227–229.

Founda, D., & Santamouris, M. Does the urban heat island increases during heat waves? Submited for publication, 2017.

Founda, D., Pierros, F., Petrakis, M., Zerefos, C., 2015. Inter-decadal variations and trends of the Urban Heat Island in Athens (Greece) and its response to heat waves. Atmos. Res. 161–162, 1–13. Available from: https://doi.org/10.1016/j.atmosres.2015.030.016.

Franco, G., Sanstad, A.H., 2007. Climate change and electricity demand in California. Climatic Change 87, 5883–5891.

Freitas, E., Christopher, M. Rozoff, William, R. Cotton, Silva Dias, Pedro L., 2007. Interactions of an urban heat island and sea-breeze circulations during winter over the metropolitan area of São Paulo, Brazil. Bound. Layer Meteorol 45.

Fung, W.Y., Lam, K.S., Hung, W.T., Pang, S.W., Lee, Y.L., 2006. Impact of urban temperature on energy consumption of Hong Kong. Energy 31, 2287–2301.

Garnaut, R., 2011. The Garnaut Climate Change Review—Update 2011: Australia in the Global Response to Climate Change Summary, Garnaut Review. ACT, Canberra.

Giannakopoulos, C., Psiloglou, B.E., 2006. Trends in energy load demand for Athens, Greece: Weather and non-weather related factors. Climate Research 31 (1), 97–108.

Giannopoulou, K., Livada, I., Santamouris, M., Saliari, M., Assimakopoulos, M., Caouris, Y., 2013. The influence of air temperature and humidity on human thermal comfort over the greater Athens area. Sustainable Cities Soc. 10, 184–194.

Goggins, W.B., Chan, E.Y.Y., Ng, E., Ren, C., Chen, L., 2012. Effect modification of the association between short-term meteorological factors and mortality by urban heat islands in Hong Kong. PLoS ONE 7 (6), pe38551.

Hajat, S., Armstrong, Ben, Baccini, Michela, Biggeri, Annibale, Luigi Bisanti, Russo, Antonio, et al., 2006. Bettina Menne, Tom Kosatsky, Impact of high temperatures on mortality is there an added heat wave effect? Epidemiology 17 (6), 632–638.

Hansen, A., Bi, P., Nitschke, M., Ryan, P., Pisaniello, D., Tucker, G., 2008. The effect of heat waves on mental health in a temperate Australian city. Environ. Health Perspect. 116(10), 1369–1375.

Hayhoe, K., Robson, M., Rogula, J., Auffhammer, M., Miller, N., VanDorn, J., et al., 2010. An integrated framework for quantifying and valuing climate change impacts on urban energy and infrastructure: A Chicago case study. Journal of Great Lakes Research 36, 94–105.

Hedquist, Brent C., Braze, Anthony J., 2014. Seasonal variability of temperatures and outdoor human comfort in Phoenix, Arizona, U.S.A. Build. Environ. 72, 377–388.

Howard, L., 1833. 'The climate of London', Vols. I-III, London.

Katavoutas, G., Georgiou, G.K., Asimakopoulos, D.N., 2015. Studying the urban thermal environment under a human-biometeorological point of view: The case of a large coastal metropolitan city, Athens. Atmos. Res. 152, 82–92.

Khan, S.M., Simpson, R.W., 2001. Effect of a heat island on the meteorology of complex urban airshed. Bound. Layer Meteorol. 100, 487–506.

Kim, Y., Baik, Jong-Jin, 2002. Maximum urban heat island intensity in Seoul. J. Appl. Meteorol. 41, 102.

Kovats, R.S., Hajat, S., Wilkinson, P., 2004. Contrasting patterns of mortality and hospital admissions during heat waves in London, UK. Occup. Environ. Med. 61(11) (2004), 893–898.

Krüger, Eduardo, Drach, Patricia, Emmanuel, Rohinton, Corbella, Oscar, 2012. Urban heat island and differences in outdoor comfort levels in Glasgow, UK. Theor. Appl. Climatol. Available from: https://doi.org/10.1007/s00704-012-0724-9.

Kubota, T., Ossen, Dilshan Remaz, 2014. Spatial characteristics of urban heat island in Johor Bahru city, Malaysia. Available from: <http://www.shibaura-it.ac.jp/about/hybrid_twinning/pdf/01_arch_02.pdf>.

Lai, Li-Wei, Cheng, Wan-Li, 2009. Air quality influences by urban heat island coupled with synoptic weather patterns. Sci. Total Environ. 407, 2724–2732.

Li, D., Bou-Zeid, E., 2013. Synergistic interactions between urban heat islands and heat waves: The impact in cities is larger than the sum of its parts. Journal of Applied Meteorology and Climatology 52, 2052.

Li, D., Sun, T., Liu, M., Yang, L., Wang, L., Gao, Z., 2015. Contrasting responses of urban and rural surface energy budgets to heat waves explain synergies between urban heat islands and heat waves. Environ. Res. Lett. 10, 054009.

Linares, C., Dıaz, J., 2007. Impact of high temperatures on hospital admissions: Comparative analysis with previous studies about mortality (Madrid). Eur. J. Public Health 18 (3), 317–322.

Lomas, K.J., Kane, T., 2013. Summertime temperatures and thermal comfort in UK homes. Build. Res. Inf. 41 (3), 259–280.

Loughnan, M.E., Neville, N., Tapper, N.J., 2010. The effect of summer temperature, age and socioeconomic circumstance on Acute Myocardial Infarction admissions in Melbourne, Australia. Int. J. Health Geogr. 9 (41).

McCarthy, M.P., Best, M.J., Betts, R.A., 2010. Climate change in cities due to global warming and urban effects. Geophys. Res. Lett. 37, L09705. Available from: https://doi.org/10.1029/2010GL042845.

Mihalakakou, P., Flocas, H.A., Santamouris, M., Helmis, C.G., 2002. Application of neural networks to the simulation of the heat island over Athens, Greece, using synoptic types as a predictor. J. Appl. Meteorol. 41, 5,519–5,527.

Mirasgedis, S., Sarafidis, Y., Georgopoulou, E., Lalas, D.P., Moschovits, M., Karagiannis, F., et al., 2006. Models for mid-term electricity demand forecasting incorporating weather influences. Energy 31, 208–227.

Mohan, M., Yukihiro, Kikegawa, Gurjar, B.R., Bhati, Shweta, Anurag, Kandya, Koichi, Ogawa, 2012. Urban heat island assessment for a tropical urban Air shed in India. Atmos. Clim. Sci. 2, 127–138.

Morris, C.J.G., Simmonds, I., 2000. Associations between varying magnitudes of the urban heat island and the synoptic climatology in Melbourne Australia. Int. J. Climatol. 20, 1931–1954.

Morris, C.J.G., Simmonds, I., Plummer, N., 2001. Quantification of the influences of wind and cloud on the nocturnal urban heat island of a large city. J. Appl. Meteorol. 171.

Oke, T.R., 1997. Urban climates and global change. In: Perry, A., Thompson, R. (Eds.), Applied Climatology: Principles and Practice. Routledge, London, pp. 273–287.

Oke, T.R., Johnson, G.T., Steyn, D.G., Watson, I.D., 1991. Simulation of surface urban heat islands under 'ideal' conditions at night - Part 2: Diagnosis and causation. Boundary Layer Meteorology 56, 339–358.

Orosa, José A., Costa, Ángel M., Rodríguez-Fernández, Ángel, Roshan, Gholamreza, 2014. Effect of climate change on outdoor thermal comfort in humid climates. J. Environ. Health Sci. Eng. 12 (1), 46. Available from: https://doi.org/10.1186/2052-336X-12-46.

Papanastasiou, D.K., Melas, D., Kambezidis, H.D., 2015. Air quality and thermal comfort levels under extreme hot weather. Atmos. Res. 152, 4–13.

Papanikolaou, N.M., Livada, I., Santamouris, M., Niachou, K, 2008. The influence of wind speed on heat island phenomenon in Athens, Greece. Int. J. Vent. 6, 4.

Pardo, A., Meneu, V., Valor, E., 2002. Temperature and seasonality influences on Spanish electricity load. Energy Economics 24 (1), 55–70.

Park, H.C., 1986. Features of the heat island in Seoul and its surrounding cities. Atmos. Environ. 20, 1859–1866.

Parkpoom, S.J., Harrison, G.P., 2008. Analyzing the impact of climate change on future electricity demand in Thailand. IEEE Transactions on Power Systems 23, 1441–1448.

Pirard, P., Vandentorren, S., Pascal, M., et al., 2005. Summary of the mortality impact assessment of the 2003 heat wave in France. Euro Surveill. 10 (7), 153–156.

Rosenfeld, A.H., Akbari, H., Bretz, S., Fishman, B.L., Kurn, D.M., Sailor, D., et al., 1995. Mitigation of urban heat islands: Materials, utility programs, updates. Energy and Buildings 22 (3), 255–265.

Ruth, M., Lin, A.C., 2006. Regional energy demand and adaptations to climate change: Methodology and application to the state of Maryland, USA. Energy Policy 34, 2820–2833.

Sailor, D.J., Munoz, J.R., 1997. Sensitivity of electricity and natural gas consumption to climate in the U.S.A.—methodology and results for eight states. Energy 22, 987–998.

Sakka, A., Santamouris, M., Livada, I., Nicol, F., Wilson, M., 2012. On the thermal performance of low income housing during heat waves. Energy Build. 49, 69–77.

Santamouris, M., 2014. On the energy impact of urban heat island and global warming on buildings. Energy and Buildings 82, 100–113.

Santamouris, M., 2015a. Analyzing the heat island magnitude and characteristics in one hundred Asian and Australian cities and regions. Science of the Total Environment 512– 513, 582–598.

Santamouris, M., 2015b. Regulating the damaged thermostat of the cities – status, impacts and mitigation strategies. Energy and Buildings, Energy and Buildings 91, 43–56.

Santamouris, M., 2016a. Innovating to zero the building sector in Europe: Minimising the energy consumption, eradication of the energy poverty and mitigating the local climate change. Solar Energy 128, 61–94.

Santamouris, M., 2016b. Cooling of buildings. Past, present and future. Energy and Buildings 128, 617–638.

Santamouris, M., Paraponiaris, K., Mihalakakou, G., 2007. Estimating the ecological footprint of the heat island effect over Athens, Greece. Clim. Change 80, 265–276.

Santamouris, M., Cartalis, C., Synnefa, A., Kolokotsa, D., 2015. On the impact of urban heat island and global warming on the power demand and electricity consumption of buildings–A review. Energy and Buildings 98, 119–124. Available from: https://doi.org/10.1016/j.enbuild.2014.09.052.

Santamouris, M., Ding, L., Fiorito, F., Oldfield, P., Paul Osmond, Paolini, R., et al., December 2016. Passive and active cooling for the outdoor built environment – Analysis and assessment of the cooling potential of mitigation technologies using performance data from 220 large scale projects. Solar Energy In Press, Corrected Proof, Available online 19.

Sarrat, C., Lemonsu, A., Masson, V., Guedalia, D., 2006. Impact of urban heat island on regional atmospheric pollution. Atmos. Environ. 40, 1743–1758.

Segal, M., Shafir, H., Mandel, M., Alpert, P., Balmor, Y., 1992. Climatic-related evaluations of the summer peak-hours' electric load in Israel. Journal of Applied Meteorology 31, 1492–1498.

Stathopoulou, E., Mihalakakou, G., Santamouris, M., Bagiorgas, H.S., 2008. Impact of temperature on tropospheric ozone concentration levels in urban environments. J. Earth Syst. Sci. 117 (3), 227–236.

Summerfield, A.J., Lowe, R.J., Bruhns, H.R., Caeiro, J.A., Steadman, J.P., Oreszczyn, T., 2007. Milton Keynes energy park revisited: Changes in internal temperatures and energy usage. Energy Build. 39, 783–791.

Thorsson, S., Lindberg, F., Björklund, J., Holmer, B., Rayner, D., 2011. Potential changes in outdoor thermal comfort conditions in Gothenburg, Sweden due to climate change: The influence of urban geometry. Int. J. Climatol. 31 (2), 324–335.

United Nations, 2014. World urbanisation projects.

United Nations Habitat, Annual report, 2014.

United Nations, Human Development Report, 2013.

Unwin, D.J., 1980. The synoptic climatology of Birmingham's urban heat Island, 1965–1974. Weather 35, 43–50.

Wangpattarapong, K., Maneewan, S., Ketjoy, N., Rakwichian, W., 2008. The impacts of climatic and economic factors on residential electricity consumption of Bangkok Metropolis. Energy and Buildings 40 (Jan. (8)), 1419–1425.

Wilkinson, P., Armstrong, B., Fletcher, T., Landon, M., Mckee, M., Pattenden, S., et al., 2001. Cold Comfort: The Social and Environmental Determinants of Excess Winter Deaths in England. The Policy Press, Bristol, pp. 1986–1996.

Wong, S.L., Wan, K.K.W., Li, D.H.W., Lam, J.C., 2010. Impact of climate change on residential building envelope cooling loads in subtropical climates. Energy and Buildings 42 (Nov. (11)), 2098–2103.

World Health Organisation, 2007. Large Analysis and Review of European Housing and Health Status (LARES). WHO Regional Office for Europe, DK-2100, Copenhagen, Denmark.

Wright, A.J., Young, A.N., Natarajan, S., 2005. Dwelling temperatures and comfort during the August 2003 heat wave. Build. Serv. Eng. Res. Technol. 26 (4), 285–300.

Yabe, Evaluation of energy saving effect for the long-term maximum power forecast (title only in original language), in: Proceedings of national convention of the Institute of Electrical Engineers of Japan (IEEJ), 2005.

Yoshikado, Hiroshi, Tsuchida, Makoto, 2009. High levels of winter air pollution under the influence of the urban heat island a long the shore of Tokyo Bay. J. Appl. Meteorol. Climatol. 35, 1804.

FURTHER READING

Tan, J., Zheng, Y., Tang, X., Guo, C., Li, L., Song, G., et al., 2010. The urban heat island and its impact on heat waves and human health in Shanghai. Int. J. Biometeorol. 54 (1), 75–84. Available from: https://doi.org/10.1007/s00484-009-0256-x.

Chapter 4

Energy Poverty and Urban Vulnerability

DEFINITIONS OF ENERGY POVERTY AND ACTUAL STATUS IN THE WORLD

According to Wikipedia (2018), "energy poverty" is defined as

> *The lack of access to modern energy services. It refers to the situation of large numbers of people in developing countries and some people in developed countries whose well-being is negatively affected by very low consumption of energy, use of dirty or polluting fuels, and excessive time spent collecting fuel to meet basic needs.*

However, energy poverty is not universally defined. Several organizations and states use numerous definitions and approaches to define and measure it. According to Barnes (2018), there are four different approaches to defining energy poverty. The first approach is based on the definition of a "minimum amount of physical energy necessary for basic needs such as cooking and lighting." However, it is quite difficult, if not impossible, to define in an exact way the required energy for cooking or for heating, given the serious differences between families and countries.

The second approach is based on the "type and amount of energy that is used for those at the poverty line." Such an approach allows a much easier measurement and monitoring of the magnitude of energy poverty. Several countries have defined thresholds over which households are considered as "energy poor." According to Barnes (2018), such an approach "is not so useful for those that might want to track the impact of energy sector reform," while "Tracking energy poverty with this method would be no more than tracking general poverty trends."

The third approach is based on a pre-defined threshold percentage of income that households spend on energy. Above this threshold, households are considered to be "energy poor." The threshold varies from between 5%–20%, depending on the local conditions, but it is widely accepted that above 10%, households seems to face serious problems in satisfying their energy needs. It is evident that the selection of an income percentage is quite

Minimizing Energy Consumption, Energy Poverty and Global and Local Climate Change in the Built Environment: Innovating to Zero. DOI: https://doi.org/10.1016/B978-0-12-811417-9.00004-0

arbitrary, and such an approach may not reflect the real dimensions and characteristics of energy poverty in a country.

The fourth approach is based on "The income point below which energy use and or expenditures remains the same, implying this is the bare minimum energy needs." Such a definition considers the magnitude of the required energy as it relates to family income. The approach considers as threshold a level where energy consumption does not increasing considerably as a function of increased income. Although such an approach requires the input of considerable data to quantify the magnitude of energy poverty, its major advantage is that it takes into account the ways that people consume energy under the specific conditions of a country.

Apart from the magnitude of the consumed energy and its relation to household income, several other indicators are used to evaluate, monitor and report the conditions of energy poverty. According to Eurostat (2018), all parameters defined in Table 4.1 are used in a statistical way to quantify levels of energy poverty and its characteristics, in terms of percentage of persons in the total population.

TABLE 4.1 Indicators Used to Report the Magnitude and the Characteristics of Energy Poverty

- The **inability to keep the home adequately warm**
- The **inability to afford paying for one week annual holiday away from home**
- The **inability to afford a meal with meat, chicken, fish (or vegetarian equivalent) every second day**
- The **inability to face unexpected financial expenses**
- The indicator on **arrears (mortgage or rent, utility bills or hire purchase),** expressing the enforced inability to pay their mortgage or rent, utility bills or hire purchase on time due to financial difficulties
- The indicator on **arrears on mortgage or rent payments**, expressing the enforced inability to pay their mortgage or rent on time due to financial difficulties.
- The indicator on **arrears on utility bills**, expressing the enforced inability to pay their utility bills on time due to financial difficulties
- The indicator **on arrears on hire purchase instalments or other loan payments** expressing the enforced inability to pay for hire purchase instalments or other loan payments on time due to financial difficulties
- The indicator on **inability to make ends meet**, based on the following groups of the subjective non-monetary indicator defining the ability to make ends meet:
  - Households making ends meet with great difficulty (EM_GD)
  - Households making ends meet with difficulty (EM_D)
  - Households making ends meet with some difficulty (EM_SD)
  - Households making ends meet fairly easily (EM_FE)
  - Households making ends meet easily (EM_E)
  - Households making ends meet very easily (EM_VE)

Source: Eurostat, 2018.

The total number of energy poor in the world cannot be defined in a precise way given the lack of data and the fuzziness of the definition. According to IEA (2017), it is estimated that almost 200 million people in developed countries, or the 15% of the total global population, suffer from energy poverty. Fig. 4.1 shows the estimated percentage of energy poverty in most of the countries in the so-called "developed world."

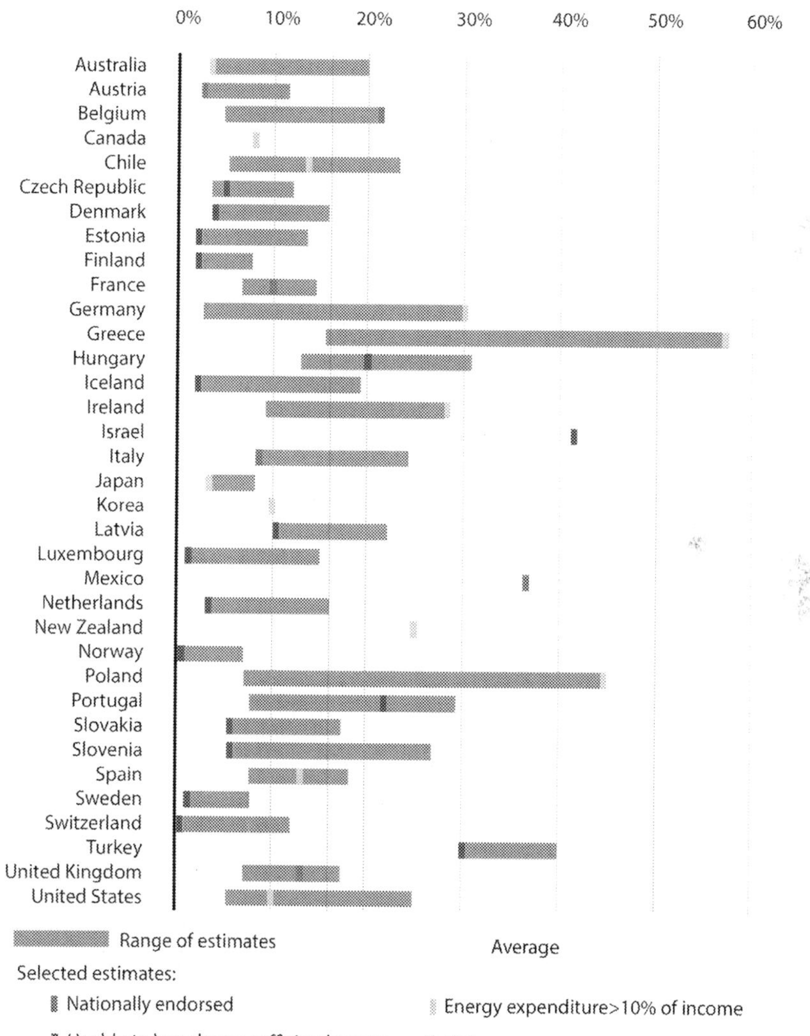

FIGURE 4.1 Levels of energy poverty in various countries in the developed world. Source: *Based on IEA data from World Energy Outlook 2017 © OECD/IEA 2017 http://www.worldenergyoutlook.org/weo2017/ Licence: www.iea.org/t&c.*

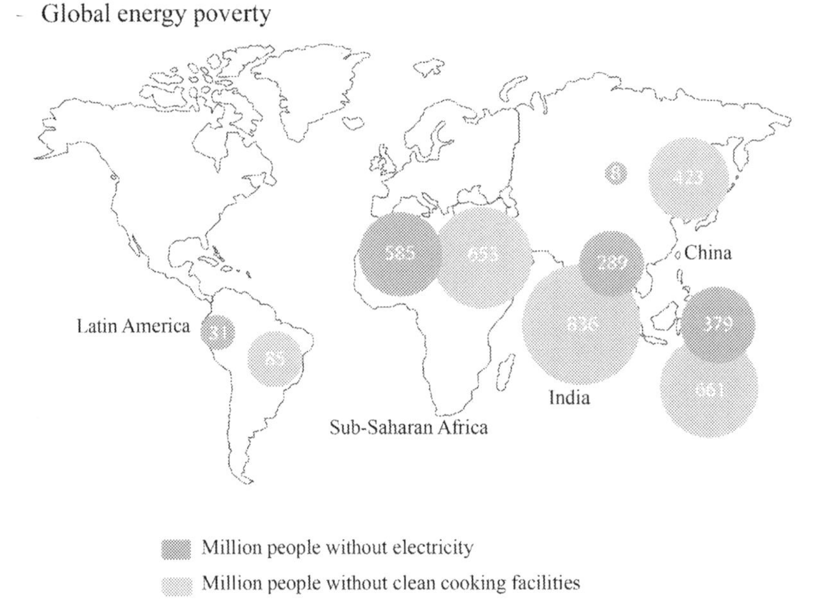

FIGURE 4.2 Million of people live in developing countries without electricity or clean cooking facilities. Source: *Based on IEA data from World Energy Outlook 2017 © OECD/IEA 2017 http://www.worldenergyoutlook.org/weo2017/ Licence: www.iea.org/t&c.*

As shown, the maximum levels of energy poverty are present in Greece, Poland, Turkey, Hungary and Germany. However, as discussed below, the specific levels of energy poverty may change when different definitions of energy poverty are used.

As concerns the so-called "developing world," estimations by the International Energy Agency (IEA) (2011) suggest that the total number of people without electricity exceeds 1.3 billion, while there are about 2.5 billion people without clean cooking facilities (Fig. 4.2).

ENERGY POVERTY IN THE DEVELOPED WORLD – FACTS AND STATISTICS

Energy poverty in the developed world is an important problem. Although there is a lack of exact data on the number of energy poor in developed countries, it is estimated to exceeds 200 million people. Most of the energy poor are in Europe and the United States.

In the United States of America, there are approximately 48 million people living below the poverty line who potentially suffer from energy poverty (Gridmate 2018). According to the United States Census Bureau (2014), national authorities and programs such as Low Income Home Energy Assistance Program (LIHEAP) provided energy assistance to about

6.9 million households in 2014. According to the Climate Change Challenge Organisation (2018), there are about 15.9 million people who are energy poor in the United States. Energy poverty affects a significant percentage of people in specific states of the United States. According to the United States Census, about 4.6 million people in Texas live below the poverty line and many of them face serious problems to cover their energy needs. Most of these households require energy assistance from the state: in a year period, about 182,505 households received assistance from LIHEAP. According to the Consumers Energy Alliance (2017), more than 900,000 homes were disconnected from the electricity grid in Texas in a single summer period because of unpaid utility bills. This figure is almost three times higher than the corresponding figure from a decade ago. In California, about 714,000 homes were disconnected from the electricity grid in a single one year period as the average price for residential power rose by 18% between 2007 and 2016.

A high percentage of the energy poor in the United States (36%) have an income higher than the existing Federal Poverty Guideline, an additional 5% are considered to be ineligible to receive benefits from the Low Income Home Energy Assistance program (Climate Change Challenge, 2018). Statistical data show that the median income level of the energy poor in the USA was USD$4330, while5% had income above USD$15,000. The variation in the energy expenditure between the energy poor and the higher income population groups is given in Table 4.2 (Climate Change Challenge, 2018). The energy poor presented a median energy expenditure close to

TABLE 4.2 Average Energy Units Consumed Per Square Foot of Living Space in the United States

Average Units Consumed Per Square Foot of Living Space		
Households In Fuel Poverty	Households Not Low-Income	Type of Home
101	82	Mobile home
73	51	Single-family home detached
71	48	Single-family home attached
86	65	2–4 Unit apartment
71	55	5 or More unit apartment
77	54	U.S. average

Source: http://www.climatechangechallenge.org/Resource%20Centre/Fuel-Poverty/USA_Fuel_Poverty.htm.

USD$975, while 25%, spent more than USD$1330. Additionally, 65% of the energy poor were 65 years old or older.

According to official European statistical data (Dol and Haffner, 2010), the percentage of people living in low income households in European countries varies from 10% to 25% (Fig. 4.3). The majority of this low income population exists in countries with the lowest national GDPs in Europe, such as Bulgaria and Romania. Fewer people in this low income population live in Europe's wealthier countries.

Specific information on the percentage of people living in European countries who are unable to keep their home warm is provided by European statistical services (Thomson and Snell, 2013). As shown in Fig. 4.4, the European average is close to 12%, while in some countries, such as Portugal, Bulgaria and Cyprus, the percentage of those unable to keep their home warm may exceed 30%. Fig. 4.5 shows the percentage of households in arrears on utility bills in the last 12 months in various European countries (Thompson and Snell, 2013). While the average percentage in Europe who are unable to keep their home warm is between 7% and 8%, in some countries, such as Bulgaria and Romania, this number may exceed 20% or even 30%.

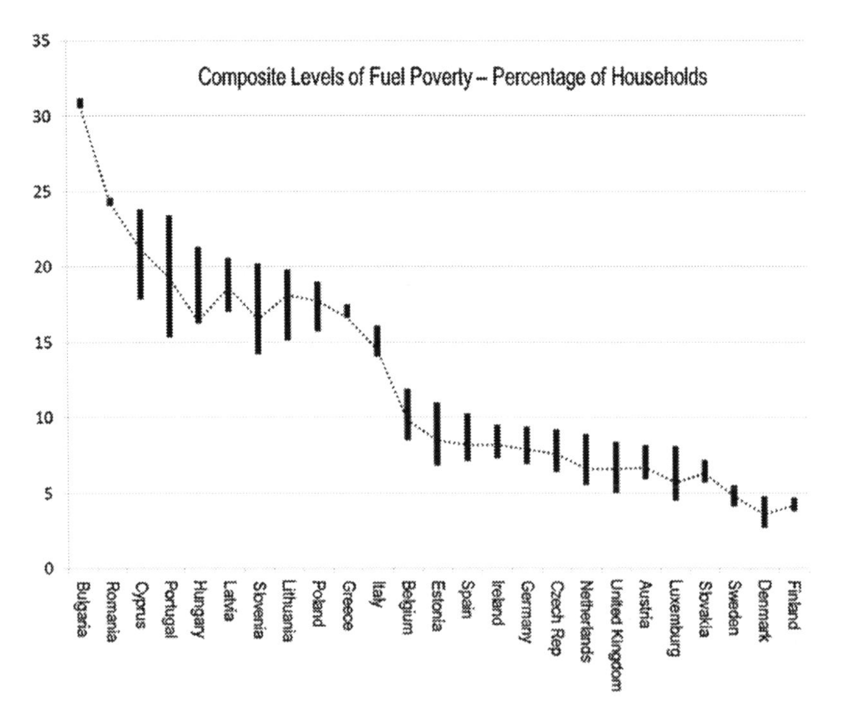

FIGURE 4.3 Calculated levels of fuel poverty in Europe for 2010. Source: *Santamouris and Kolokotsa (2015).*

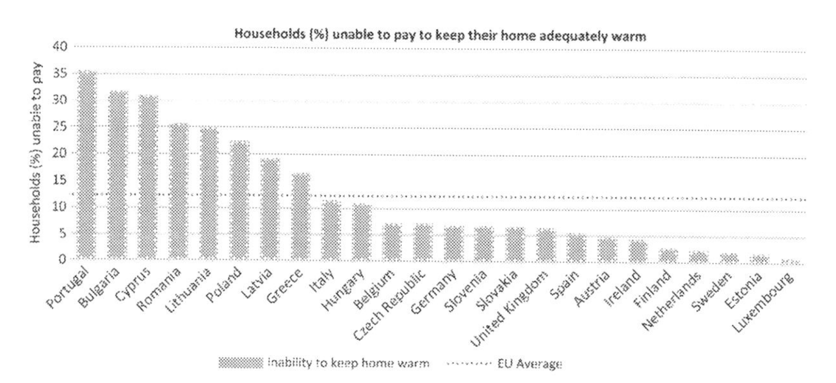

FIGURE 4.4 Households unable to pay to keep their home adequately warm. Source: *Thomson and Snell (2013).*

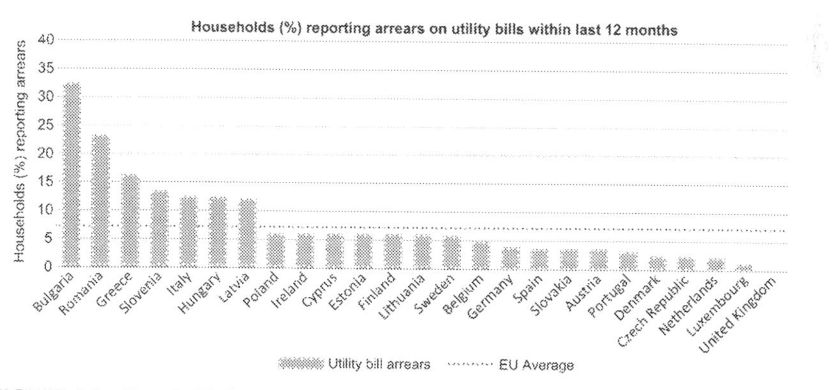

FIGURE 4.5 Households in arrears on utility bills in the last 12 months by country. Source: *Thomson and Snell (2013).*

To provide a more concrete and integrated way to estimate energy poverty in Europe, several indicators may be combined to create a common indicator. Thompson and Snell (2013), proposed to combine three specific indicators into an integrated index to account for energy poverty:

1. The percentage of households not able to keep their home adequately warm.
2. The percentage of households with arrears in utility bills over the last 12 months.
3. The percentage of households living in houses with a leaking roof, damp walls/floors/foundations.

Based on the three specific indicators, four scenarios have been developed (Table 4.3). In the first scenario, the indicator related to the ability to keep homes adequately warm is assigned a weight of 0.5, while the

TABLE 4.3 Composite Levels of Fuel Poverty Across EU25 (% of Households)

Country	Scenario 1	Scenario 2	Scenario 3	Scenario 4
Austria	6.2	5.9	8.2	6.7
Belgium	9.2	8.5	11.9	9.8
Bulgaria	31.1	31.2	30.5	30.6
Cyprus	23.8	17.8	22.8	21.2
Czech Republic	7.5	6.4	9.2	7.6
Denmark	2.7	3.3	4.8	3.6
Estonia	6.8	8.1	11	8.5
Finland	3.8	4.7	4.3	4.2
Germany	7.6	6.9	9.4	7.9
Greece	16.8	16.6	17.5	16.8
Hungary	16.2	16.4	21.3	17.8
Ireland	7.3	8.2	9.5	8.2
Italy	13.8	14	16.1	14.5
Latvia	18.9	17	20.6	18.6
Lithuania	19.9	15.1	19.8	18.1
Luxembourg	4.5	4.6	8.1	5.7
Netherlands	5.5	5.5	8.9	6.6
Poland	19	15.7	19	17.7
Portugal	23.4	15.3	19.5	19.2
Romania	24.6	24	24.4	24.1
Slovakia	6.4	5.7	7.2	6.3
Slovenia	14.2	15.7	20.2	16.5
Spain	7.6	7.1	10.3	8.2
Sweden	4.1	5	5.5	4.8
United Kingdom	6.5	5	8.4	6.6

Source: Thompson and Snell (2013).

remaining two indicators have a weight of 0.25 respectively. In the next scenario, the second indicator, referring to energy arrears, is assigned a weight equal to 0.5, while the other two indicators have a weight of 0.25. In the third scenario, the third indicator, referring to houses with damp and so on, has a weight equal to 0.5 and the two other indicators is at 0.25. Finally, in the last scenario, all three indicators are assigned an equal weight.

Despite the differences in the absolute predicted value, the four proposed combinations reflect the same image in quantitative terms. It is evident that countries with a low national GDP present at the same time a high percentage of energy poverty, while in the richest countries, energy poverty is quite limited.

ENERGY POVERTY CHARACTERISTICS IN THE DEVELOPING WORLD

Energy poverty is a serious problem in developing countries. Energy is a significant engine to fight poverty and improve life quality in the developing and poor zones of the planet. According to the World Bank (Albouy and Nadifi, 1999), by 2020 about 70% of the world's population will live in cities and almost 60% of them will live below poverty level. To fight against poverty and improve the living conditions of vulnerable populations, additional energy must be spent to satisfy the basic needs of people and to promote and improve their quality of life. According to the IEA (2010), there are about 1.4 billion people in the world who lack access to electricity. Almost 85% of them live in rural areas. To satisfy demand, thousand of new megawatts of electricity infrastructure must be installed, especially in the next few decades. The cost of the required additional power capacity is estimated to be very high. Albouy et al. estimate that the cost of generating this new power will exceed USD$2 trillion of dollars over the next 30 years. The IEA (2010) estimates that achieving universal access to modern energy services by 2030 will require almost USD$756 billion of additional investments, or USD$36 billion per year, for the period 2010–2030.

In recent years, important progress has been made to improve infrastructure and provide electricity in less developed countries (Energypedia, 2018). Fig. 4.6 provides information on the evolution of the number of people in less developed countries who are already connected to the electricity grid. As stated by the IEA," Without additional dedicated policies, by 2030 the number of people drops, but only to 1.2 billion. Some 15% of the world's population will still lack access, the majority in sub-Saharan Africa"

Although there is a serious shortage of electricity capacity in the developing world, poor countries pay more than average for energy. According to Serageldim and Brown (1995), citizens of poor countries pay almost 12% of their income to buy energy services. This is almost five times higher than in the OECD countries, and it is evident that it places a heavy burden on the citizens of less developed countries.

In less developed countries, the ability to connect to the grid is a strong function of family income. As shown in Figs. 4.7 and 4.8, the higher the

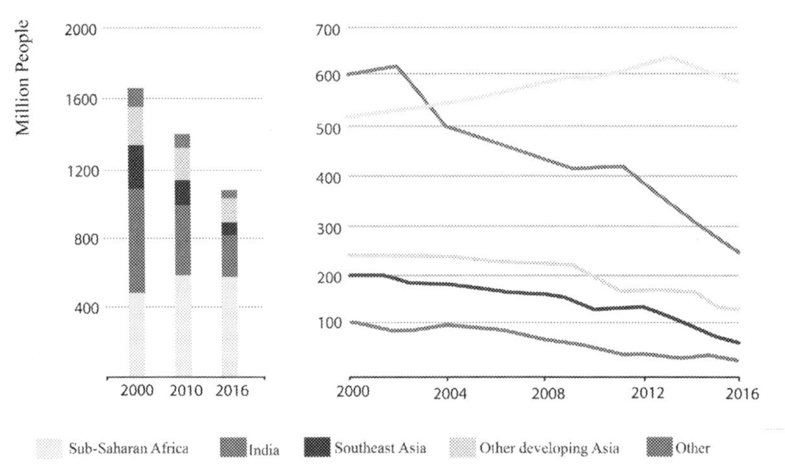

Progress on electricity access is being made in all parts of the world, led by developing countries in Asia, in particular India

FIGURE 4.6 Growth in the number of people being connected to the electricity grid in less developed countries. Source: *Energypedia (2018).*

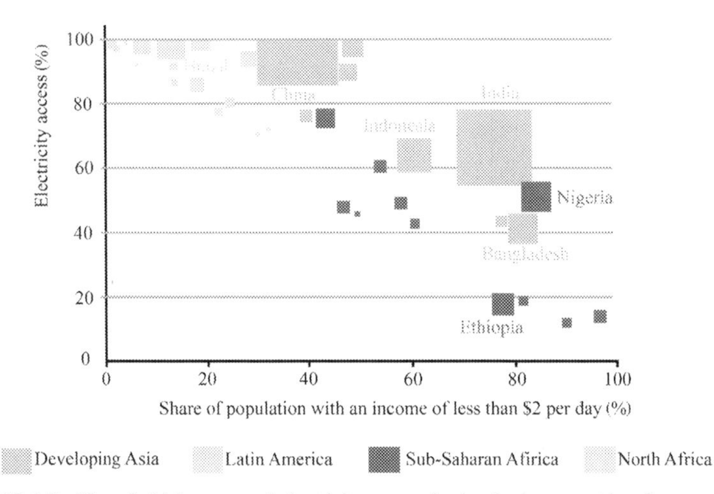

FIGURE 4.7 Household income and electricity access in developing countries. Source: *Based on IEA data from World Energy Outlook 2010 © OECD/IEA 2017 http://www.worldenergyoutlook.org/weo2010/ Licence: www.iea.org/t&c.*

family income in a country, the higher the percentage of the population connected to the grid.

Energy imports are one of the more important sources of foreign economic debt in developing countries and constitute a significant economic factor in increasing poverty. As reported by Birol (2002), "in over 30 countries, energy imports exceed 10% of the value of all exports," while "in about 20 countries, payments for oil imports exceed those for debt servicing."

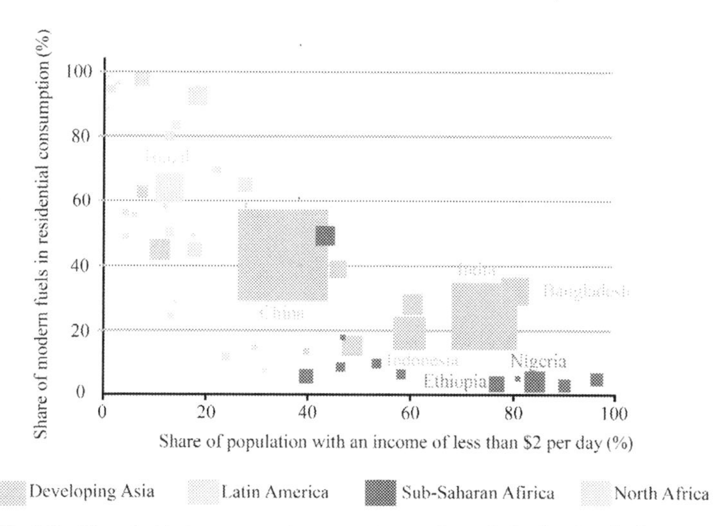

FIGURE 4.8 Household income and access to modern fuels in developing countries. Source: *Based on IEA data from World Energy Outlook 2010 © OECD/IEA 2017 http://www.worldenergyoutlook.org/weo2010/ Licence: www.iea.org/t&c.*

Developing countries rely heavily on the use of inappropriate and polluting fuels for heating, cooking and even lighting (Table 4.4) (IEA, 2010). IEA data indicate that the number of people using biomass is expected to rise from 2.7 billion in 2010 to 2.8 billion in 2030 (IEA, 2010).

Burning inappropriate fuels usually tremendously increases the concentration of harmful indoor pollutants and creates serious indoor air quality problems. According to the estimations of the WHO, by the year 2030 indoor air pollution from the use of biomass in low efficiency stoves will results in over 1.5 million premature deaths per year, or about 4,000 deaths per day. This is a greater than estimates for premature deaths from malaria, tuberculosis or HIV/AIDS (IEA, 2010). Unfortunately, in poor cities where the per capita income is below 7USD$50, only 70% of the population are connected to the grid, and in many of these cities the grid supplies electricity for just for a few hours per day. It is widely accepted that the type of fuel used by families for heating and cooking is a function of their income. Poor households use polluting fuels such as kerosene, paraffin, wood, and so on, while wealthier families may use cleaner fuels, such as gas and electricity (Smith, 1990, 1994). This is known as the "energy ladder." According to Waddams (2001), families in developing countries are able to switch over to non-polluting fuels once their income reaches USD$1000–1500. In many of the less developed countries, the cost of energy services is quite high and disproportional to household income. As reported by Barnes et al. (1994), fees for connecting to the electricity grid may cost USD$600, far beyond the economic capacity of the local poor population.

TABLE 4.4 Number of People Without Access to Electricity and Who Rely on the Use of Traditional Biomass, 2009 (Million)

	Number of People Lacking Access to Electricity	Number of People Relying on the Traditional Use of Biomass for Cooking
Africa	587	657
Sub-Saharan Africa	585	653
Developing Asia	799	1937
China	8	423
India	404	855
Other Asia	387	659
Latin America	31	85
Developing countries[a]	1438	2679
World[b]	1441	2679

[a]*Includes Middle East countries.*
[b]*Includes OECD and transition economies.*
Note: The World Energy Outlook *maintains a database on electricity access and reliance on the traditional use of biomass, which is updated annually. Further details of the IEA's energy poverty analysis are available at www.worldenergyoutlook.org/development.asp.*
Source: Based on IEA data from World Energy Outlook 2010 © OECD/IEA 2017 http://www.worldenergyoutlook.org/weo2010/ Licence: www.iea.org/t&c.

The specific problems related to the use of polluting fuels are well reflected in Table 4.5. As shown, citizens in 45 cities in less developed countries consume less energy, but using more polluting fuels such as wood and other biomass (Barnes et al., 1994). As already mentioned, the final cost of energy for each household in less developed countries is a strong function of family income. Low income families, defined as the poorest 20% of the population, may spend almost 15%–22% of their income on energy (Fig. 4.9). As shown, the lower the income, the highest the relative energy expenditures and the wider the use of polluting fuels for heating and cooking.

ANALYSIS OF INDOOR CLIMATIC CONDITIONS IN LOW INCOME HOUSEHOLDS: ENERGY CONSUMPTION, ENVIRONMENTAL QUALITY AND VULNERABILITY

Low income homes suffer from poor indoor environmental quality. Thermal comfort is not at the proper levels suggested as appropriate by international standards and regulations, indoor temperature may be much lower or higher than the thresholds, ventilation rates may be below the accepted levels, and

TABLE 4.5 Fuel Use in 45 Cities by Ease of Access to Electricity

Access to Electricity in City	Average Monthly Household Income (United States Dollars)	Average Population (Thousands)	Wood	Charcoal	Kerosene	LPG	Electricity
Percentage of households using fuel							
Very difficult	33	23	56.4	73.4	57.6	26.6	21.1
Difficult	67	124	72.3	33.5	65.2	21.8	42.8
Easy	62	514	24.1	62.7	50.4	21.6	47.7
Very Easy	77	1153	22.1	34.5	42.6	47.8	90.5
Fuel use (kgs of oil equivalent per capita per month)							
Very difficult	33	23	1.31	10.09	0.35	1.49	0.24
Difficult	67	174	7.27	2.54	0.46	0.91	1.24
Easy	62	514	2.83	7.20	.1.10	0.50	2.00
Very Easy	77	1.1.53	1.71	1.75	1.75	2.00	2.79

Source: Waddams, 2001.

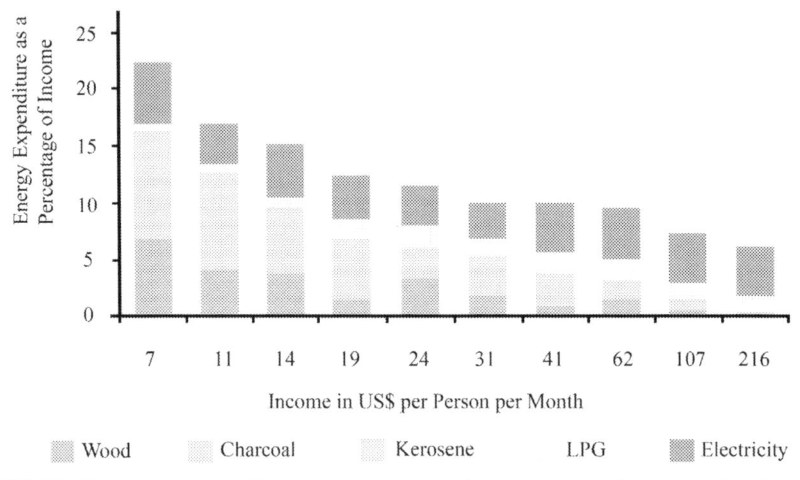

FIGURE 4.9 Energy expenditure as a percentage of income. Average income deciles for 45 cities in 12 countries. Source: *World Bank (1995), Barnes et al. (1994).*

lighting characteristics and the magnitude of other environmental parameters may not satisfy the minimum standards (Kolokotsa and Santamouris, 2015).

Indoor Thermal Comfort During the Summer Period

There is a considerable mass of information and knowledge available concerning prevailing indoor comfort conditions in low income and vulnerable homes during the summer season. Several experimental surveys carried out in warm and less warm climates have collected data on this topic. A list of the existing data in Europe, the corresponding conditions and the main conclusions is given in Table 4.6 (Kolokotsa and Santamouris, 2015).

Indoor comfort conditions during the summer period are defined by international standards and regulations. CIBSE suggests that an appropriate indoor temperature for the warm period is close to 24°C, and 25°C for the living rooms and the bedroom of houses. The maximum suggested indoor temperatures are 28°C and 26°C correspondingly (CIBSE, 2006). Alternatively, adaptive thermal comfort standards, mainly proposed for naturally ventilated houses, suggest significantly higher indoor temperatures as the result of the dynamic interaction between the residents and their environment (De Dear and Brager, 1998, CIBSE, 2006).

In a survey performed by the World Health Organization (2007), data on indoor comfort conditions were collected in numerous European countries, including France, Germany, Slovakia, Hungary, Portugal, Italy, Lithuania and Switzerland. About 9% of the population participating in the survey

TABLE 4.6 Experimental Studies on Overheating Problems in the Homes of Various Income Groups in Europe

Reference	Country	Year	Number of Houses	Results
Anon (2007)	Frances, Germany, Slovakia, Hungary, Portugal, Italy, Switzerland, Lithuania	2002–2003	3373	Almost 9% of the people believe that their house had a permanent heat related problem during the summer period, while 13% declared that overheating may happen sometimes.
Wright et al. (2005)	UK	2003	5 in London and 4 in Manchester	Indoor temperatures maintained at about 5 K higher temperature than the ambient temperature. In a house in Manchester the bedroom temperature was around 36.0°C and the living room 30°C, while in a flat in London temperature has risen up to 37.9 C
Summerfield et al. (2007)	UK	2005–2006	15	Indoor temperatures were maintained above the ambient temperature while maximum indoor temperatures around 29°C are recorded for ambient temperatures close to 27°C
Wingfield et al. (2008)	UK	2006	4	Measured indoor temperature was peaking above 30°C
Firth et al. (2007)	UK	2006		The average maximum indoor temperature was close to 29.5°C while the corresponding maximum ambient temperature was 31.9 C. During the 6 days of the monitoring and in particular between 11 pm to 8 a.m. in some of the buildings, indoor temperature exceeded the threshold of 26 C for about 45 h.
Mavrogianni et al. (2010)	UK	2009	36	Almost 15 of the 36 monitored houses presented indoor temperatures higher than 26°C.
Firth and Wright (2008); Beizaee et al. (2013).	UK	2007	62	The average proportion of time with indoor temperatures over 25°C, varied between 0.6% to 9.1%, while the corresponding proportion of time over 28°C was between 0.1% to 4%.

(Continued)

TABLE 4.6 (Continued)

Reference	Country	Year	Number of Houses	Results
Lomas and Kane (2013)	UK	2009	268	Almost 28% of the living rooms and 88% of the bedrooms were severely overheated according to the CIBSE criteria. The average recorded temperature for the period between 08:00 and 22:00 was 22.2 C and the average minimum and maximums were 19°C and 25.2 C. The absolute maximum recorded temperature was 35.0 C
Yohanis and Mondol (2010)	UK	2004–2005	25	For 80% of the dwelling, the average indoor temperatures during the summer ranged between 20°C and 23°C.
Sakka et al. (2012)	Greece	2007	50	The average indoor minimum temperatures were always higher than 28°C while indoor temperatures as high as 40°C was recorded. The average indoor temperature during the period of the heat waves was almost 4.2 K higher than the temperatures recorded during the normal summer period. For almost 85% of the time during the heat waves period indoor temperatures were above 30°C. Spells of about 216 continuous hours above 30°C are recorded while in very hot houses spells of six continuous days above 33°C are found.
Zavadskas et al. (2008)	Lithuania		1	Indoor temperatures during the summer period varied between 23°C–25°C.

Source: Kolokotsa and Santamouris (2015).

reported the existence of permanent overheating problems, while in about 13% of cases overheating happened occasionally.

In Lithuania, Zavadskas et al. (2008) performed measurements in a low income home and found that indoor temperatures during the summer period varied from 23–25°C. In the UK, Wright et al. (2005) carried out measurements during the 2003 heat wave period in five non air conditioned low income houses in London and in another four in Manchester. They reported very high levels of the average maximum indoor temperature. Indoor temperatures in London were close to 37.4°C and 32.1 in Manchester. The absolute maximum recorded temperatures were 36.0°C and 37.9°C in Manchester and London respectively. Indoor temperatures were kept almost 5 K, higher than the ambient one. Wingfield et al. (2008) performed measurements of the indoor thermal comfort levels in four houses in Stamford Brook, UK, during the summer period and reported maximum indoor temperatures close to 30°C. Summerfield et al. (2007), reported the results of an energy and environmental monitoring of 15 low energy houses in Milton Keynes in the UK. Indoor temperatures were found to be above the ambient one, while the maximum recorded indoor temperature was 29°C and the ambient temperature was 27°C. Mavrogianni et al. (2010) monitored 36 houses in London during the summer of 2009. Almost 15 of the 36 dwellings presented indoor temperatures above the threshold of 26°C for about 45 hours. Firth and Wright (2008) monitored about 62 dwellings in Leicester during the 2006 heatwave. The average maximum indoor temperature was 29.5°C, and the maximum ambient temperature was close to 31.9°C. The same authors, Firth and Wright (2008), have reported data from the monitoring of 207 dwellings across the UK during the summer of 2007. The maximum measured temperature was 28.9°C, while indoor temperatures exceeded the threshold of 25°C between 0.6% and 9.1%, and 29°C between 0.1% and 4% throughout the whole monitoring period. Lomas and Kane (2013) also performed measurements in 268 dwellings in Leicester, UK, during the summer of 2009. About 28% of the living rooms and 88% of the bedrooms were seriously overheated. The average maximum recorded temperature was close to 25.2°C, while the absolute maximum was close to 35°C. Finally, Yohanis and Mondol (2010) presented experimental data on the indoor temperature of 25 dwellings in Northern Ireland. For almost 80% of the dwellings, the average indoor temperature varied from 20–23°C.

In Athens, Greece, where the ambient summer temperature is much higher than in the UK, Sakka et al. performed an extended monitoring of fifty low income dwellings during the heat wave period of 2007. During the experiments, the maximum outdoor temperature reached 44.8°C, while the average daily ambient temperature was close to 34.9°C. As reported, the mean indoor temperature was always above 28°C, while the absolute maximum temperature reached 40°C. For almost 85% of the time the indoor temperature exceeded 30°C. The average indoor temperature during the heat

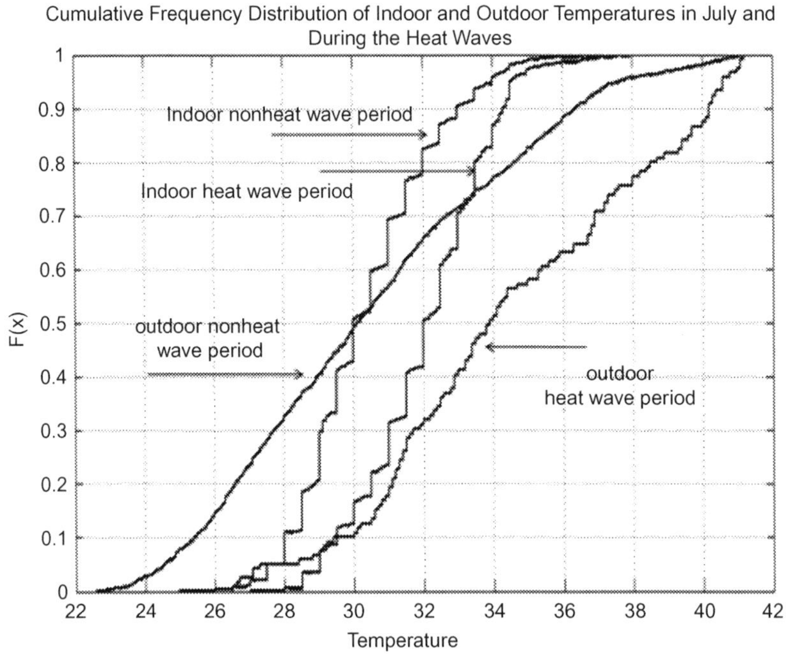

FIGURE 4.10 Cumulative frequency distribution of indoor and outdoor temperatures during the heat wave period and the rest of July. Source: *Sakka et al. (2012).*

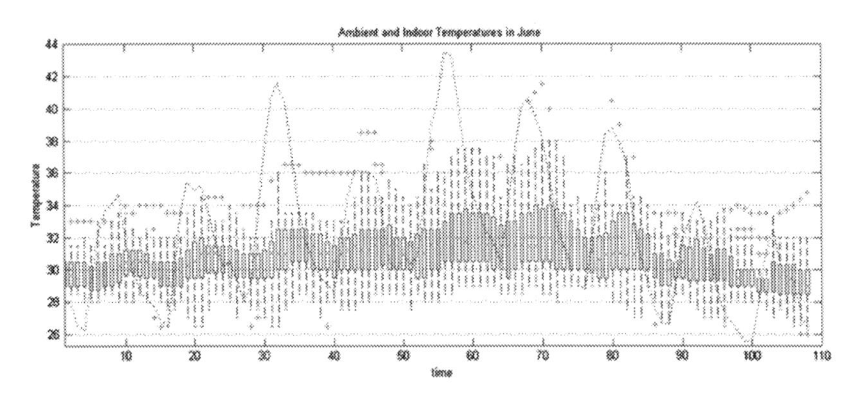

FIGURE 4.11 Box plot of the temporal variation of indoor temperatures during the period 22–30 June and the evolution of the ambient temperature. Time step = 2 h. Source: *Sakka et al. (2012).*

wave period was about 4.2°C higher than the corresponding indoor temperature before the heat wave. Spells of six consecutive days with continuous indoor temperature above 33°C were measured in many of the houses monitored (Figs. 4.10–4.13).

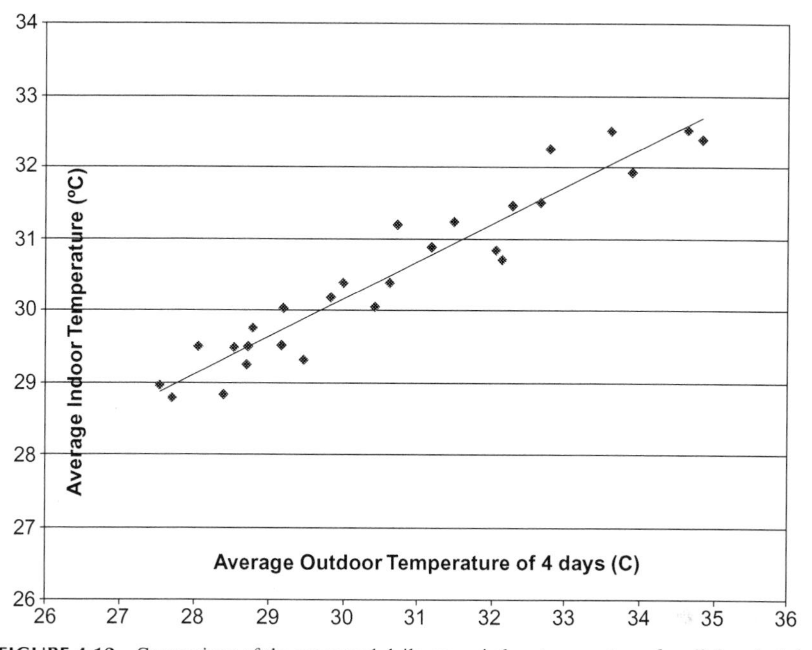

FIGURE 4.12 Comparison of the measured daily mean indoor temperatures for all days in July against the predicted results, using a time sequence of the present day plus the three immediate previous days. Source: *Sakka et al. (2012).*

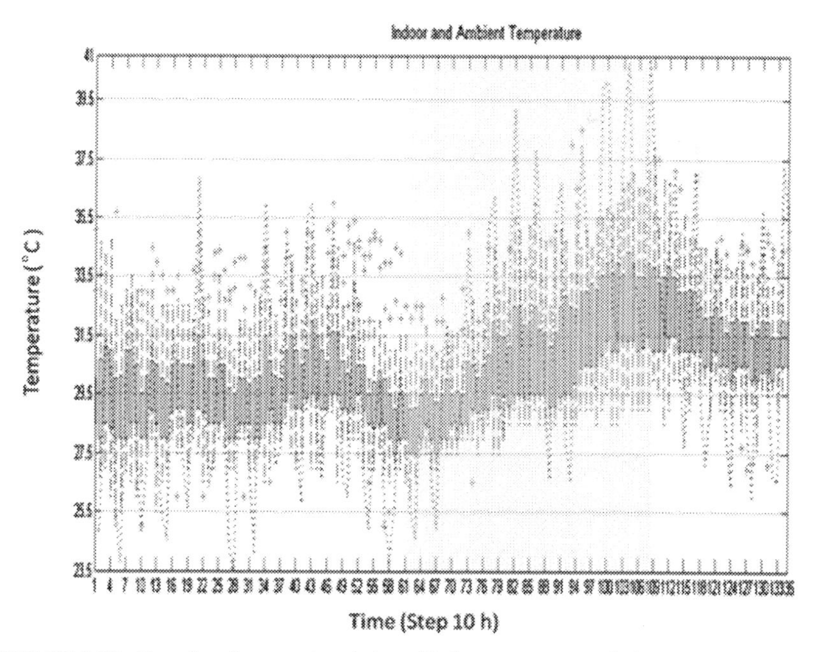

FIGURE 4.13 Box plot of temporal variation of indoor temperatures during the whole of July and the evolution of the ambient temperature. Time step = 5 h. Source: *Sakka et al. (2012).*

It is obvious that the magnitude of indoor overheating is much higher in warm climates than in cold ones, but for both climatic conditions the corresponding maximum indoor temperatures exceed the thresholds for comfort and health. However, it should be pointed out that the thermal experience and the adaptation of the people in warm climates is completely different than in cold climates as the local population is more acclimatized in high ambient temperatures. According to Baccini et al. (2008), the health threat is not linearly correlated to the ambient temperature. Instead, it very much depends on the physiological characteristics of the local population.

Discomfort Problems of Low Income Households in Winter

Heating homes is necessary when ambient temperatures are quite low. Even in warm climates, heating may be required during the winter period. Inadequate heating of houses, especially in cold climates, is associated with low or very low indoor temperatures and cold houses. Cold dwellings pose serious social, health and economic problems for many developed and underdeveloped countries. According to existing international standards, indoor comfort temperatures during the winter period should range between 18°C to 21°C. CIBSE suggests a minimum allowed indoor temperature for bedrooms close to 16°C, while for the rest of the spaces, the recommended temperature is 18°C (Peeters et al., 2009). The World Health Organization (2007), proposes a threshold temperature of 20°C as a benchmark indoor temperature for vulnerable populations, while Boardman (1991) suggests a minimum temperature of 18°C. According to Healy and Clinch (2002), the proper indoor temperature for health reasons is close to 21°C for vulnerable populations and 18°C for sedentary activities and for healthy residents.

Indoor winter temperatures in low income homes are well-studied. Several experimental and theoretical works have been carried out around the world, although most of the experimental data are from the UK. Table 4.7 summarizes the main studies that have investigated the levels of indoor temperatures in low income dwellings (Kolokotsa and Santamouris, 2015).

Important information on the levels of indoor temperature in low income houses in warm climates is provided by Santamouris et al. (2015). Fifty low income dwellings in Athens, Greece, were monitored during the winter period of 2013. The climate was very mild, with a mean outdoor ambient temperature in January of around 10.6°C, and minimum and maximum temperatures close to 0.9°C and 19.0°C respectively. Three days of very low temperatures occurred during the monitoring period with an average ambient temperature of around 2.5°C, 3.9°C and 6.8°C. Analysis of the data shows that indoor temperatures were much lower than the set threshold for comfort and hygienic purposes (Fig. 4.14). In parallel, there is a very clear and strong correlation between the levels of the indoor temperature and the corresponding household income (Fig. 4.15).

TABLE 4.7 A Full List of Studies Aiming to Investigate the Levels of Indoor Temperatures in Low Income Homes

Reference	Country	Year	Number of Houses	Results
Anon (2007)	France, Germany, Slovakia, Hungary, Portugal, Italy, Switzerland, Lithuania	2002–2003	3373	Almost 10% of the people believe that their house had a permanent heat related problem during the winter period, while 14% declared that overheating may happen sometimes.
Alevizos et al. (2013)	Greece	2012–2013	50	The average indoor temperatures varied between 11.7 and 21.1°C Minimum temperatures ranged between 5.2 and 18.8°C and in most of the buildings, minimum temperatures were below 15°C About 5% of the sample, indoor temperature was at freezing levels. During the specific cold days, the average median temperature for all houses was close to 14.7°C, while the average and absolute minimum were 13.2 and 5.5°C respectively. For almost 25% of the houses the average indoor temperature was below 12°C.
Hunt and Gidman (1982)	UK	1978	100	The average dwelling temperature was close to 15.8°C, while the average temperature of the living rooms, kitchen and bedroom were 18.3°C, 16.7°C and 15.2°C. Minimum recorded indoor temperatures were as low as 6°C while for the 15% of the dwellings, indoor temperatures were lower than 12°C.
Burholt and Windle (2006)	UK	2005	421	Almost one fifth of the low income population faced serious difficulties to keep warm their house and had to use additional clothes While almost 90% of the residents living in houses with roof insulation don't needed any extra clothes, the corresponding percentage for houses with non-insulated roof was 61%.

(*Continued*)

TABLE 4.7 (Continued)

Reference	Country	Year	Number of Houses	Results
Oreszczyn et al. (2006)	UK	2001–2002 and 2002–2003	1600	The median standardized temperature of the living room during the day time was 19°C while the corresponding temperature of the bedroom during the night time was 17.1°C
Summerfield et al. (2007)	UK	2005–2006	14	Winter temperatures were kept quite high ranging between 16°C the minimum to 22°C.
Shortt and Rugkåsa (2007)	UK	2000–2002	54	They performed a pre and post assessment of the comfort levels and is found that while the degree of dissatisfaction concerning the quality of the indoor thermal environment before the energy measures was very high, people reported highly satisfied after the energy rehabilitation. Comfort levels and health related indices have also improved considerably.
Hong et al. (2009)	UK	2001–2002 and 2002–2003	2500	While the average temperature in the reference buildings prior to any intervention was close to 16.4°C, the temperature in insulated dwellings was 17.6°C, in centrally heated houses was 18.3°C while in insulated and insulated and centrally heated dwellings it was 19.2°C. In parallel, the neutral temperature has increased from 18.7°C in the reference houses to 18.75°C in the insulated ones, 19.06°C in the centrally heated houses and 19.32°C in the centrally heated and insulated buildings.
Yohanis and Mondol (2010)	UK	2004–2005	25	For about 80% of the dwellings the average indoor temperature during the winter period was between 15°C and 20°C. The temperature in the living room of the 12% of the more warm houses ranged between 21.5°C and 24.7°C, while the corresponding temperature for the colder houses ranged between 13 C to 16.5°C.

Critchley et al. (2007)	UK	2001–2002	888	In 220 houses the mean living room and bedroom temperatures were below 18°C and 16°C and the houses were characterized as cold houses
Zavadskas et al. (2008)	Lithuania	2007	1	Indoor temperatures varied between 20°C–24°C.
Holgersson and Norlén (1984)	Sweden	1982	144	The average temperature in single family houses ranged between 17.8°C and 23.2°C, while in multifamily dwellings varied between 20.2°C to 238°C.
Kavgic et al. (2012)	Serbia	2009–2010	96	The average temperature in the living rooms was 22.8°C. During the evening, the living rooms temperature in individually centrally heated buildings, was for the 22% of the time below 18°C, while the night bedroom temperatures were for the 37% of the time below 18 C.

Source: Kolokotsa and Santamouris (2015).

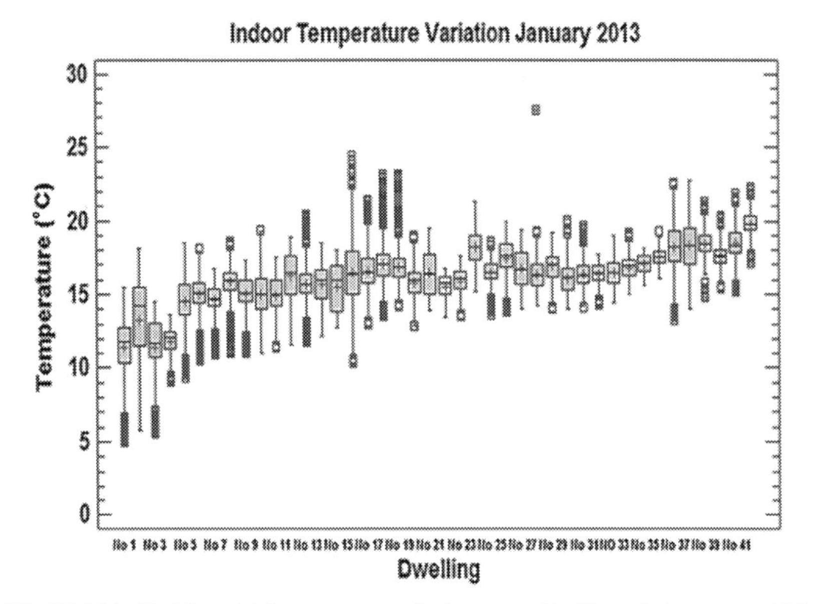

FIGURE 4.14 Variation of indoor temperature in the surveyed buildings during January 2013. Source: *Santamouris et al. (2014).*

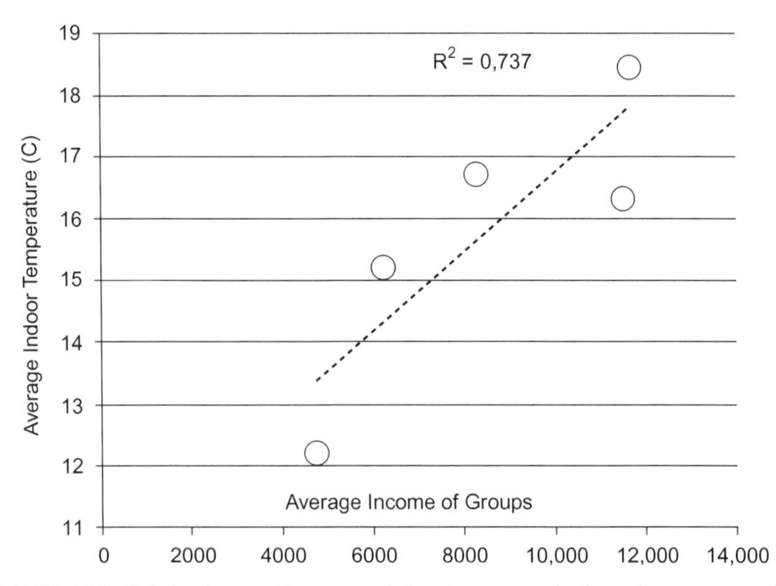

FIGURE 4.15 Relation between the average indoor temperature in the various groups against the corresponding annual income. Source: *Santamouris et al. (2014).*

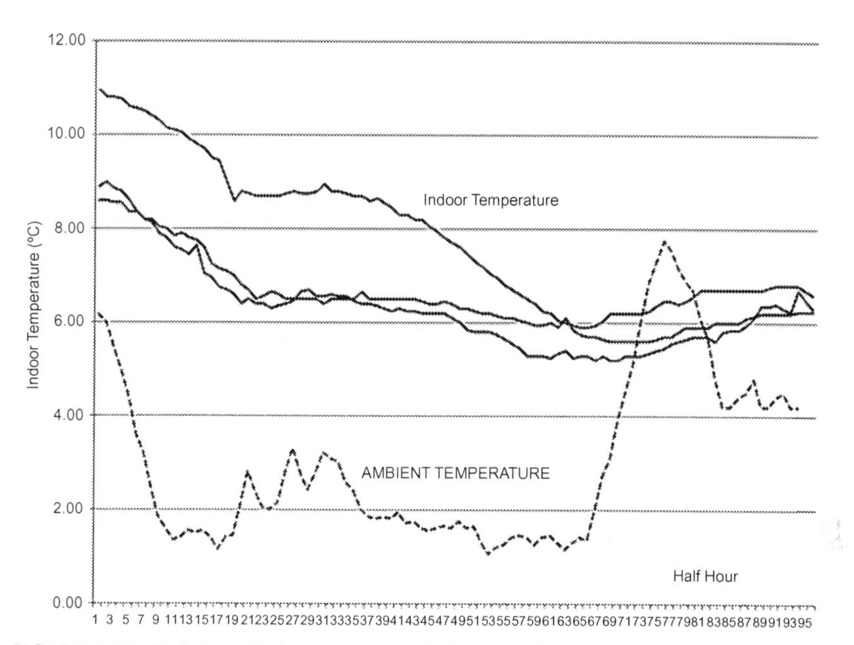

FIGURE 4.16 Variation of indoor temperature in three buildings of Group An and the ambient temperature for the 8th and 9th of January. Source: *Santamouris et al. (2014)*.

The mean indoor temperature in the selected dwellings varied between 11.7°C and 21.1°C, while most of the houses were far below 18°C. The minimum indoor temperature for most of the dwellings was below 15°C, while the absolute minimum recorded indoor temperature was 5.2°C (Fig. 4.16).

A similar study was carried out in Cyprus to identify the levels of indoor temperature in low income houses (Pignatta et al., 2016). Measurements were performed for one complete year starting from December 2013 in 38 low income households in different parts of the country. Questionnaires were used to assess health and social conditions in the households. The range of the recorded temperature in all monitored households is shown in Figs. 4.17 and 4.18. As shown, low income households in Cyprus use to live with quite low indoor temperatures, despite the warm climate in the country. The average indoor temperature during the winter period varies from 16–19°C, about 2–3°C lower than the temperatures recommended in existing comfort standards.

Low temperatures were attributed to the poor conditions of the building envelope and the improper use of HVAC (Heating Ventilation and Air Conditioning), systems. In fact, only 30% of the monitored houses were insulated, while only 53% of the houses had double glazed windows.

Based on the collected monitoring data and the questionnaires, households were classified into three clusters. The first cluster, A, contains all

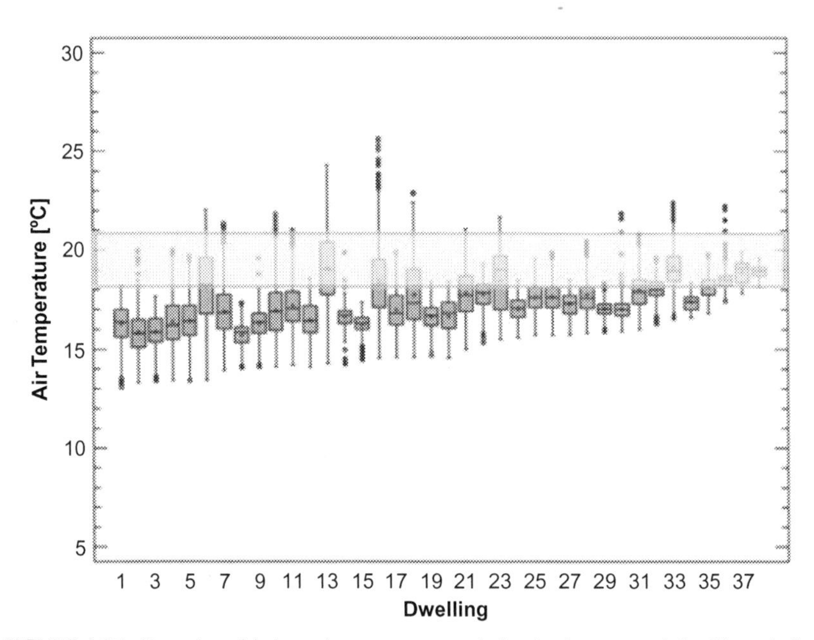

FIGURE 4.17 Box plot of indoor air temperature variation in the surveyed dwellings during January 2014. The winter thermal comfort conditions are within the horizontal colored band. Source: *Pignatta et al. (2017).*

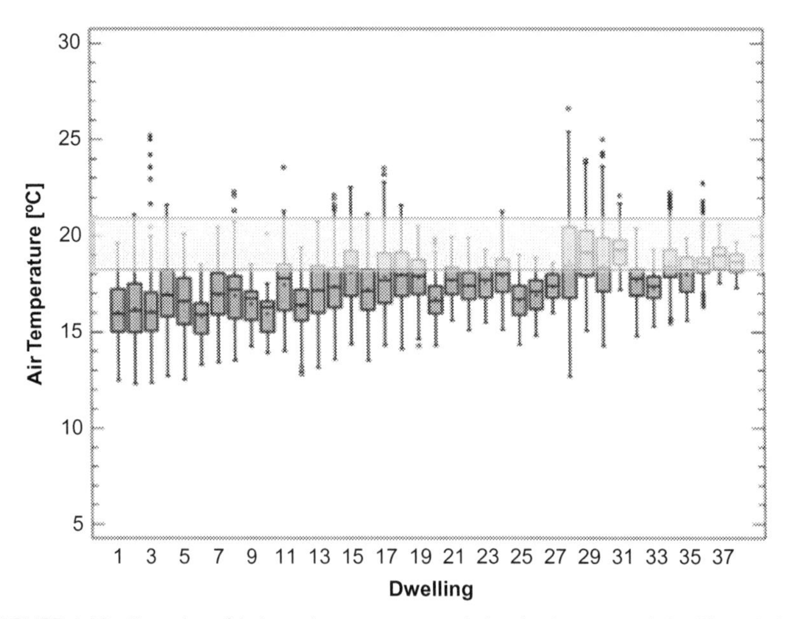

FIGURE 4.18 Box plot of indoor air temperature variation in the surveyed dwellings during February 2014. The winter thermal comfort conditions are within the horizontal colored band. Source: *Pignatta et al. (2017).*

TABLE 4.8 General Characteristics of Households in the Three Defined Clusters

	Cluster A	Cluster B	Cluster C
General Identity of Households			
No. of households [−]	12	15	11
Average persons per household [−]	2.17	2.07	2.64
Average occupiers' age [years]	41	44	40
Percentage of employed people [%]	38.46	9.68	3.45
Percentage of unemployed or non-active people [%]	30.78	51.61	37.93
Percentage of retired people [%]	15.38	25.81	27.59
Percentage of students [%]	15.38	12.90	31.03
Average annual income[a] (2013) [€]	9800	9907	10,057
Average annual income per person (2013)[€]	4516	4786	3809
Average education score (0–3)[b]	2.31	2.26	2.28
Percentage of population with tertiary education [%]	65.38	61.29	58.62
Percentage of population with upper secondary education [%]	15.38	12.90	24.14
Percentage of population with low education [%]	3.85	16.13	3.45
Percentage of population with compulsory education [%]	15.38	9.68	13.79

[a] *Taxes included.*
[b] *0 = compulsory education, 1 = low school, 2 = upper secondary school, 3 = tertiary education.*
Source: Pignatta et al. (2017).

houses presenting lower indoor temperatures, while the third cluster, C, contains households with the highest indoor temperatures. The characteristics of the households and the dwellings belonging to the three clusters are given in Tables 4.8–4.9.

Fig. 4.19 presents the variation of the indoor temperature in representative buildings of the three clusters during the coldest day of the monitoring period. It is clear that the average indoor temperature in Cluster A buildings is quite low, and much lower that the threshold levels. Similar conclusions may be drawn for the rest of the clusters, although in Clusters B and C the indoor temperature is significantly higher than in Cluster A.

TABLE 4.9 General Characteristics of the Dwellings in the Three Defined Clusters

	Cluster A	Cluster B	Cluster C
General characteristics of the dwellings			
Average surface of the dwellings [m^2]	92	103	109
Average year of construction [−]	1982	1976	1991
Percentage of detached or semi-detached houses [%]	66.67	73.33	63.64
Percentage of apartments [%]	33.33	26.67	36.36
Percentage of rented dwellings [%]	41.67	40.00	18.18
Percentage of insulated dwellings [%]	50.00	20.00	18.18
Percentage of dwellings with double glazing windows [%]	66.67	53.33	36.36
Percentage of insulated dwellings with double glazing [%]	50.00	20.00	18.18
Percentage of dwellings with an installed A/C [%]	25.00	26.67	18.18
Percentage of dwellings with a central heating system [%]	33.33	33.33	45.45
Percentage of dwellings with another auxiliary heating systems (mobile) [%]	58.33	60.00	54.55

Source: Pignatta et al. (2017).

The levels of the recorded indoor temperature in the three cluster buildings are given in Fig. 4.20. It is clear that indoor temperatures in cluster A and B buildings were much lower than the suggested comfort temperature throughout the whole winter period.

A very strong correlation between household income and the average indoor temperature for each cluster is found (Fig. 4.21). As expected, the lower the income, the lower the average indoor temperature during the winter period.

Cold homes are a serious problem for Northern Europe, and more particularly for Ireland and the United Kingdom. For many years, both countries designed and implemented specific policies to improve the energy performance of low income houses and protect their dwellers. The application of the policies could be considered successful: according to the Department of Energy and Climate Change of the UK, the average indoor temperature in

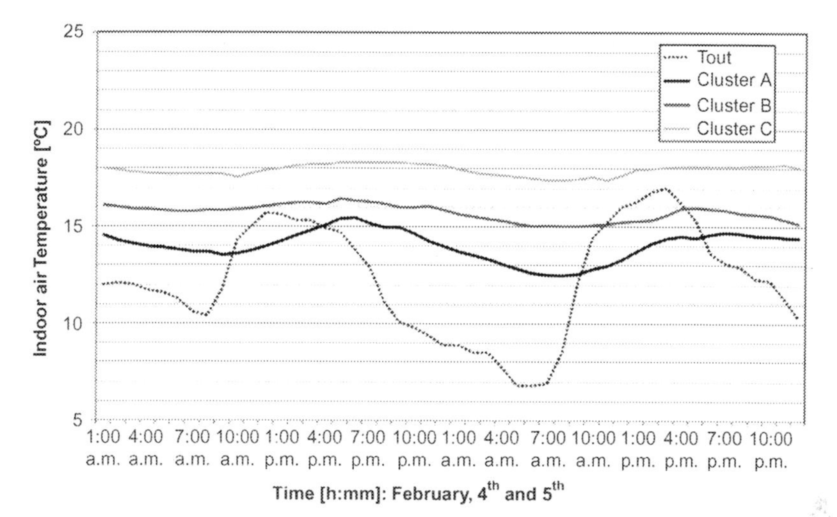

FIGURE 4.19 Daily variation of indoor air temperature in three dwellings of clusters A, B, and C and the outdoor air temperature measured in Limassol New Port during the coldest days of monitoring (February 4th and 5th 2014). Source: *Pignatta et al. (2017).*

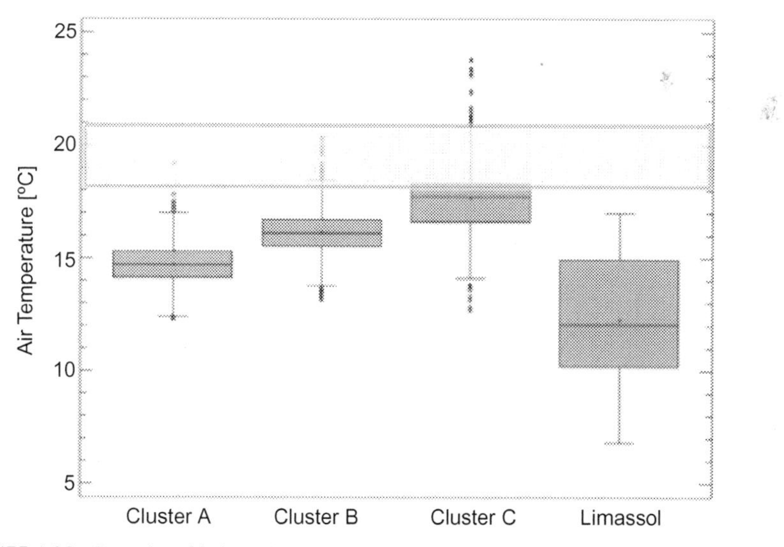

FIGURE 4.20 Box plot of indoor air temperature variation in the three defined clusters of surveyed dwellings and the outdoor air temperatures provided by the meteorological station of Limassol (i.e., January–February 2014). The winter thermal comfort conditions are within the horizontal colored band. Source: *Pignatta et al. (2017).*

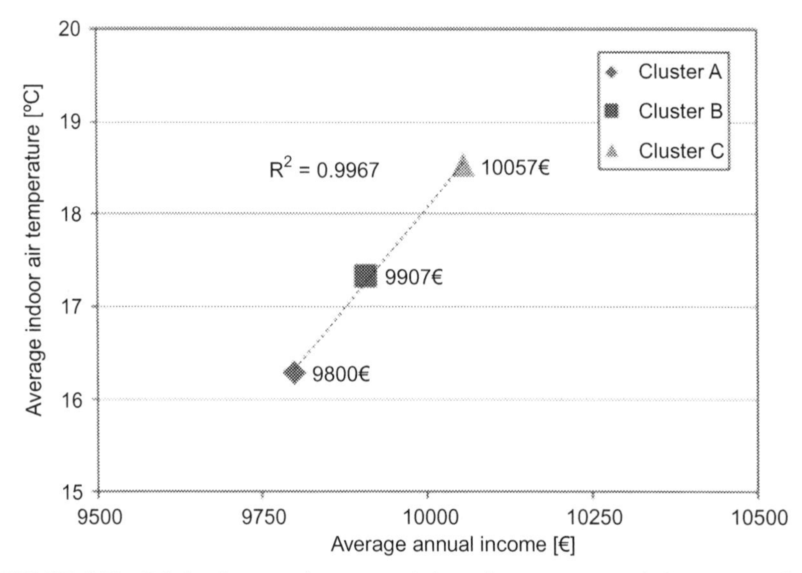

FIGURE 4.21 Relation between the average indoor air temperature and the corresponding average annual income (2013) for the three clusters A, B, and C. Source: *Pignatta et al. (2017).*

houses with central heating increased by 3.6°C and is now close to 17.3°C, compared to 13.7°C in 1970. Moreover, the average indoor temperature in non centrally heated homes increased by 3.6°C and is now close to 14.8°C compared to 11.2°C in the 1970s.

Numerous experimental studies in the UK have attempted to measure the levels of indoor winter temperature in low income houses. Hunt and Gidman (1982) measured the indoor temperature in about 100 dwellings during the winter of 1978. The average indoor temperature was found to be close to 15.8°C, while the temperature in the different rooms varied from 15.2–18.3°C. The minimum recorded temperature was 6°C, while the average outdoor temperature was 6.6 ± 2.2°C. For almost 15% of the houses the indoor temperature was below 12°C. Oreszczyn et al. (2006) monitored almost 1600 low income houses during the period 2001–2003. The median indoor living room temperature was 19°C, while the average night temperature in the bedrooms was close to 17.1°C. Summerfield et al. (2007) measured the indoor temperature in 14 low energy houses in Mylton Keynes. The average indoor temperature ranged from 16–22°C. Burholt and Windle (2006) monitored about 421 old low income dwellings in Wales. They found that about 20% of the residents faced serious problems keeping their home warm. Most of them would wear additional clothes to keep themselves warm. It is observed that almost 90% of the residents living in roof insulated homes did not needed to wear any additional clothes, while in non roof insulated buildings the corresponding percentage was close to 60%. In another study performed by Shortt and Rugkasa (2007) in Northern Ireland, 54 low income houses that had been rehabilitated to improve their energy performance were monitored to assess the levels of indoor temperature during the winter period. Measurements were performed before and after the rehabilitation and it was found that the indoor comfort levels improved considerably after the energy retrofitting of the dwellings. Hong et al. (2009) reported the results of an extended analysis of experimental data from 2500 energy-rehabilitated low income houses in the UK. Retrofitting of the dwellings was performed according to Warm Front National Program standards. Indoor temperatures were measured before and after the energy retrofitting. The average indoor temperature in insulated buildings increased by 1.3 K and after the energy retrofitting it was close to 17.6°C. In centrally heated dwellings it was 18.3°C, while in insulated and centrally heated homes it reached 19.2°C. Hutchinson et al. (2006) further analysed the monitoring results of the previous study, reporting that dwellings in the most deprived quartile presented an indoor temperature below 16°C for almost 37% of the time. The corresponding percentage in the wealthier quartile was just 10%. Yohanis and Mondol (2010) monitored 25 houses in Northern Ireland during the winter period. As reported, the average winter indoor temperature for almost 80% of the dwellings varied from 15–20°C, while in the colder houses it ranged from 13–16.5°C.

In Northern Europe, Holgersson and Norlen (1984) monitored 144 houses in Sweden during the spring period. They found that the average indoor temperature varied from 17.8–23.8°C. Finally, Kavgic et al. (2012) performed detailed measurements in 96 dwellings in Belgrade, Serbia, during the winter time. While the average indoor temperature was at acceptable levels, it was found that the night temperature in the bedrooms was below 18° for almost 37% of the houses.

Based on the above studies, it is obvious that most of the low income houses suffer from serious discomfort problems during the winter period. Indoor temperatures may be very low, even below 10°C, and this is a serious threat for the health of the dwellers. It is evident that energy retrofitting houses can greatly improve their environmental performance and indoor comfort, and protect the health of low income households.

Lighting and Visual Problems in Low Income Houses

Studies have shown that the lack of proper indoor lighting levels may have a serious impact on human mental health. A 2007 WHO study revealed that improper indoor daylight levels and poor quality of view are associated with chronic anxiety, depression and other mental problems. In particular, the study found that poor daylight conditions and a bad view out of the window increased the chance of experiencing depression by 60% and 40% correspondingly.

Several studies have shown that low daylight levels in houses may create important psychological and physiological problems. As shown by Edwards and Tortellini (2002), a proper level of daylight increases productivity, decreases stress and is beneficial for health. A 2011 study by Rybkowska

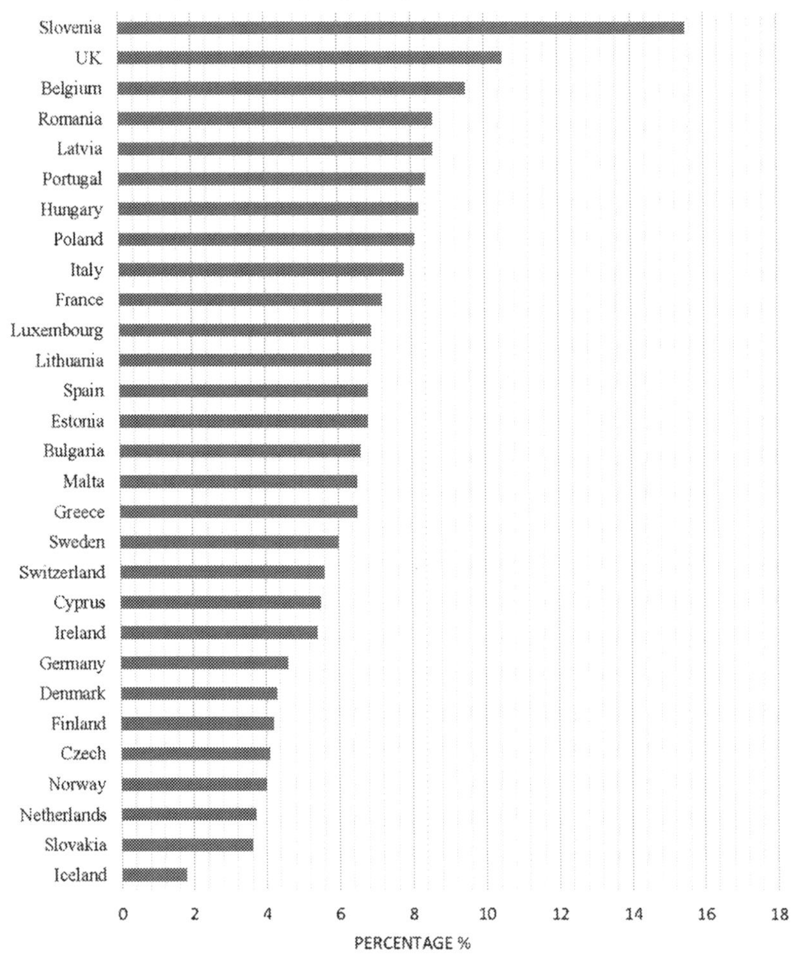

FIGURE 4.22 Percentage of houses reporting problems of darkness at home. Source: *Rybkowska and Schneider (2011).*

and Schneider of potential darkness problems in European homes shows that the average percentage of dark homes in Europe is close to 6.8% (Fig. 4.22). Several countries, such as Slovenia and the UK, present much higher percentages, close to 15.5% and 10.6% respectively.

Ventilation and Indoor Air Quality Problems in Low Income Houses

Adequate ventilation of dwellings contributes to achieving excellent thermal comfort conditions and proper indoor air quality levels. Several studies have shown that inadequate ventilation or the use of non-ventilated heating systems are associated with a higher frequency of asthma and respiratory problems (World Health Organization, 2007). According to Wargocki et al. (2002), review of recent literature on to the impact of ventilation on health, non-adequate ventilation levels are associated with infections, sick building symptoms, asthma, inflammation, allergies and short term sick leave. The use of non-ventilated combustion sources or non-ventilated heating systems, such as imperfect ovens and braziers, may be the source of harmful indoor air quality problems (Kovacevic, 2004). Such a problem is very common in less developed counties and affects mainly women and children. Kovacevic (2004) reported similar problems in Serbia and Montenegro. In about 46% of the low-income houses they inspected, solid fuels are used for heating, cooking and other uses, resulting in the women who would feed the ovens with solid fuels being exposed to very high concentrations of indoor pollutants.

Households living in polluted environments face serious health threats. Indoor and outdoor pollutants greatly affect the quality of life of households and may have a serious impact on their residents' health. It is well accepted that exposure to NOx, Ozone VOCs and particulates produced by combustion processes may cause serious human health problems, ranging from simple irritations to death. Santamouris et al. (2007) measured indoor environmental conditions in several dwellings in Greece and reports that the concentration of TVOCs in low income houses was much higher than the corresponding concentration in high income groups (Fig. 4.23).

Highly industrialized urban zones and areas with a very high production of anthropogenic heat usually present higher levels of outdoor pollution. Most of the population living in such urban zones are low income (Mielck, 2002; Stahtopoulou et al., 2009). Kovats and Ebi (2006) have demonstrated that the level air pollution perceived by the low and medium income population is considerably higher than that of the high income group. Rybkowska and Schneider (2011) calculated the percentage of the European population in various countries suffering from high pollution levels in residential areas. They found the mean value in Europe to be close to 17%, however in particular countries, such as Malta and Lithuania, the corresponding percentage may exceed 30% of the total number of households (Fig. 4.24).

A serious problem in cold homes is the presence of mold and damp. Low indoor temperatures combined with high indoor humidity levels may create mold and damp, which increases the size of fungal and allergic mites and their populations. Chronic exposure to damp in houses may cause important health problems such as asthma, allergies and respiratory problems (Sun and Sundell, 2013). Healy and Clinch (2002) have estimated the percentage of houses in various European countries suffering from the presence of damp spores (Fig. 4.25). As shown, a13% of European homes suffer from damp problems, while in some countries the percentage may exceed 20%. These figures are concerningly high.

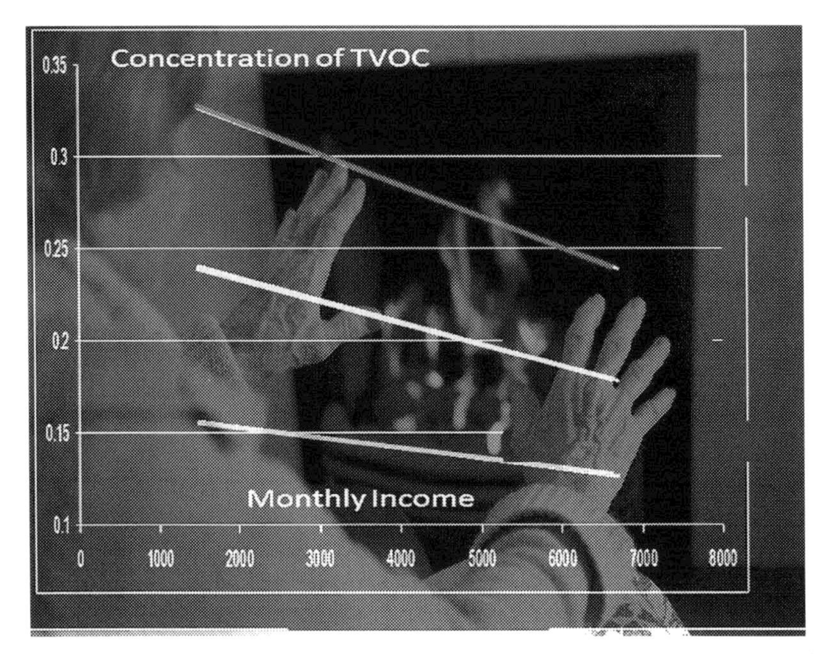

FIGURE 4.23 Concentration of TVOCs measured in the houses of several income groups in Greece. Colored lines correspond to measurements performed in different urban zones. Source: *Santamouris et al. (2007).*

As reported by Santamouris and Kolokotsa (2015), fungi grow only when relative humidity is higher than 60%. Also, mites undergo maximum growth at humidity levels around 80%, while growth is minimum below 50%. Molds may be the source of important allergic symptoms in homes. According to Hardin et al. (2003), almost 5% of the individuals living in these homes may develop allergic symptoms from mold, while it is found that there is a very strong relation between mold, damp, indoor condensation and the severity of illness. WHO (2007) reported that residents exposed to mold have much worse health than those living in houses without mold Residents living in homes with mold and damp presenting high rates of problems related to anxiety and depression, asthma and bronchitis, arthritis, migraine, cold and diarrhea (World Health Organization, 2007). Up to 60% may develop depression. Very similar results regarding the impact of mold, damp and condensation on health are also reported by Harrington et al. (2003), Platt et al. (1989), Hopton and Hunt (1996), Revie (1998) and Khanom (2000).

When low income houses are rehabilitated to improve their energy performance, post assessment surveys have shown that the problems of damp, mold and condensation are drastically reduced, and the health symptoms and hygienic problems are equally reduced. Shortt and Rugkasa (2007) studied 54 rehabilitated low-income homes in Northern Ireland. Almost 60% of

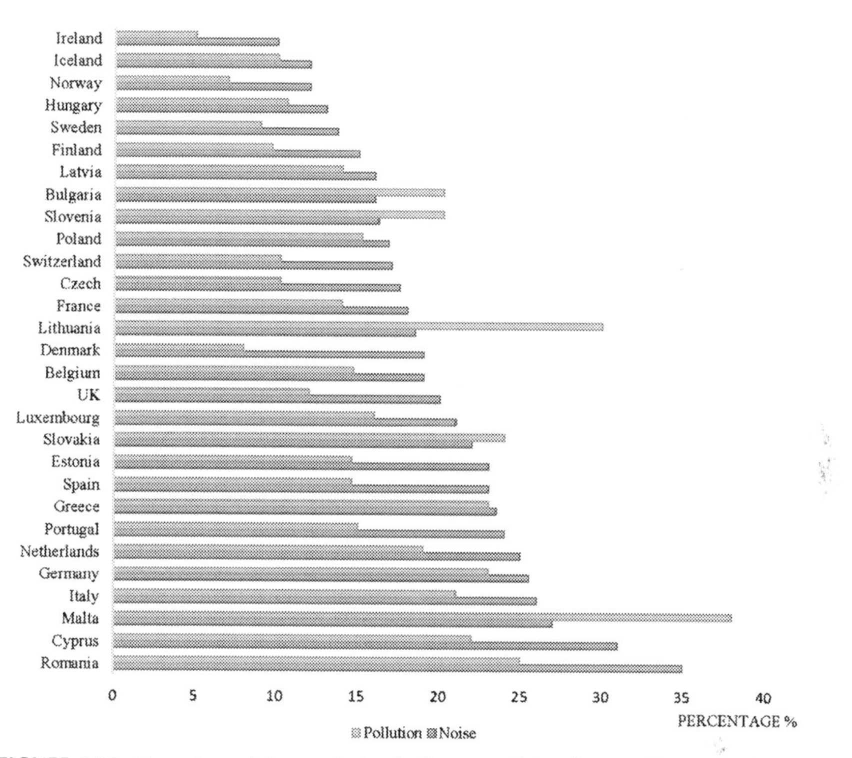

FIGURE 4.24 Percentage of the population in Europe suffering from problems of noise and pollution in residential areas. Source: *Rybkowska and Schneider (2011)*.

them had reported problems with damp and mold, but after rehabilitation the problem completely disappeared, and the number of illnesses per household was reduced from 1.43 to 0.91. In parallel, the number of visits to health services was reduced from 2.5 to 1.3 per head in a three-month period.

Numerous studies have shown that low income populations are exposed to significantly higher concentrations of pollutants than high-income groups. Low-income households live in the proximity of traffic and polluting industrial sources, tobacco smoke environments, and under improper building conditions that are associated with the presence of damp (Chuang et al., 1999; Shapiro et al., 1999). Healy and Clinch (2004) report that almost 50% of the population experiencing problems with damp were living under energy poverty conditions.

Exposure to Noise Pollution

It is well known that exposure to noise pollution is a serious problem and a significant annoyance for human beings. Kryter (1985) has shown that

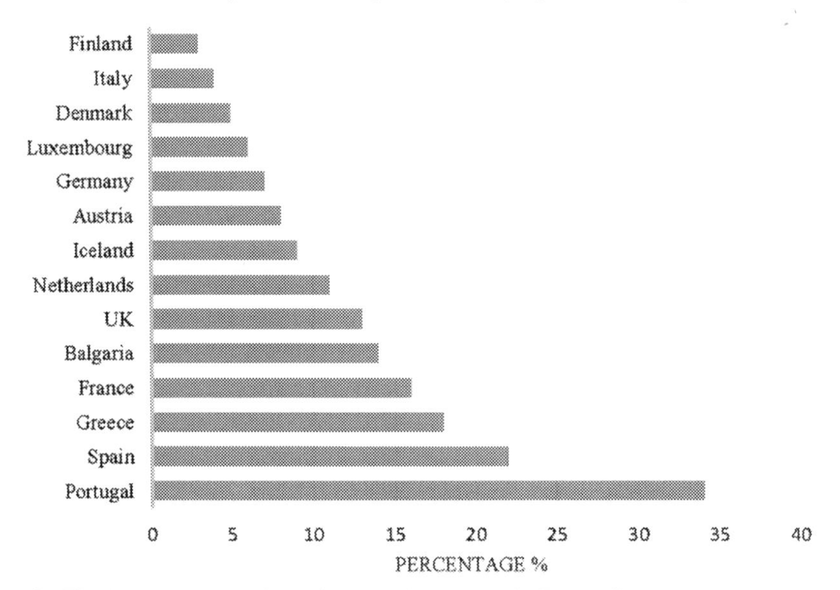

FIGURE 4.25 Percentage of dwellings with damp spores in Europe. Source: *Healy and Clinch (2002).*

people living in very noisy environments suffer from stress and sleep disturbance, become irritable and may develop the habit of shouting. Barreiro et al. (2005) notes that urban families are willing to spend almost four euros per year and per decibel to decrease exposure to noise. Low income families are usually more exposed to noise problems than the higher income population (Kohlhuber et al., 2006). Fig. 4.24 shows the percentage of the European population exposed to noise problems (Rybkowska and Schneider, 2011). While the average figure in Europe is 22%, in several countries, including Romania, Cyprus, Malta, Italy, Germany and the Netherlands, the percentage of the population submitting complaints may be much higher, perhaps up to 35%. Bilger and Carrieri (2013) found that increased level of noise in the neighborhood has a negative impact on the global health of residents. Thus exposure to high noise levels is another important and serious environmental problem that low income populations must face.

ANALYSIS OF THE URBAN ENVIRONMENTAL CONDITIONS IN LOW INCOME URBAN DISTRICTS: PROBLEMS AND CHARACTERISTICS

As already discussed, exposure to very high or very low ambient temperatures, environmental pollution and noise may have serious adverse effects on human health, increase the cost of the national health system and increase vulnerability and mortality rates (World Health Organization, 2007). This is the conclusion of several studies (see Keatinge et al., 2000, Basu and Samet, 2002), which show that low income households living in deprived neighborhoods are at a much higher risk of being exposed to extreme weather conditions. It is widely accepted that dwellers living with improper indoor and outdoor environmental conditions present a much higher frequency of heat-related illnesses. Several studies from the UK, France and Italy support this conclusion (Rozzini et al., 2004; Kovats and Ebi, 2006). Mc Gregor et al. (2007) also report that in poor urban zones the vulnerability of low income people is significantly increased, while Liddell and Morris (2010) conclude that children living in houses without inadequate environmental quality present an increased risk of being admitted to hospital (30%). Similarly, the World Health Organization (2007) report that residents of uninsulated houses, or houses with single glass or without a tight roof, may suffer much more frequently than those living in adequate houses. It is telling that Evans and Kantrowitz (2002) found that those who have difficulty in keeping their home at a proper temperature present almost the double risk of undergoing surgery four or more times, and are twice as likely to use outpatient services compared to those who never experience the problem. The same authors have also reported that poor environmental quality within homes costs the National Health Service of the UK around GBP£2 billion per year.

During extreme weather conditions, mortality rates within a low income population are much higher than in the rest of the population (Michelozzi et al., 1999; Klinenberg, 2002; Vandentorren et al., 2006). Despite the progress achieved recently, mortality rates during the winter period remain high due to low indoor temperatures (Clinch et al., 2000). According to Healy (2003), there are about 30,000–60,000 preventable deaths annually in the UK that can be attributed to low indoor temperatures and poor indoor environmental quality.

As already mentioned, low income populations are exposed to higher pollution levels than the rest of the population (Brown, 1995; Evans and Kantrowitz, 2002; Kohlhuber et al., 2006). As reported by Gesundheit (2009), low income children in Germany are exposed to benzene in much more that the children belonging to households with high incomes. The same author also observed similar differences regarding lead concentration in the blood of children.

The problem of poor health and inadequate indoor air quality was also observed during the monitoring of low income housing in Cyprus (Pignatta et al., 2017). As shown in Table 4.10, dwellings in low income Cluster A present significant environmental problems that present serious health problems for their inhabitants.

The documentation presented shows clearly that the low income population is exposed to hazardous indoor and outdoor pollutants much more than

TABLE 4.10 General Characteristics of Housing and Neighborhood in the Three Defined Clusters

	Cluster A	Cluster B	Cluster C
General characteristics of the housing			
Level of noise (0–1)[a]	0.33	0.26	0.18
Percentage of dwellings with problems of mold [%]	41.67	40.00	27.27
Percentage of dwellings with problems of rats [%]	30.77	13.33	27.27
Perception on the dwelling quality (−2 to 2)[b]	−0.42	0.13	0.00
General characteristics of the neighborhood			
Perception on the quality of the neighborhood (−2 to 2)[b]	1.17	0.80	0.45
Existence of green in the neighborhood (0–1)[c]	0.66	0.73	0.64
Quality of the outside view from the window (−2 to 2)[b]	0.92	0.40	−0.09

[a] *0 = not a noisy environment, 1 = noisy environment.*
[b] *−2 = very bad, −1 = bad, 0 = acceptable, 1 = good, 2 = very good.*
[c] *0 = non-existence of green, 1 = existence of green.*
Source: Pigmatta et al. (2017).

the wealthier population. This has a serious impact on their health and well-being, and increases the cost to the national health system considerably.

BUILDING QUALITY AND ITS ROLE IN ENERGY POVERTY

Existing statistical data shows that low income households live in houses of much lower quality than the wealthiest population groups. Data collected from the European Survey of Life Quality (2003) shows that there is a very strong variation in housing size as a function of income. As shown in Fig. 4.26, the average number of rooms per person for the low income population is close to 1.6, while the average value is 1.9 and 2.3 for the high income population.

A survey carried out in Greece to identify the characteristics of houses for the various income groups (Santamouris et al., 2007) shows that the low income population lives in much smaller houses than average (Fig. 4.27). As reported, the average surface of houses in the lower income population group is almost 2.5 times less than that of the higher income population.

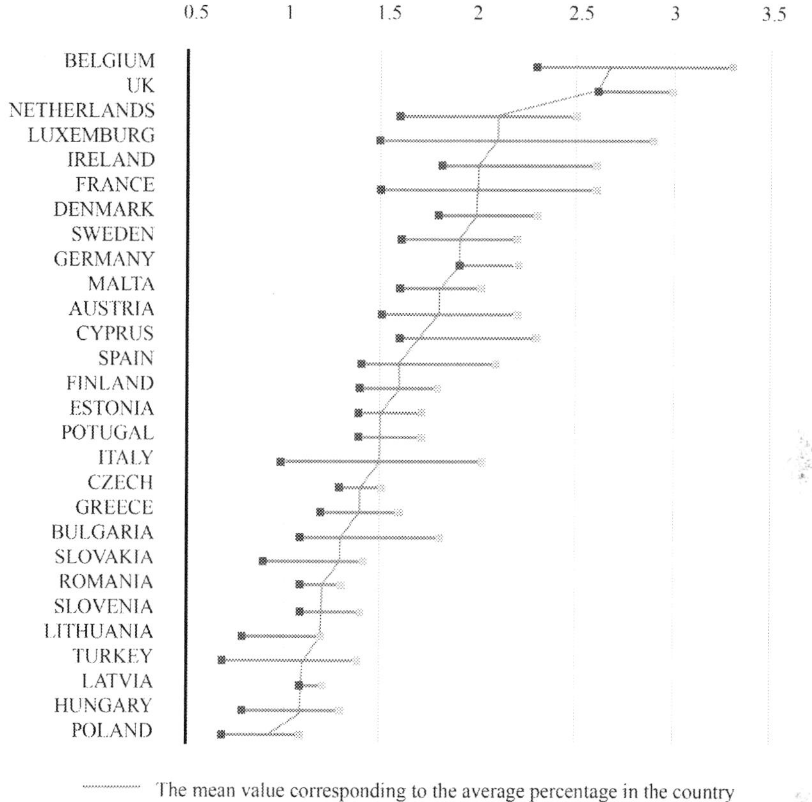

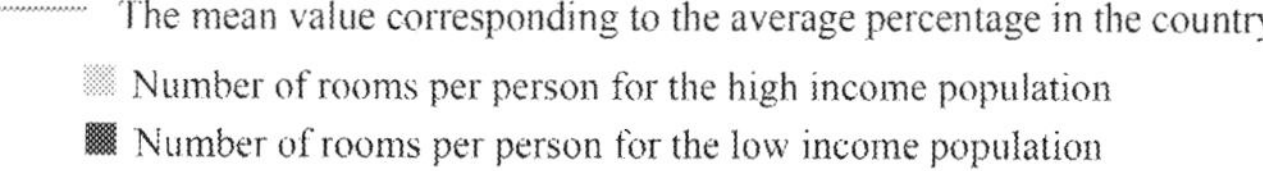

FIGURE 4.26 Average number of rooms per person by household income quartile in European countries. Source: *European Survey of Life Quality (2003).*

Concerning the age of the dwellings, it is well known that low income households tend to live in older buildings. The mean age of dwellings for the low income group is 29 years, whereas for the highest income level households the mean age of dwellings is 19 years.

Very few studies have been done on the energy and environmental characteristics of low income houses. The European Quality of Life Survey (2003) found that the percentage of the low income population living in houses with leaking floors, walls and roofs, or with the presence of rot in window frames, is much higher in lower income groups (Fig. 4.28). The problem seems to be more significant in Eastern European than in the western part of Europe, where intensive energy conservation measures have been implemented.

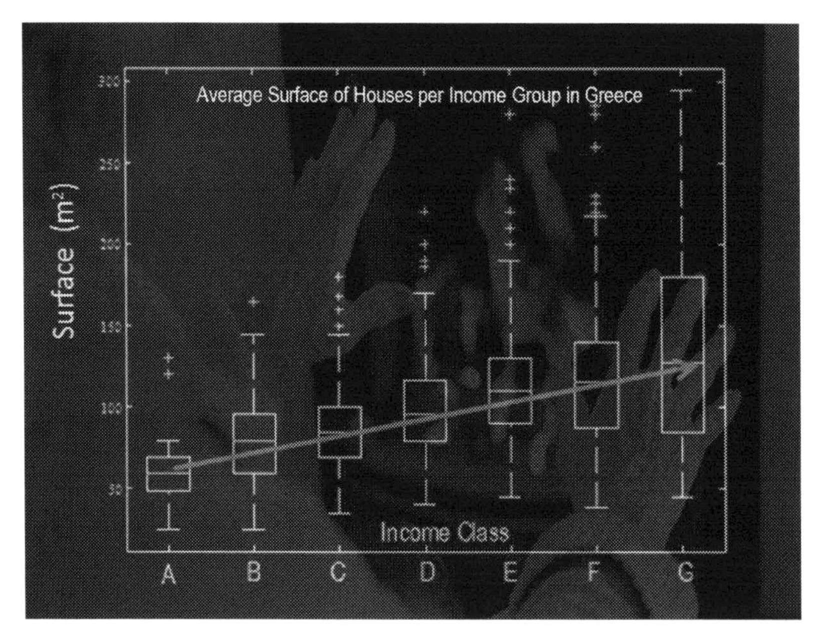

FIGURE 4.27 Variation of the size of housing in the various income groups in Greece. Source: *Santamouris et al. (2014).*

Similar distribution patterns are observed regarding the European population claiming leaking and rotting windows (Figs. 4.29–4.30) (European Quality of Life Survey, 2003). As observed, about 20%–55% of the low income households in eastern and southern European countries are living in houses with leaking windows. The corresponding percentage in the high income group is 40%–60% lower. There is a clear correlation between the level of economic development of a country and the percentage of leaking and rotting windows observed. Countries with a greater level of economic development, such as in Scandinavian, present a much lower percentage of the population living in houses with leaking or rotting windows.

Existing information on the energy conservation features installed in low income houses is quite rare. Partial data exist for selected countries around the world. In almost all countries, the percentage of energy conservation measures installed in low income houses is considerably lower compared to the corresponding figures for the high income population. In Greece, Santamouris et al. (2012) performed a large scale survey and found the percentage of dwellings with insulation, double glazing, and combined insulation and double glazing for the various income groups (Fig. 4.31).

As shown, almost 30% of the low income population lives in insulated homes, while the corresponding percentage for the higher income group exceed 70%. Furthermore, about 23% of low income households live in

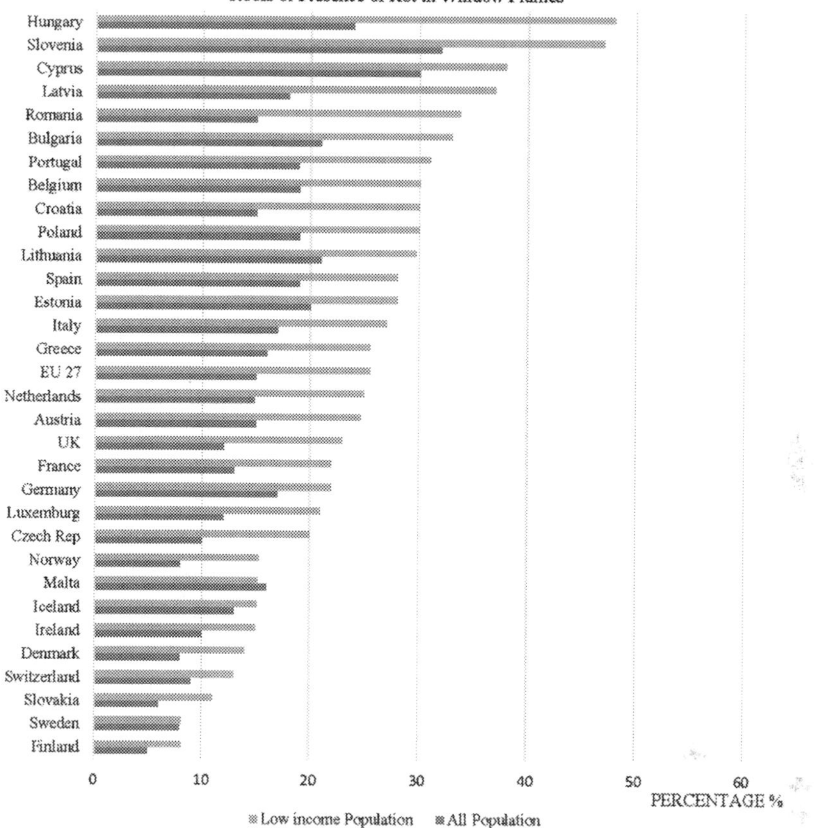

FIGURE 4.28 Percentage of European population living in houses with leaking floors, walls and roofs, or with the presence of rot in window frames. Source: *European Quality of Life Survey (2003).*

dwellings with double glazing, while for the high income group the percentage is close to 68%. Finally, only 8% of the low income families live in insulated houses with double glazing, compared with 68% of the higher income group.

In the UK, due to recently applied intensive energy saving measures, the percentage of houses with cavity insulation in the walls has increased from 25% to 34%, and the percentage of houses with double glazing has increased from 61% to 90% (Palmer and Cooper, 2011). It is unfortunate that the available data on housing conditions in the UK does not make a direct distinction between the income groups. Information on the housing conditions of the low income population may be extracted in an indirect way. According to the same authors, in 1970 the average rate of heat loss of houses in the UK

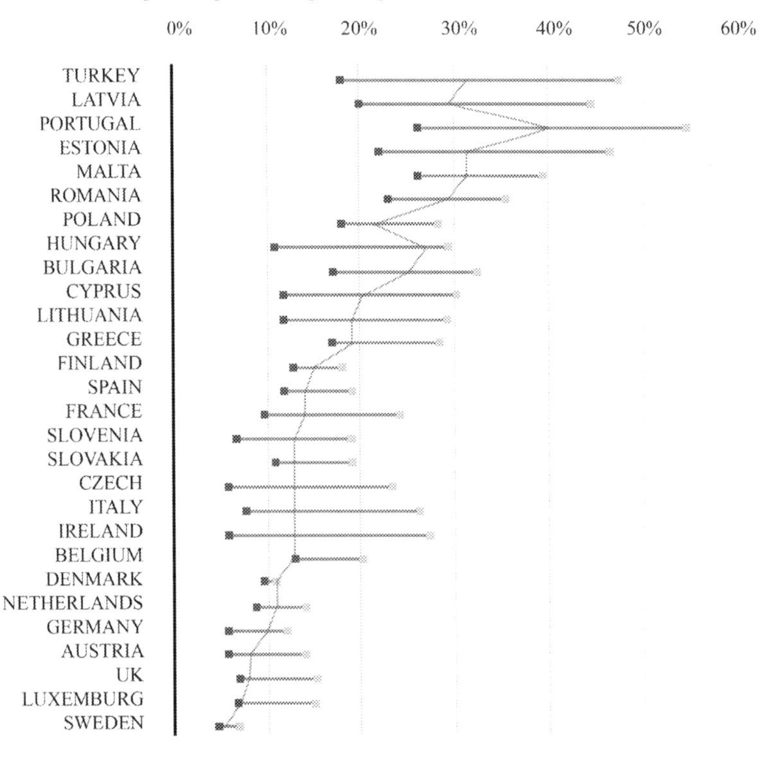

FIGURE 4.29 Percentage of the European population claiming leaking windows. The upper part of the bar corresponds to the percentage of the low income population, the mean value of the bar corresponds to the average percentage in the country, and the lower end of the bar corresponds to the percentage within the high income group. Source: *EuropeanQuality of Life Survey (2003)*.

was close to 376 W/K, while it decreased to 254 W/K in 2010. In parallel, Standard Assessment Procedure (SAP), the energy rating of homes, has improved significantly for all income groups due to the intensive energy savings measures applied by the government. As reported, the SAP score for low income population houses increased from close to 42.5 in 1996 to 52.4 in 2008. Finally, in Belgium, the percentage of insulated houses increased from 43% to 58%, (McKinsey, 2009).

Based on data provided by various sources, the percentage of insulated houses in the southern European countries varies from 30%–40% in western Europe between 60%–70%, while in the northern European countries it exceeds 90% (EACI, 2003; Erhorn Klutting et al., 2007; Datamine, 2009;

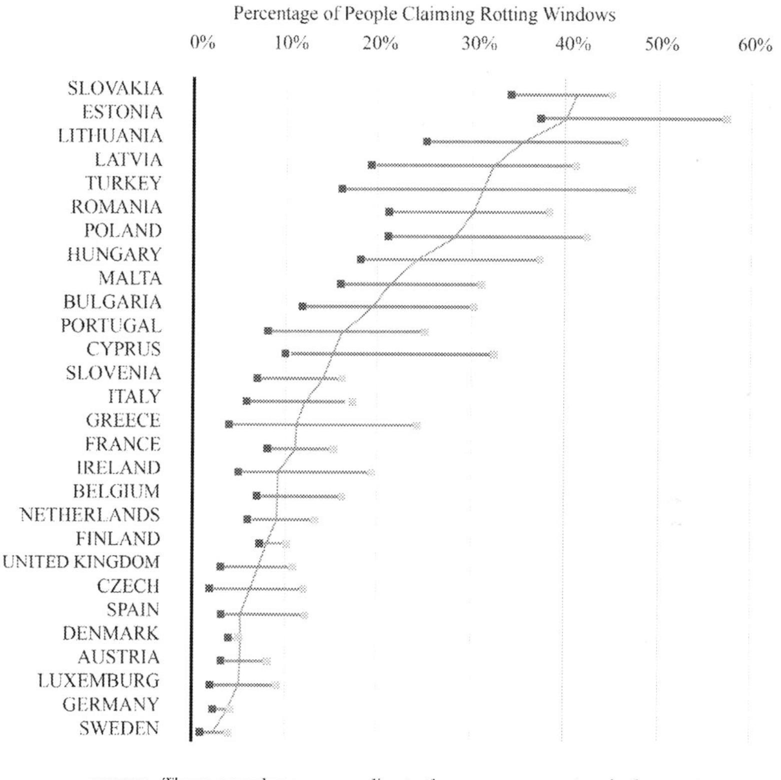

FIGURE 4.30 Percentage of the European population claiming rotting windows. The upper part of the bar corresponds to the percentage of the low income population, the mean value of the bar corresponds to the average percentage in the country, and the lower end of the bar corresponds to the percentage within the high income group. Source: *European Survey of Life Quality (2003).*

Tabula, 2013). In Eastern Europe, most of the buildings were constructed during the Soviet period and the energy savings standards are quite low (European Quality of Life Survey, 2003).

The impact of insulation on indoor ambient temperatures is quite significant. Santamouris et al. (2002) correlated nominal thermal losses through the envelope of the house against the average indoor ambient temperature in non-thermostatically controlled houses (Fig. 4.32). As shown, there is an almost linear relation between thermal losses through the envelope and the average indoor temperature. The authors also investigated the relation between the quality of insulation in the envelope and household income * Fig. 4.33). As shown, there is a very strong correlation between the two

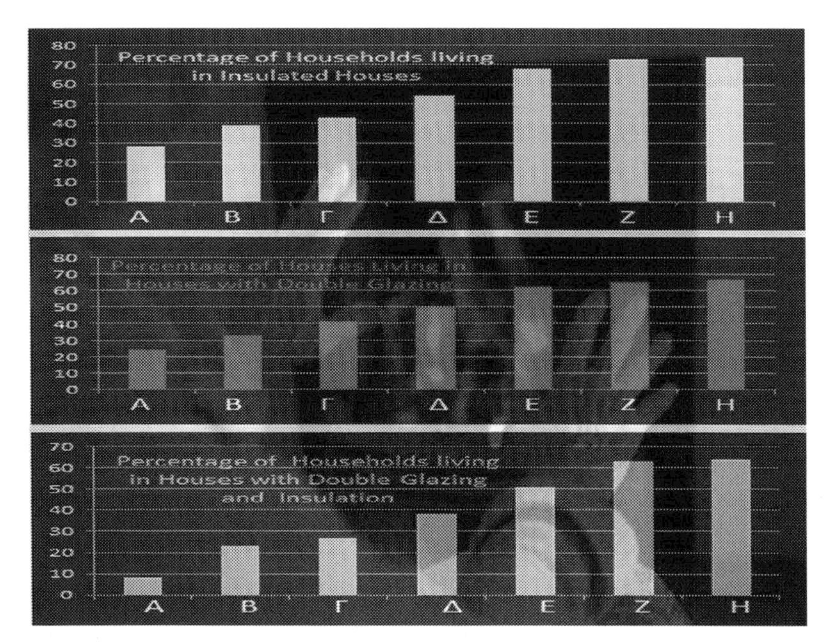

FIGURE 4.31 Percentage of insulated homes with double glazing and homes that are both insulated and have double glazing in Greece by income group. Source: *Santamouris et al. (2012).*

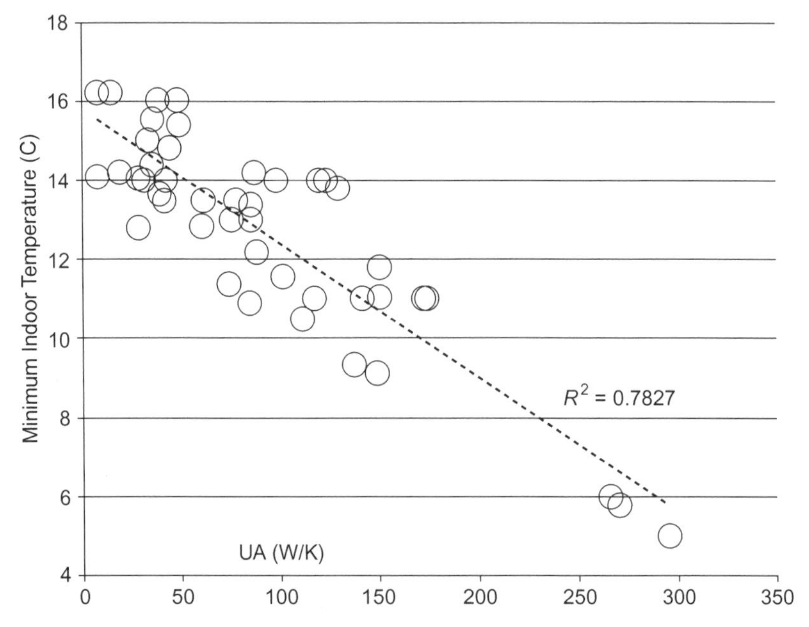

FIGURE 4.32 Variation of the minimum indoor temperature in the dwellings, as a function of the corresponding thermal losses through the envelope (UA). Source: *Santamouris et al. (2014).*

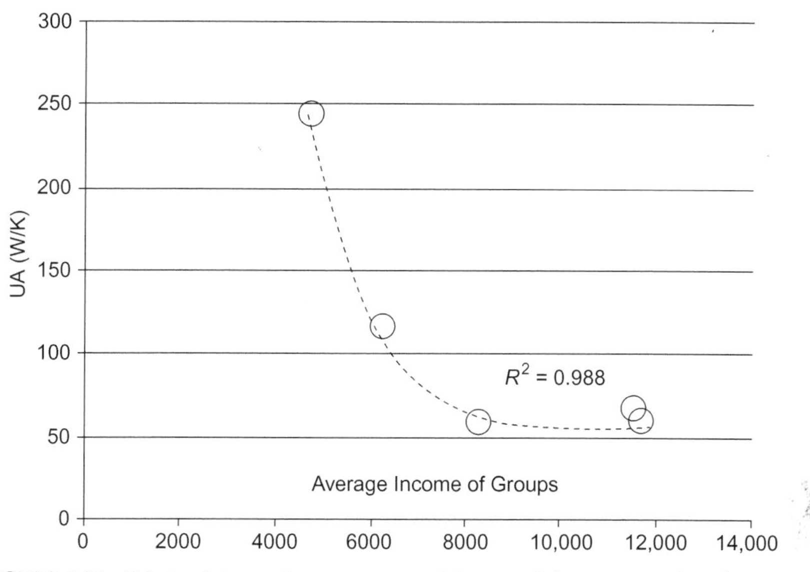

FIGURE 4.33 Relation between the average annual income of the groups against the corresponding mean UA value. Source: *Santamouris et al. (2014).*

parameters and it is clear that low income households tend to live in less insulated or non-insulated dwellings.

High infiltration rates in houses have a serious impact on energy consumption and global indoor environmental quality (Dascalaki et al., 1999). Very few studies are available of infiltration rates in low income houses. Santamouris et al. (2012) measured the infiltration rate in a significant number of dwellings occupied by several income groups (Fig. 4.34). As shown, low income households tend to live in houses presenting much higher infiltration rates than in the higher income groups. High infiltration rates have a serious impact on indoor thermal conditions and the required energy to maintain the house at the appropriate indoor temperature.

LEVELS OF ENERGY CONSUMPTION OF LOW INCOME HOUSEHOLDS

Apart from the environmental and health issues related to energy poverty, the availability and provision of energy to low income households is an issue of major concern. The present chapter investigates the levels of energy consumption of low income households in developed countries and extract the information necessary to characterize the energy status of poor populations. Energy indicators and units vary substantially between the various countries, and so a thorough comparison of the existing data is difficult, if not

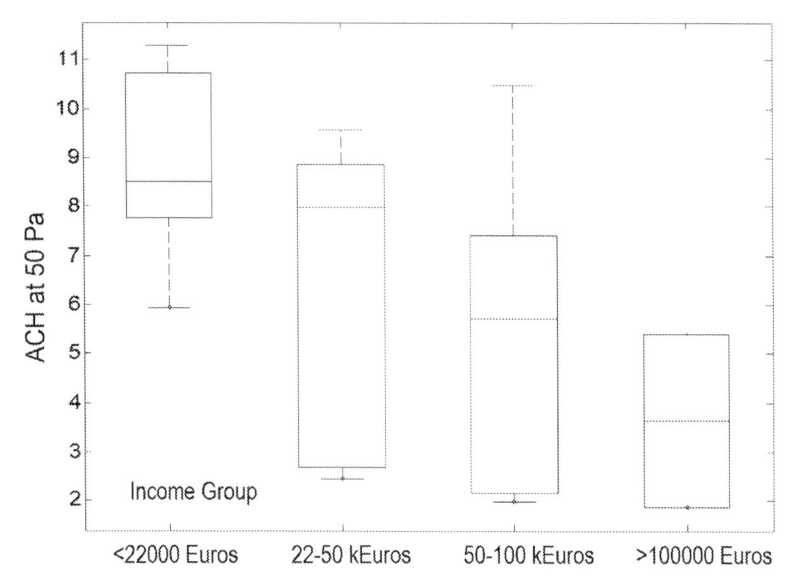

FIGURE 4.34 Measured infiltration levels in dwellings of various income groups in Greece. Source: *Santamouris et al. (2012).*

impossible. The specific information collected in each country where data is available is presented and discussed in the following paragraphs.

In the UK, Warm Front and Warm Front Plus are the governmental institutions that develop and implements energy policies related to fuel and energy poverty in low income households (Hong et al., 2009).

Data on the energy consumption of low income households has been collected through several national surveys aiming to design and implement energy saving measures in poor quality houses (Hutchinson et al., 2006). The collected data were used to calculate the energy rating of the houses, using the official energy rating scheme of the UK, SAP. SAP indicates the energy performance of houses on a scale from zero (poor performance) to 120 (excellent performance). T average SAP score in England in 2001 was close to 51. The results of the energy surveys for the cold homes were analysed using two main predictors: The SAP rating and heating costs. A direct correlation between a house's characteristics and the heating cost of the energy consumed by poor households was documented. Before the energy retrofitting performed by the Warm Front, the average SAP score of the dwellings was 41, compared to a national average of 51. About 82% of the homes considered for rehabilitation were uninsulated or partially insulated, and almost 75% of them did not have a central heating system. After the energy performance rehabilitation performed by the Warm Front, the average SAP score of the low income dwellings increased to 62 (Hong et al., 2009). In a similar study, Oreszczyn et al. (2006) assessed the performance of the houses in the Warm Front program in five urban areas of the UK for periods between 2001–2002 and 2002–2003. The research concluded that for the 26% of the houses, the power necessary to maintain the indoor temperature close to 19.3°C was 250 W/K, while for 20% of the sample the power needed to keep the houses at 17.9°C was close to 500 W/K. As expected, the study shows that once the power requirement decreases, indoor temperatures significantly increase.

Burholt and Windle (2006) studied the energy performance of low income housing occupied by elderly people in Wales. They concluded that the percentage of income spent for energy purposes was very high for all households with an income less than GBP£140 per week. About 29% of the participants in the study spent more than 10% of their income on heating. In parallel, about 38% of the households did not had central heating and 49% did not had draught proofing. The study shows that the percentage of elderly people suffering from energy poverty is much higher than the one calculated by the 10% threshold, as a considerable part of the population are not able to keep the house properly warm. Critchley et al. (2007) investigated the reasons why a high number of energy rehabilitated houses, where more efficient heating systems have been installed, operate below the threshold comfort indoor levels. The study was performed in a sample of 856 households, 96 of which presented a SAP rating below 50, while about 60% of the total sample had a SAP score above 65. It is concluded that despite the important improvement of the energy efficiency and energy performance of the houses, the owners prefer to keep thermostats at relatively low temperatures to decrease the total energy cost. Hong et al. (2006), using data from another similar study, concluded that the reduction of energy consumption in energy rehabilitated houses was not as significant as expected. This was attributed to the specific occupant's behavior, the "take back and rebound effect," and possibly to improperly installed heating systems. The Rebound effect refers to the increase of the energy consumption of households after an energy renovation. Because of the intensive energy rehabilitation program in the UK, the number of households living in energy poverty was reduced by 0.25 million from 2010 to 2011, and the total number of energy poor was close to 4.5 million.

Druckman and Jackson (2008) analysed patterns of energy consumption of various groups of households in the UK as a function of various socio-economic aspects and geographical parameters. They studied zones presenting a high level of deprivation and typical households. Their results and conclusions are summarized in Figs. 4.35 and 4.36.

As shown, the use of energy in "city living," "multicultural" and "constrained" households are 15% lower than the average consumption in the UK, demonstrating the important differences in energy consumption by the low income population.

Several studies of energy consumption within low income populations have been performed in Central Europe. Brunner et al. (2012) used the

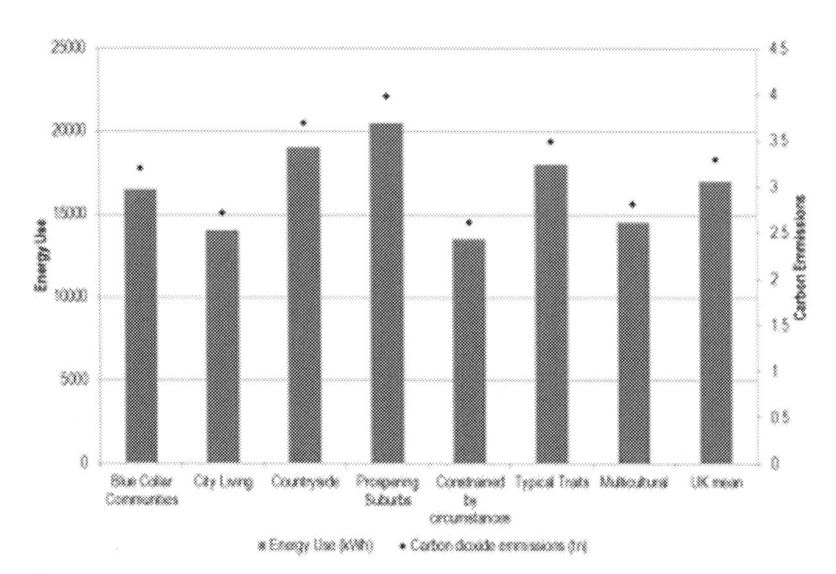

FIGURE 4.35 The consumption of fuel in UK households. Source: *Kolokotsa and Santamouris (2015).*

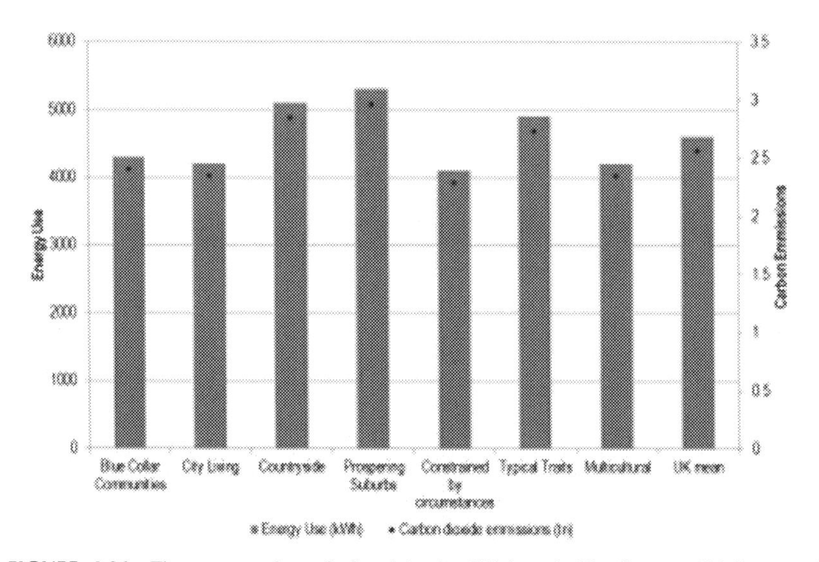

FIGURE 4.36 The consumption of electricity in UK households. Source: *Kolokotsa and Santamouris (2015).*

results of the project “Sustainable Energy Consumption and Lifestyles in Poor and At Risk of Poverty Households” in Austria, and provided information on the levels of energy poverty in Austria. The project studied the levels of energy consumption of the low income population in Vienna. Using questionnaires, they found that the participating households suffer from poor living conditions because of the inappropriate housing and unaffordable heating and lighting. Energy expenses varied from 3.6%−18.7% of their income, depending on the way the households were accustomed to live. Poor quality windows and low thermal standards of the envelope were among the major issues to consider. Until 2008, more than one million people in Austria were at risk of poverty (12.4%). Although there is a lack of an official definition of energy poverty in Austria, the latest data show that about 3% of the population, or 237,000 citizens, are not able to heat their house to adequate temperature levels. Increasing energy prices are a serious concern for the poorest quartile of the population. It is characteristic that between 2004 and 2005, the lower income quartile in Austria spent about 5.6% of their income for energy purposes, while this has increased to 8.3% between 2009 and 2010. On the contrary, the percentage of the energy expenditure of the richest quartile decreased from 3.8% in 2004−2005 to 3.3% in 2009−2010.

In another central European country, Germany, almost 3 million households have benefited from specific social assistance from national housing programs. The energy consumption of the households varies between 1800−4600 kWh/year for families with one to five members per dwelling respectively. The monthly energy bill varies from EUR€76−90 (Kolokotsa and Santamouris, 2015). Grösche (2010) reports results from a large survey of almost 6000 households in Germany that received welfare from the national system during the years 2005−2006. He compared their rent and space heating requirements against those of households that did not receive any welfare (Table 4.11).

It was found that households receiving welfare pay less rent (EUR€5.62 per square meter, compared with EUR€6.3 per square meter for the non-welfare group). Also, they present slightly higher space heating requirements (EUR€1.12 per square meter compared with EUR€1.09 per square meter for the welfare and the non-welfare groups respectively).

Herrero and Ürge-Vorsatz (2010) report that between 2005 and 2007 almost 16.7% of the population in Hungary, corresponding to 1.65 million citizens, faced problems with arrears in energy bills. In the same country, in 2007, about 18% of the population continued to present the same problem. This was the highest percentage in Europe. The average energy consumption in Hungary for a dwelling connected to the district heating network is close to 70 kWh/m^3 or 189 kWh/m^2. The cost of this consumption may represent from 5%−20% of the family's income, without considering maintenance expenses, which may rise to 60% of household income.

Santamouris et al. (2007) collected energy, social and financial data from 1110 households in Athens, Greece. They report that the average energy consumption for heating varies between 154 kWh/m^2/year and 200 kWh/m^2/year, with a median value close to 130 kWh/m^2/year. They found that the

TABLE 4.11 Average Statistics for Germany Case Studies

	2005		2006	
	Welfare	No Welfare	Welfare	No Welfare
Observations	650	4469	638	4071
Living spaces (m^2)	66	73	67	73
Monthly rent expenditures				
Total	366	450	366	456
Per m^2 (€/m^2)	5.62	630	5.62	6.33
Monthly energy expenditures				
Total	366	450	366	456
Per m^2 (€/m^2)	5.62	630	5.62	6.33
Gas price (cent/kWh)	5.58		6.57	
Heating degree days	3606		3449	
Average rent (€/m^2)	5.20		5.24	

Source: Kolokotsa and Santamouris (2015).

maximum energy consumption for heating purposes corresponds to the lower and the higher income groups (Figs. 4.37–4.38).

As observed, for low income populations cover their heating needs, they must use about 120% more energy per square meter of building and person than the high income group. Similar results have been obtained regarding the electricity energy consumption of the various income groups (Fig. 4.39).

The study also revealed that low income groups have to use almost 100% more energy per unit of surface and inhabitant to satisfy their cooling needs during the summer period than the high income households (Fig. 4.40).

The study has shown that for all the income groups, the average income fraction spent for heating and electricity is close to 2.4% and 3.1% respectively. Thus, about 5.5% of the annual income is spent for energy purposes. The fraction of the income spent by each group for heating and electricity purposes is given in Fig. 4.41. While annual heating expenditures represent almost 0.6% for the high income group, it is close to 6.2% for the poorest people. The annual cost of electricity supply is close to 5.9% and 1.1% of the income of the poorest and richest groups respectively.

A second study concerning low income households in Greece was carried out by Santamouris et al. (2013). The study was performed for the years 2010–2011 and aimed to identify the impact of the economic crisis on the energy consumption of the low income population. According to the study,

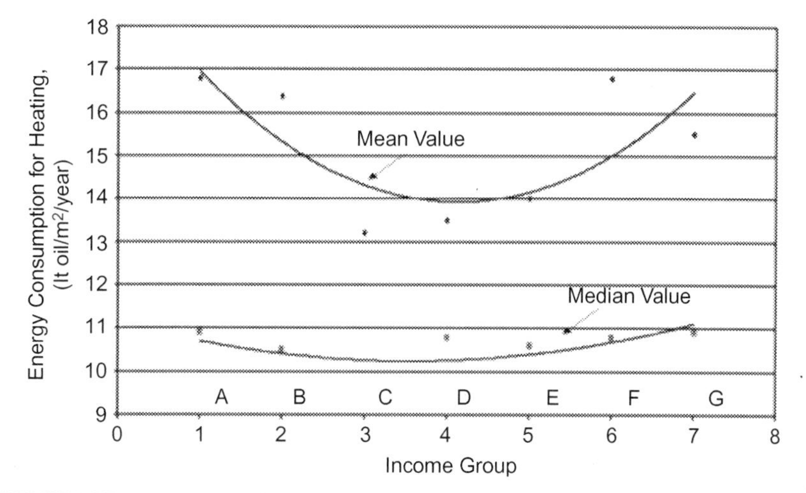

FIGURE 4.37 Mean and median primary energy consumption for heating (litters oil/m2/year), for all income groups. Source: *Santamouris et al. (2007).*

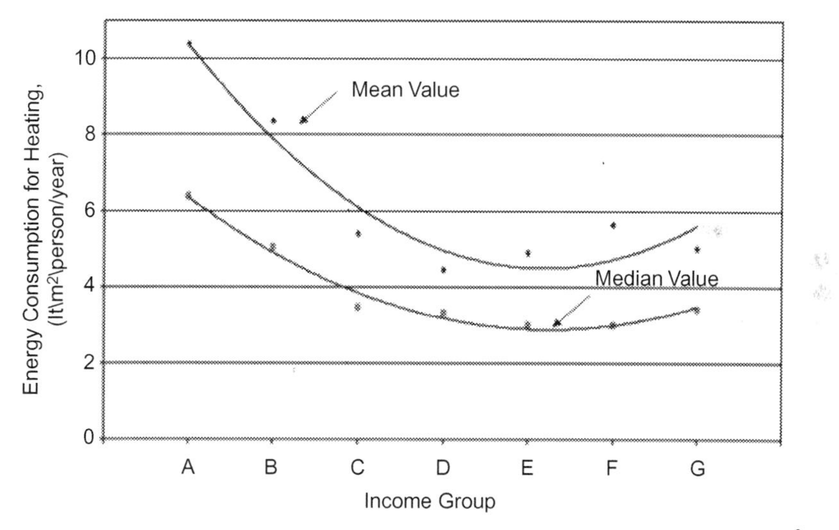

FIGURE 4.38 Mean and median primary energy consumption for heating (liters oil/person/m^2/year), for different income groups. Source: *Santamouris et al. (2007).*

the energy consumption for heating during the winter of 2010, a quite mild winter, varied from almost zero to 882.79 kWh/m^2/year, with an average consumption close to 134 kWh/m^2/year. The corresponding energy consumption for heating during the winter of 2011–2012, which was much colder than the previous one, varied from zero to 677 kWh/m^2/year, with an average close 110 kWh/m^2/year.

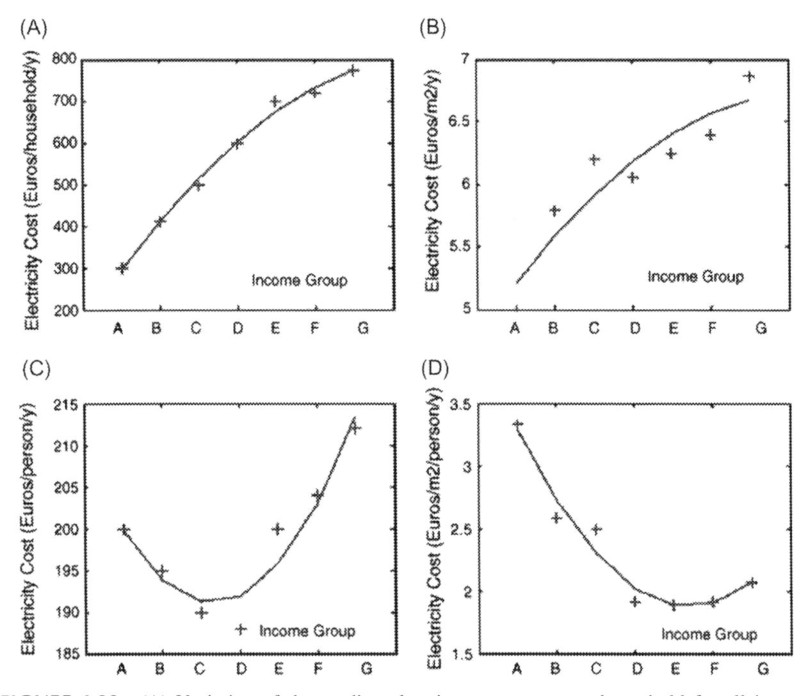

FIGURE 4.39 (A) Variation of the median electric energy cost per household for all income groups; (B) variation of the median electric energy cost per square meter for all income groups; (C) variation of the median electric energy cost per person for all income groups; (D) variation of the median electric energy cost per square meter and person for all income groups. Source: *Santamouris et al. (2007).*

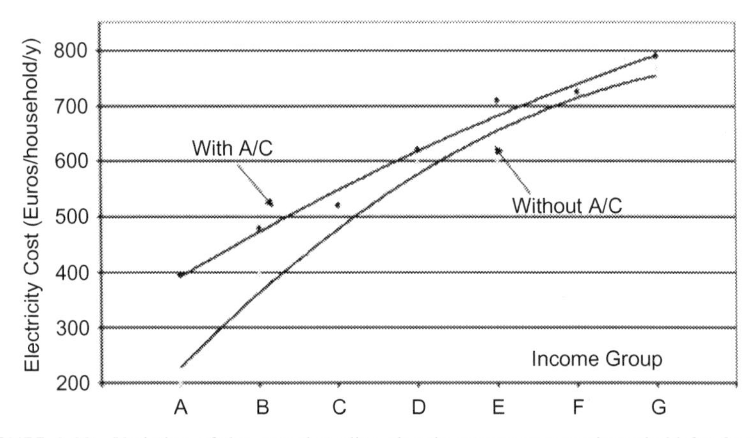

FIGURE 4.40 Variation of the annual median electric energy cost per household for families with and without air conditioning for all income groups. Source: *Santamouris et al. (2007).*

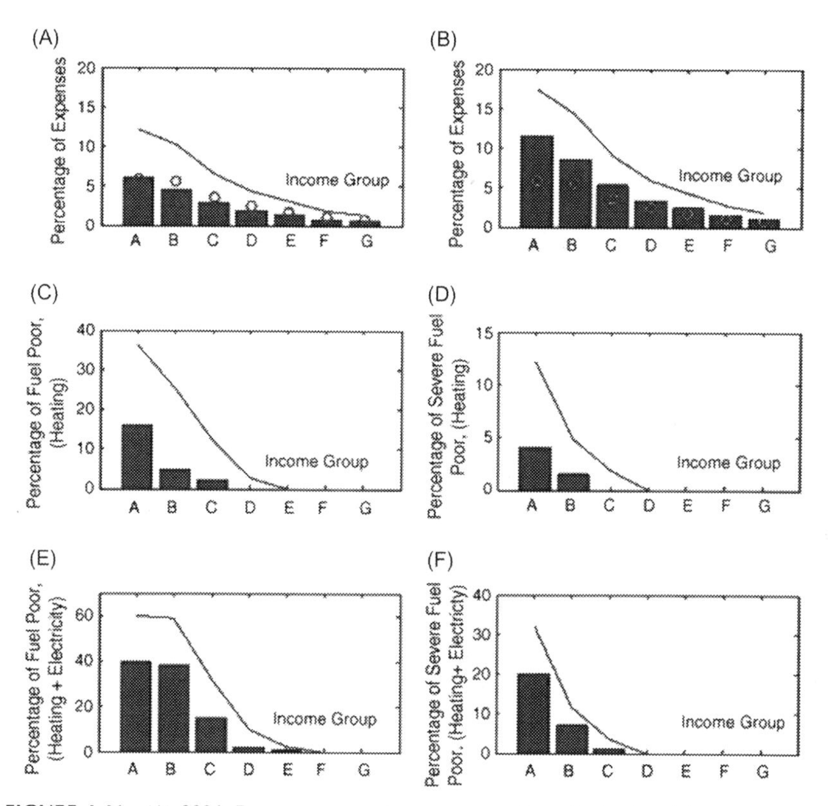

FIGURE 4.41 (A) 2004: Bar: percentage of annual expenditure for heating (o): percentage of annual expenditure for electricity line: percentage of annual expenditure for heating and electricity. (B) 2006: Bar: calculated percentage of annual expenditure for heating (o): calculated percentage of annual expenditure for electricity line: calculated percentage of annual expenditure for heating and electricity. (C) Bar: percentage of fuel poor (heating) per income group for 2004, line: calculated percentage of fuel poor (heating), for 2006. (D) Bar: percentage of severe fuel poor (heating), per income group for 2004, line: calculated percentage of severe fuel poor (heating), for 2006. (E) Bar: percentage of fuel poor (heating + electricity), per income group for 2004, line: calculated percentage of fuel poor (heating + electricity), for 2006. (F) Bar: percentage of severe fuel poor (heating + electricity), per income group for 2004, line: calculated percentage of severe fuel poor (heating + electricity), for 2006.

Despite the colder climate, households spent almost 20 kWh/m^2/year less energy, a reduction of 15% compared to the previous year. The corresponding decrease of the heating energy consumption for the low income group exceeded 50%. An analysis of the energy consumption characteristics of the low income population in Albania, Bosnia and Herzegovina, Bulgaria, Croatia, FYROM, Moldova, Romania, Serbia and Montenegro was prepared by the European Bank of Reconstruction and Development (Kolokotsa and Santamouris, 2015).

According to this study, a typical electricity consumption in Albania is around 360 kWh/month in winter and 320 kWh/month in summer. In rural areas, energy consumption is considerably lower, at 220 kWh/month, due to the extensive use of biomass as a heating source. The corresponding energy consumption in Bosnia and Herzegovina was found to be between 210 kWh/month and 260 kWh/month. In Croatia, the corresponding energy consumption varies between 173 kWh/month and 330 kWh/month, with an average consumption close 260 kWh/month. In Bulgaria, the average energy consumption is close to 195 kWh/month, while it is observed that the most vulnerable population groups spent about 16% of their income for energy purposes. In Moldova the average energy consumption at the end of the previous century was close to 100 kWh/month, but it decreased to 55 kWh/month in 2002. In Romania, a social tariff is implemented for all households consuming less than 60 kWh/month. Almost 43% of the households in the country subscribe to this plan, and about 95% of them mange to consume less than 60 kWh/month. It is characteristic that the average consumption of the households that subscribe to the specific energy plan was as low as 35 kWh/month, while the corresponding consumption of the other social groups that did not participate in the low tariff was 110 kWh/month.

In Serbia, the average energy consumption for heating is quite high, at 240 kWh/m^2/year. As reported by Kavgic et al. (2012), old and poorly maintained buildings consume almost two to three times more energy than new buildings. Electricity consumption in Serbia varies between 350 kWh/month to 490 kWh/month, with an average value close to 400 kWh/month.

Specific data on the energy consumption of low income households in the different parts of Europe (Table 4.12) may result in the following conclusions:

- In Europe, the percentage of the population spending more than 20% of their income on heating varies significantly. This is because of the significant social and economic differences between the different parts of Europe. As shown in Fig. 4.42, in some European regions almost 46% of the population suffer from energy poverty, while in other zones the corresponding percentage may not exceed 1% (Ezratty and Ormandy, 2011).
- Using data from the MURE-ODYSSEE project and data base, the relationship between energy consumption and the income level is established (Fig. 4.43). It is evident that energy use increases with income. However, there is a very high dispersion, and even for low income levels energy consumption varies significantly.
- The available consumption data show that although there is a strong relationship between income and the energy consumption, other factors, such as the quality and the characteristics of the buildings, determine the final energy consumption of the dwellings. This is clear when energy consumption data from similar income groups living under similar climatic conditions are compared (Lapillonne et al., 2012).

TABLE 4.12 The Energy Consumption of Low Income Populations in Various European Countries

Country	Energy Data	Comments	Reference
UK	SAP rating and heating costs.	The mean SAP rating increased from 51 to 62 due to Warm Front	Hong et al. (2009).
UK	Energy demand in W/K for various cities and households	The 26% of the sample was less than 250 W/K to maintain indoor temperature 19.3°C and for the 20% of the sample was more than 500 W/K to maintain indoor temperature 17.9°C	Oreszczyn et al. (2006)
UK	SAP rating	Almost 11% had a SAP rating below 50 and 60% had a SAP higher than 65.	Critchley et al. (2007)
UK	kWh/year	The energy use for "city living," "multicultural" and "constrained" households is almost 15% lower than the UK mean consumption.	Druckman and Jackson (2008).
Northern Ireland	Fuel bill	Fuel bills are an estimated 27% higher than in the rest of the UK	Shortt and Rugkåsa (2007)
Austria	Percentage of energy expenses vs total income.	The range of energy expenses is between 3.6%–18.7% depending on the inhabitants' way of living. The lowest income quartile used 5.6% and 8.3% of its budget for energy.	Brunner et al. (2012), Burnner and Spitzer (2011).
Germany	kWh/year	1800–4600 kWh for 1 to 5 people per dwelling respectively with a monthly bill among €76–90. The households receiving welfare had lower rent (5.62 and 6.30 €/m^2 for welfare and non-welfare recipients respectively) but higher space heating requirements (1.12 and 1.09 €/m^2 for welfare and non-welfare recipients respectively).	Grösche (2010)
Hungary	Percentage of energy expenses vs total income.	The average dwelling connected to a district heating system annual heating demand is 70 kWh/m^3 (189 kWh/m^2). 5% to 20% of the households' income, and can rise up to 60% of their income.	
Albania	kWh/month	360 kWh per month in winter and 320 kWh per month in summer	Anon (2003)

(*Continued*)

TABLE 4.12 (Continued)

Country	Energy Data	Comments	Reference
Bosnia and Herzegovina	kWh/month	average winter monthly consumption was 255 kWh per household and average summer monthly consumption was 211 kWh per household	Anon (2003)
Bulgaria	kWh/month	Average monthly consumption in Bulgaria is 195 kWh.	Anon (2003)
Moldova	kWh/month	In 1997, the average family consumed approximately 100 kWh/month, however in 2002 average consumption had fallen to approximately 55 kWh/month.	Anon (2003)
Romania	kWh/month	It has been found that households subscribing to the social tariff consume on average 35 kWh per month, whereas households not receiving the social tariff consume on average 110 kWh/month.	Anon (2003)
Serbia and Montenegro	kWh/month	Electricity consumption is approximately 400 kWh per household. Regional average household consumption varies between approximately 350 kWh/month in Western Serbia to 490 kWh in Belgrade.	Anon (2003)
Greece	kWh/m^2	Mean energy consumption for heating 154 kWh/m^2/year and 200 kWh/m^2/year. Moreover the highest energy consumption was noticed for the very low and very high income groups.	Santamouris et al. (2007a)
Greece	kWh/m^2	Energy consumption varied from 0.035 to 882.79 kWh/m^2 for 2010. Energy consumption 2011 varied from 0.035 to 676.80 kWh/m^2 with an average of 109.6 and a median of 88.05 kWh/m^2.	Santamouris et al. (2013)

Source: Kolokotsa and Santamouris (2015)

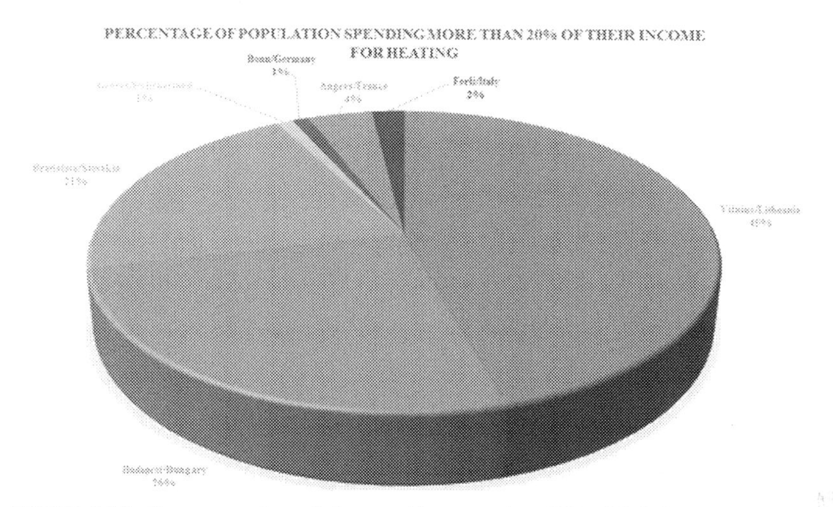

FIGURE 4.42 Percentage of population spending more than 20% of their income on heating. Source: *Kolokotsa and Santamouris (2015).*

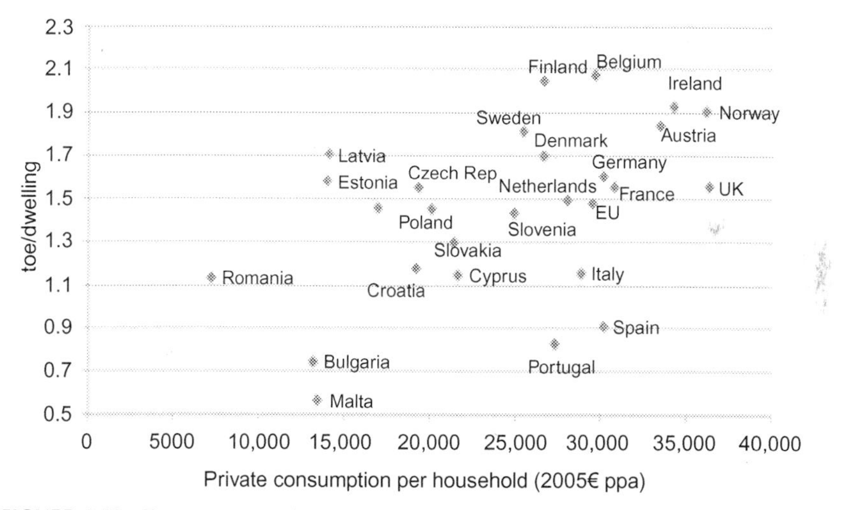

FIGURE 4.43 Energy consumption per dwelling vs income per household in 2005. Source: *Kolokotsa and Santamouris (2015).*

– Important experience has been gained in various states through the implementation of policies and programs to eradicate energy poverty. For example, the Warm Front in the UK provides welfare, but at the same time offers significant incentives to improve the energy and environmental performance of low income houses. It is important to transfer the existing experience and improve how policies are implemented in the rest of the developed countries.

SOCIAL AND HEALTH STATUS OF LOW INCOME HOUSEHOLDS

It is well accepted that energy poverty is associated with inappropriate indoor temperatures, either during the warm or cold periods of the year. As already mentioned, several studies have shown that indoor temperatures in cold houses during the winter period, or in warm houses during the summer period, are far from meeting internationally accepted temperature levels for comfort, while in many cases are out of the recommended sanitary levels for human beings. As a matter of fact, energy poverty and inappropriate indoor temperatures are often associated to serious health problems (Thomson et al., 2001).

Studies aiming to investigate the impact of very low or very high indoor temperatures on human health are quite limited, and the issue has only quite recently gained attention. Although there exists serious research aiming to identify the impact of very high or very low ambient temperatures on human mortality and morbidity (Baccini et al., 2008), very few studies have focused on the health impact of non-appropriate indoor temperatures (Linares and Diaz, 2008). Recent research has found that exposure to high ambient temperatures causes serious respiratory, cardiovascular and cerebrovascular disorders (Knowlton et al., 2004). Other research has shown that exposure to very high temperatures may increase the risk of thrombosis as the viscosity of the blood decreases, while it may cause problems of thermoregulation in the aged population and impaired kidney function (Bates, 2005; Flynn et al., 2005).

The impact of high ambient temperatures on human mortality is a quite well-researched area. However, few studies are available about the variability of the human morbidity during hot weather conditions. Several recent studies indicate that during heat waves the number of hospital admissions related to heat stroke, heat exhaustion, neurological mental illness and renal diseases increases considerably (Santamouris and Kolokotsa, 2015). The rate of increase is usually much higher for the low income and vulnerable population. In particular, Kovats and Ebi (2006) report that during the 2003 heat wave in the UK, adults living in social institutions presented a much higher rate of heat-related health problems than the rest of the population. Similar findings are reported by Rozzini et al. (2005) for Italy and France for the same heat wave period.

The impact of low indoor temperatures on human health is better documented than the corresponding impact of very high temperatures. Several studies have shown that exposure to low indoor temperatures is the source of serious respiratory deceases, increased blood pressure, more frequent accidents, risk of stroke, bronchitis, pneumonia, allergies, cold throat illness, asthma, mental deficiencies and arthritis (Shortt and Rugkasa, 2007; Santamouris and Kolokotsa, 2015). The WHO (2007) performed a large-scale health survey in eight European countries aiming to identify the links between house quality and health problems. The study concluded that residents of cold homes, or of homes with a poor envelope quality that is uninsulated and single glazed, suffer much more frequently from respiratory problems than those living in well-protected houses with proper indoor environmental conditions. In parallel, it is found that about 16.8% of residents living in poor quality houses declared that they have bad health conditions, while the corresponding percentage for residents of proper houses was 4.7% (World Health Organization, 2007). Similar results are also reported by DETR in the UK (2000). They add that people living in cold houses needed more frequent medical treatment. It is characteristic that residents of cold houses visited health services twice as often as residents living in properly heated homes (Evans et al., 2000), and children living in cold homes presented an increased risk, 30%, of visiting hospitals or health centers (Liddel, 2008).

Vulnerable people and aged residents who are older than 65 years present a higher sensitivity to very low or very high indoor temperatures. Santamouris and Kolokotsa (2015) report that the aged population present an increased rate of respiratory problems and are exposed to significant physiological stress. This is because the aged population presents a diminished cold-induced thermoregulation that may decrease the temperature of the body and result in death. Additionally, several research studies have clearly shown that low indoor temperatures are associated with poor mental health. As reported by Thomson and Snell (2013) and Harrington et al. (2005), very low indoor temperatures have a serious and damaging psychological effects, including stress, social isolation, depression and constraints on mobility.

Apart from the serious health issues associated with cold and warm homes, poor indoor environmental conditions have a significant impact on the economics of health services. As reported by Evans et al. (2000), homes of poor environmental quality cost the National Health System (NHS) of the UK almost GBP£2 billion per year, while for Northern Ireland alone the cost approaches GBP£21 million per year.

Similar results are reported by Pignatta et al. (2017) for low income households in Cyprus (Table 4.13).

TABLE 4.13 Health General Conditions and Problems in the Three Defined Clusters

	Cluster A	Cluster B	Cluster C
Health general conditions and problems			
Self-reported health level (−2 to 2)[a]	0.96	0.19	0.24
Percentage of population with at least a serious disease [%]	15.38	16.13	31.03
Percentage of population with depression problem [%]	3.85	12.90	13.79
Percentage of population with allergies [%]	3.85	9.68	3.45
Percentage of smokers [%]	42.31	38.71	31.03

[a] *−2 = very bad, −1 = bad, 0 = acceptable, 1 = good, 2 = very good.*
Source: Pignatta et al. (2017).

REFERENCES

Albouy, Yves, Nadifi, Nadia, 1999. Impact of Power Sector Reform on the Poor: A Review of Issues and the Literature. World Bank, Energy, Mining, and Telecommunications Department, Washington, D.C.

Baccini, M., Biggeri, A., Accetta, G., Kosatsky, T., Katsouyanni, K., Analitis, A., et al., 2008. Heat effects on mortality in 15 European cities. Epidemiology 19, 711–719.

D. Barnes: The Concept of Energy Poverty, http://www.energyfordevelopment.com/2010/06/energy-poverty.html, 2018.

Barnes, Douglas F., Dowd, Jeffrey, Qian, Liu, Krutilla, Kerry, Hyde, William, 1994. Urban Energy Transitions, Poverty, and the Environment: Understanding the Role of Urban Household Energy in Developing Countries. Draft Report, Industry and Energy Department. World Bank, Washington DC.

Barreiro, J., Sanchez, M., Viladrich-Grau, M., 2005. How much are people willing to pay for silence? A contingent valuation study. Appl. Econ. 37, 1233–1246.

Basu, R., Samet, J.M., 2002. Relation between elevated ambient temperature and mortality: A review of the epidemiologic evidence. Epidemiol. Rev. 24, 190–202.

Bates, D.V., 2005. Why do older patients die in a heatwave? QJM: Monthly Journal of The Association of Physicians 98 (11), 840–841.

Beizaee, a, Lomas, K.J., Firth, S.K., 2013. National survey of summertime temperatures and overheating risk in English homes. Build. Environ 65, 1–17. Available from: https://doi.org/10.1016/j.buildenv.2013.03.011.

Bilger, M., Carrieri, V., 2013. Health in the cities: when the neighborhood matters more than income. J. Health Econ. 32, 1–11. Available from: https://doi.org/10.1016/j.jhealeco.2012.09.010.

Birol : Energy and Poverty, World Outlook 2002, Energy Forum, World Bank, June 4–5 2002, Washington.

Boardman, B., 1991. Fuel Poverty: From Cold Homes to Affordable Warmth. Belhaven Press, London.

Brown, P., 1995. Race, class, and environmental health: A review and systematization of the literature. Environ. Res. 69, 15–30. Available from: https://doi.org/10.1006/enrs.1995.1021.

Brunner, K.-M., Spitzer, M., Christanell, A., 2012. Experiencing fuel poverty. Coping strategies of low-income households in Vienna/Austria. Energy Policy 49, 53–59. Available from: https://doi.org/10.1016/j.enpol.2011.11.076.

Burholt, V., Windle, G., 2006. Keeping warm? Self-reported housing and home energy efficiency factors impacting on older people heating homes in North Wales. Energy Policy 34, 1198–1208. Available from: https://doi.org/10.1016/j.enpol.2004.09.009.

CIBSE, 2006. Guide A, Environmental Design. 7th (ed.) Consumers Energy Alliance, 2017: https://consumerenergyalliance.org/2017/10/americans-living-energy-poverty/

United States Census Bureau; Campaign for Home Energy Assistance, "Investing in LIHEAP Why Energy Assistance Is More Important Than Ever," March 2014.

Chuang, J.C., Callahan, P.J., Lyu, C.W., Wilson, N.K., 1999. Polycyclic aromatic hydrocarbon exposures of children in low-income families. J. Expo. Anal. Environ. Epidemiol. 9, 85–98.

Climate Change Challenge : 2018 http://www.climatechangechallenge.org/Resource%20Centre/Fuel-Poverty/USA_Fuel_Poverty.htm

Clinch, J. Peter, Healy, John, D., 2000. Housing standards and excess winter mortality. J. Epidemiol. Commun. Health 54, 719–720.

Critchley, R., Gilbertson, J., Grimsley, M., Green, G., Street, H., 2007. Living in cold homes after heating improvements: Evidence from warm-front, England's home energy efficiency scheme. Appl. Energy 84, 147–158. Available from: https://doi.org/10.1016/j.apenergy.2006.06.001.

Dascalaki, E., Santamouris, M., Bruant, M., Balaras, C.A., Bossaer, A., Ducarne, D., et al., 1999. Modelling large openings with COMIS. Journal of Energy and Buildings 30, 105–115.

Datamine, 2009. Datamine Project: Collecting DATA from Energy Certification to Monitor Performance Indicators for New and Existing buildings. EACI. European Commission, Brussels, Brussels.

De Dear, R., Brager, G.S., 1998. Developing an adaptive model of thermal comfort and reference. ASHRAE Trans. 104, 1–18.

K. Dol and M. Haffner, "Housing Statistics in the European Union," 2010.

Druckman, A., Jackson, T., 2008. Household energy consumption in the UK: A highly geographically and socio-economically disaggregated model. Energy Policy 36, 3167–3182.

EACI : "Towards an European Building Code," EACI, European Commission, Brussels, Brussels, 2003.

Edwards, L., Torcellini, P., 2002. A Literature Review of the Effects of Natural Light on Building Occupants.

Energypedia: https://energypedia.info/wiki/Energy_Poverty, 2018.

Erhorn-Kluttig, H., Erhorn, H., De Boer, J., 2007. Applying the Energy Performance of Buildings Directive (EPBD) to existing buildings. Fraunhofer Institut Bauphysik 34.

European Quality of Life Survey," Ireland, 2003.

Eurostat 2018: http://ec.europa.eu/eurostat/statistics-explained/index.php/EU_statistics_on_income_and_living_conditions_(EU-SILC)_methodology_-_economic_strain#Description

Evans, G., Kantrowitz, E., 2002. Socioeconomic status and health: The potential role of environmental risk exposure. Annu. Rev. Public Health 23, 303–331. Available from: https://doi.org/10.1146/annurev.publhealth.23.112001.112349.

Evans, J., Hyndman, S., Stewart-Brown, S., Smith, D., Petersen, S., 2000. An epidemiological study of the relative importance of damp housing in relation to adult health. Journal of Epidemiology and Community Health 54 (9), 677–686.

Ezratty, V., Ormandy, D., 2011. The Health Benefits of Energy Efficiency. IEA workshop on evaluating the co-benefits of low-income weatherisation, Dublin.

Firth, S., Benson, P., Wright, A.J., 2007. The 2006 Heatwave: Its Effect on the Thermal Comfort of Dwellings. Third Annual Meeting, Network for Comfort and Energy Use in Buildings, Windsor, UK.

Firth, S.K., Wright, A.J., 2008. Investigating the Thermal Characteristics of English Dwellings: Summer Temperatures. 27–29.

Flynn, A., McGreevy, C., Mulkerrin, E.C., 2005. Why do older patients die in a heatwave? QJM—Monthly Journal of the Association of Physicians 98 (3), 227–229.

Gesundheit, U., 2009. Federal Environment Agency Umwelt, Gesundheit und soziale Lage [Environment, Health and Social Status] (Berlin).

Grösche, P., 2010. Housing, energy cost, and the poor: Counteracting effects in Germany's housing allowance program. Energy Policy 38, 93–98. Available from: https://doi.org/10.1016/j.enpol.2009.08.056.

Hardin, B., Kelman, B., Saxon, A., 2003. Adverse human health effects associated with molds in the indoor environment. Journal of Occupational and Environmental Medicine 45 (5), 470–478 (American College of Occupational and Environmental Medicine).

Harrington, B.E., Heyman, B., Merleau-Ponty, N., Stockton, H., Ritchie, N., Heyman, A., 2005. Keeping warm and staying well: Findings from the qualitative arm of the Warm Homes Project. Health & Social Care in the Community 13 (3), 259–267.

Healy, J., Clinch, J., 2004. Quantifying the severity of fuel poverty, its relationship with poor housing and reasons for non-investment in energy-saving measures in Ireland. Energy Policy 32, 207–220.

Healy, J.D., 2003. Excess winter mortality in Europe: A cross country analysis identifying key risk factors. Epidemiol. Commun. Health 2003 (57), 784–789.

Healy, J.D., Clinch, J.P., 2002. Fuel Poverty In Europe: A Cross-Country Analysis Using A New Composite Measurement. Environmental Studies Research Series Working Papers 2002.

Herrero, S.T., Ürge-Vorsatz, P.D., 2010. Fuel Poverty in Hungary: A First Assessment.

Holgersson, M., Norlén, U., 1984. Domestic indoor temperatures in Sweden. Build. Environ. 19, 121–131.

Hong, S.H., Oreszczyn, T., Ridley, I., 2006. The impact of energy efficient refurbishment on the space heating fuel consumption in English dwellings. Energy Build. 38, 1171–1181.

Hong, S.H., Gilbertson, J., Oreszczyn, T., Green, G., Ridley, I., Green, C., et al., 2009. A field study of thermal comfort in low-income dwellings in England before and after energy efficient refurbishment. Build. Environ 44, 1228–1236. Available from: https://doi.org/10.1016/j.buildenv.2008.09.003.

Hopton, J., Hunt, S., 1996. The health effects of improvements to housing: A longitudinal study. Housing Studies 11 (2), 271–286.

Hunt, D.R.G., Gidman, M.I., 1982. A national field survey of house temperatures. Build. Environ. 17, 107–124. Available from: https://doi.org/10.1016/0360-1323(82)90048-8.

Hutchinson, E.J., Wilkinson, P., Hong, S.H., Oreszczyn, T., 2006. Can we improve the identification of cold homes for targeted home energy efficiency improvements? the Warm Front Study Group Appl. Energy 83, 1198–1209. Available from: https://doi.org/10.1016/j.apenergy.2006.01.007.

International Energy Agency, World Energy Outlook, 2010.

International Energy Agency, World Energy Outlook, 2011.

International Energy Agency 2017: Energy Access Outlook. Gridmate: https://www.gridmates.com/blog_posts/37.

Kavgic, M., Summerfield, A., Mumovic, D., Stevanovic, Z.M.Z.M.Z., Turanjanin, V., 2012. Characteristics of indoor temperatures over winter for Belgrade urban dwellings: Indications of thermal comfort and space heating energy demand. Energy Build. 47, 506–514. Available from: https://doi.org/10.1016/j.enbuild.2011.12.027.

Keatinge, W.R., Donaldson, G.C., Cordioloi, E., Martinelli, M., Kunst, A.E., Mackenbach, J.P., et al., 2000. Heat-related mortality in warm and cold regions of Europe: Observational study. Br. Med. J. 321, 670–673.

Khanom, L., 2000. In: Rudge, J., Nicol, F. (Eds.), Cutting the Cost of Cold: Affordable warmth for healthier homes. Taylor & Francis, London.

Klinenberg, 2002. Heat Wave: A Social Autopsy of Disaster in Chicago. University of Chicago Press.

Knowlton, K., Rosenzweig, C., Goldberg, R., Lynn, B., Gaffin, S., Hogrefe, C., et al., 2004. Evaluating globalclimate change impacts on local health across a diverse urban region. Epidemiology 15 (4), S100.

Kohlhuber, M., Mielck, A., Weiland, S.K., Bolte, G., 2006. Social inequality in perceived environmental exposures in relation to housing conditions in Germany. Environmental Research 101 (2), 246–255.

Kolokotsa, D., Santamouris, M., 2015. Review of the indoor environmental quality and energy consumption studies for low income households in Europe. Science of the Total Environment 536, 316–330.

A. Kovacevi ˇc, ´Stuck in the Past: Energy, Environment and Poverty in Serbia and Montenegro, United Nations Development Programme, Belgrade, 2004

Kovats, R.S., Ebi, L.E., 2006. Heatwaves and public health in Europe. Eur. J. Pub. Health 16 (6), 592–599.

Kryter, K.D., 1985. The Effects of Noise on Man. Academic Press.

Lapillonne, B., Sebi, C., Pollier, K., Mairet, N., 2012. Energy Efficiency Buildings in the EU Trends in Lessons From the ODYSSEE MURE Project.

Liddell, C., 2008. The impact of Fuel Poverty on Children. University of Ulster, Belfast.

Liddell, C., Morris, C., 2010. Fuel poverty and human health: A review of recent evidence. Energy Policy 38, 2987–2997. Available from: https://doi.org/10.1016/j.enpol.2010.01.037.

Linares, C., Díaz, J., 2008. Impact of high temperatures on hospital admissions: Comparative analysis with previous studies about mortality (Madrid). European Journal of Public Health 18 (3), 317–322.

Lomas, K.J., Kane, T., 2013. Summertime temperatures and thermal comfort in UK homes. Build. Res. Inf. 41, 259–280. Available from: https://doi.org/10.1080/09613218.2013.757886.

Mavrogianni, A., Davies, M., Wilkinson, P., Pathan, A., 2010. LONDON HOUSING AND climate CHANGE: Impact on comfort and health — preliminary results of a summer overheating study. Open House Int. 35, 49–59.

McGregor, Glenn, Mark Pelling, Tanja Wolf, Simon Gosling, 2007. The social impacts of heat waves, Science Report – SC20061/SR6, Published by: Environment Agency, Rio House, Waterside Drive, Aztec West, Almondsbury, Bristol, BS32 4UD.

McKinsey, "Unlocking Energy Efficiency in the United States ECONOMY," 2009.

Michelozzi, P., Perucci, C.A., Forastiere, F., Fusco, D., Ancona, C., Dell'Orco, V., 1999. Inequality in health: Socio-economic differentials in mortality in Rome, 1990–95. J. Epidemiol. Commun. Health 53 (11), 687–693.

Mielck, J. Heinrich, Social inequalities and distribution of the environmental burden among the population (environmental justice), Gesundheitswesen 64 (7) (2002) 405–416 (Bundesverband der Arzte des Offentlichen Gesundheitsdienstes(Germany)).

Oreszczyn, T., Hong, S.H., Ridley, I., Wilkinson, P., 2006. Determinants of winter indoor temperatures in low income households in EnglandWarm Front Study Grp Energy Build. 38, 245–252. Available from: https://doi.org/10.1016/j.enbuild.2005.06.006.

J. Palmer and I. Cooper, "Great Britain's Housing Energy Fact File." Department of Energy and Climate Change, 2011.

Peeters, L., de Dear, R., Hensen, J., D'haeseleer, W., 2009. Thermal comfort in residential buildings: Comfort values and scales for building energy simulation. Appl. Energy 86, 772–780. Available from: https://doi.org/10.1016/j.apenergy.2008.07.01.

Pignatta, G., Chatzinikola, C., Artopoulos, G., Papanicolas, C.N., Serghides, D.K., Santamouris, M., 2017. Analysis of the indoor thermal quality in low income Cypriot households during winter. Energy and Buildings 152, 766–775.

Platt, S.D., Martin, C.J., Hunt, S.M., Lewis, C.W., 1989. Damp housing, mould growth, and symptomatic health state. BMJ 298 (6689), 1673–1678 (Clinical research ed.).

C. Revie, Fuel Poverty, Housing and Health in Paisley, Glasgow, Energy Action Scotland, 1998.

Rozzini, R., Zanetti, E., Trabucchi, M., 2004. Elevated temperature and nursing home mortality during the 2003 European heat wave [2]. Journal of the American Medical Directors Association 5 (2), 138–139.

A. Rybkowska, M. Schneider, Housing Conditions in Europe in 2009, Eurostat, 2011.

Sakka, A., Santamouris, M., Livada, I., Nicol, F., Wilson, M., 2012. On the thermal performance of low income housing during heat waves. Energy Build. 49, 69–77.

Santamouris, M., Kolokotsa, D., 2015. On the impact of Urban overheating and extreme climatic conditions on housing energy comfort and environmental quality of vulnerable population in Europe. Energy and Buildings 98, 125–133. Available from: https://doi.org/10.1016/j.enbuild.2014.08.050.

Santamouris, M., Kapsis, K., Korres, D., Livada, I., Pavlou, C., Assimakopoulos, M.N., 2007. On the relation between the energy and social characteristics of the residential sector. Energy and Buildings. 39, 893–905.

Santamouris, M., Alevizos, S.M., Aslanoglou, L., Mantzios, D., Milonas, P., Sarelli, I., et al., 2014. Freezing the poor—Indoor environmental quality in low and very low income households during the winter period in Athens. Energy and Buildings 70, 61–70.

Serageldim, R. Barrett and Joan Martin Brown (Editors): The Business of Sustainable Cities, Environmentally Sustainable Development Proceedings, Series No 7. The World Bank, Washington, DC, 1995.

Shapiro, G.G., Wighton, T.G., Chinn, T., Zuckrman, J., Eliassen, A.H., Picciano, J.F., et al.,1999. House Dust Mite Avoidance for Children With Asthma in Homes of Lowincome Families vol. 103.

Shortt, N., Rugkåsa, J., 2007. 'The walls were so damp and cold' fuel poverty and ill health in Northern Ireland: Results from a housing intervention. Health & Place 13, 99–110 (Mar (1)).

Smith (1990): 'Dialectics of improved stoves', in L. Kristoferson and others (editors), Bioenergy, Contribution to Environmentally Sustainable Development, Stockholm Environment Institute, Stockholm.

Smith, K.R. (1994) Workshop on the Energy-Environment Nexus: Indian Issues and Global Impacts (Center for the Advanced Study of India, Univ. of Pennsylvania, Philadelphia).

Stathopoulou, M., Synnefa, A., Cartalis, C., Santamouris, M., Karlessi, T., Akbari, H., 2009. A surface heat island study of Athens using high-resolution satellite imagery and measurements of the optical and thermal properties of commonly used building and paving materials. International Journal of Sustainable Energy 3, 59–76.

Summerfield, A.J., Lowe, R.J., Bruhns, H.R., Caeiro, J.A., Steadman, J.P., Oreszczyn, T., 2007. Milton Keynes Energy Park revisited: Changes in internal temperatures and energy usage. Energy Build. 39, 783–791. Available from: https://doi.org/10.1016/j.enbuild.2007.02.012.

Sun, Y., Sundell, J., 2013. On associations between housing characteristics, dampness and asthma and allergies among children in Northeast Texas. Indoor Built Environ. 22, 678–684. Available from: http://dx.doi.org/10.1177/1420326X13480373.

"Tabula, Typology Approach for Building stock Energy Assessment," EACI, European Commission, Brussels, Brussels, 2013.

Thomson, H., Snell, C., 2013. Quantifying the prevalence of fuel poverty across the European Union. Energy Policy 52, 563–572.

Thomson, H., Petticrew, M., Morrison, D., 2001. Health effects of housing improvement: systematic review of intervention studies. BMJ 323 (no. 7306), 187–190 (Clinical research ed.).

Vandentorren, S., Bretin, P., Zeghnoun, A., Mandereau-Bruno, L., Croisier, A., Cochet, C., et al., 2006. August 2003 heat wave in France: Risk factors for death of elderly people living at home. Eur. J. Pub. Health 16 (6), 583–591.

Waddams, Price, 2001. Better energy services, better energy sectors—and links with the poor. The World Bank.

Wargocki, P., Sundell, J., Bischof, W., Brundrett, G., Fanger, P.O., Gyntelberg, F., et al., 2002. Ventilation and health in non-industrial indoor environments: report from a European multidisciplinary scientific consensus meeting (EUROVEN). Indoor Air 12 (2), 113–128.

Wikipedia: Energy Poverty, 2018.

Wingfield, J., Bell, M., Miles-Shenton, D., South, T., Lowe, B., 2008. Evaluating the Impact of an Enhanced Energy Performance Standard on Load-bearing Masonry Domestic Construction.

World Bank. Energy Issues.: Consequences of Energy Policies for the Urban Poor, 1995.

World Health Organization, Large Analysis and Review of European Housing and Health Status (LARES), World Health Organization, DK-2100 Copenhagen, Denmark, 2007.

Wright, A.J., Young, A.N., Natarajan, S., 2005. Dwelling temperatures and comfort during the August 2003 heat wave. Build. Serv. Eng. Res. Technol. 26, 285–300. Available from: https://doi.org/10.1191/0143624405bt136oa.

Yohanis, Y.G., Mondol, J.D., 2010. Annual variations of temperature in a sample of UK dwellings. Appl. Energy 87, 681–690.

Zavadskas, E., Raslanas, S., Kaklauskas, A.A., 2008. The selection of effective retrofit scenarios for panel houses in urban neighborhoods based on expected energy savings and increase in market value: The Vilnius case. Energy Build. 40, 573–587. Available from: https://doi.org/10.1016/j.enbuild.2007.04.015.

FURTHER READING

Barnes, D., Halpern, Jonathan, 2002. The role of energy subsidies. World Bank.

Chapter 5

Defining the Synergies Between Energy Consumption–Local Climate Change and Energy Poverty

EXPLAINING THE SYNERGIES BETWEEN ENERGY CONSUMPTION, ENERGY POVERTY AND LOCAL CLIMATE CHANGE

As explained in the first chapter of this book, energy consumption in the building sector, local and global climate change and energy poverty are very much interrelated. The present chapter aims to identify the synergies between them and their impacts. It will also comprehensively analyze their interrelations and document all complex trade offs between them.

Analysis of the Synergies Between the Energy Consumption of Buildings and Local Climate Change

The synergies between the energy consumption of buildings and local and global climate change are quite evident and well understood. A decrease in energy consumption will cause lower greenhouse gas emissions and result in less waste heat release, thus reducing both global and local overheating problems. In parallel, an increase in the ambient temperature, either at a local or global level, will cause a serious increase in the energy consumption of buildings, especially for cooling purposes, and will potentially decrease the corresponding heating demand. Although the relation between the energy consumption of buildings and global and local climate change are evident, the specific synergies and impacts are quite complex and are associated with detailed positive and negative influences. Fig. 5.1 presents a detailed and comprehensive map of the specific interconnections and influences. Energy consumption by the building sector influences global and local climate change through three precise links, which are analyzed below. In parallel,

Minimizing Energy Consumption, Energy Poverty and Global and Local Climate Change in the Built Environment: Innovating to Zero. DOI: https://doi.org/10.1016/B978-0-12-811417-9.00005-2

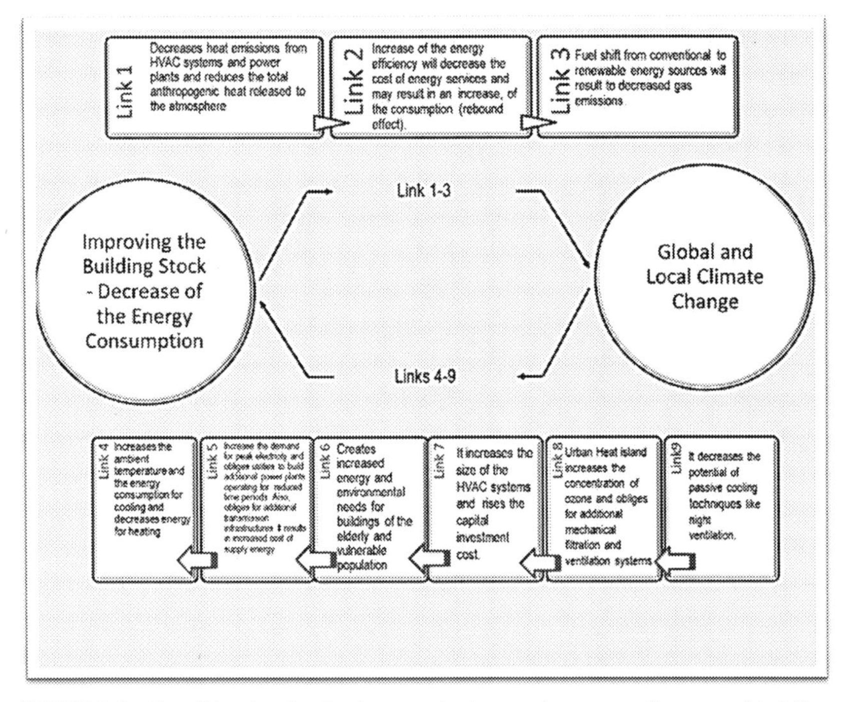

FIGURE 5.1 Causalities that develop between the improved energy performance of buildings and global and local climate change. *Adopted from Santamouris, M., 2016. Innovating to zero the building sector in Europe: Minimising the energy consumption, eradication of the energy poverty and mitigating the local climate change, Solar Energy 128, April 2016, Pages 61–94.*

climate change may impact the energy consumption of buildings through six specific ways, as will also be discussed.

Link 1: Low Energy Consumption Decreases Heat Emissions from the HVAC Systems and Power Plants, and Reduces the Total Anthropogenic Heat and CO_2 Emissions Released into the Atmosphere

Efficient technologies must be used to decrease the energy consumption of the building sector. Given that new buildings present a small fraction of the total building stock, emphasis should be given to the retrofitting of existing buildings. A decrease in energy consumption results in lower emission of greenhouse gases and waste heat, and is associated with a reduction in the amplitude of local and global climate change. According to existing statistics, by 2020 CO_2 emissions in Europe may decrease between 5% to 35%, proportional to retrofitting the building stock and the decarbonization of the energy supply industry (BPIE, 2011).

When a slow decarbonization rate is considered for the energy supply sector, the expected reduction of carbon dioxide emissions may reach 54 MT CO_2 per year or 5%, while under a "business as usual" scenario the expected decrease may reach 161 MT CO_2 or 16%. Finally, when a fast decarbonisation scenario is considered, the reduction of CO_2 emissions could be close to 286 MT CO_2 or 35%. For 2050, when a slow decarbonization is considered, the reduction of emissions may vary between 18% to 73% depending on the degree of retrofitting of the building stock. On the contrary, under a fast decarbonization scenario the expected decrease may range from 72% to 91% (BPIE, 2011).

A decrease in energy consumption by the building sector helps significantly to decrease the magnitude of the Urban Heat Island (UHI) and the strength of local climate change. Lower energy consumption is associated with a decrease in waste and anthropogenic heat from heating and cooling systems and results in a lower UHI intensity. The anthropogenic heat released in cities depends on the characteristics of the energy sources and the local climate. Data from London, UK, reveals that the magnitude of anthropogenic heat released varies between 2380 Wh/m^2 on a sunny day, to 4960 Wh/m^2 on a cold winter's day (Hamilton et al., 2009). Simulations carried out by McCarthy (2009) for several southern European locations shows that when anthropogenic heat increases from 15 W/m^2 to 45 W/m^2, the local ambient temperature may increase by 0.5°C, while Bohnenstengel et al. (2014) estimate that the intensity of the UHI in London may increase up to 1.5°C due to the additional anthropogenic heat.

Link 2: An Increase in the Energy Efficiency of Buildings will Decrease the Cost of Energy Services and may Result in an Increase in Energy Consumption (Rebound Effect)

It is well known that energy consumption may increase when the energy efficiency of buildings is seriously improved and the energy cost decreases. This is known as the "rebound effect" and may result in a serious reduction of expected energy gains (Wigley, 1997). Important research on the rebound effect has been carried out over the last few years, with many countries paying considerable attention to its impact. It should be mentioned that the House of Lords in the UK has expressed serious concerns about the real and important impact of the rebound effect, and considers it to be one of the major reason for the discrepancy between expected and achieved energy conservation in buildings.

The rebound effect and its impact is defined as the ratio between reduced energy savings and expected energy savings. It has been calculated for various European countries (Grepperud and Rasmussen, 2004; Hanley et al., 2009; Turner, 2009). In Italy, Cellura et al. (2013) found that the impact of the rebound effect varies from 0.91%–13.7%. Other studies have shown that in the residential sector, the impact of the rebound effect varies from 10% to 30% for space heating to 5%–12% for lighting (Greening et al., 2000).

Link 3: Fuel Shift From Conventional to Renewable Energy Sources Will Result in Decreased Gas Emissions

Decarbonization of the energy supply sector is mainly achieved using renewable energy sources installed at the building or central level. It may significantly reduce emissions of greenhouse gases and decrease the amplitude of global and local climate change.

Grid-connected photovoltaics may play a major role facilitating a more significant penetration of renewables in the supply sector. In Europe, the total capacity of grid-connected photovoltaics was close to 81 GW in 2013 (EPIA, 2014). Among them, about 2.4 GW were installed in residential homes and 2.96 GW in commercial buildings (EPIA, 2014). It is expected that by 2018, the capacity of roof top photovoltaics installed in Europe may vary from 5.9 GW to 12.4 GW (EPIA, 2014).

Defaix et al (2012) estimate that the technical potential of building-integrated photovoltaics in Europe-27 is close to 951 GWp and may supply 840 TWh of electricity per year, which is equal to 22% of the electricity consumption foreseen by 2030. If we take into account that the intensity of the grid GHG emissions in Europe is 289 gCO2e/KWH (IEA, 2011), the potential decrease of the greenhouse gases should reach 10% of the total GHC emissions in Europe for 2010.

Global and local climate change affects the energy consumption of buildings in the following ways.

Link 4: Higher Ambient Temperatures Increase Energy Consumption for Cooling and Decrease Energy Consumption for Heating

It is evident that an increase in the ambient temperature will result in a decrease of the energy load necessary for heating purposes and an increase of the corresponding cooling load. Several scientific investigations aiming to identify the energy impact of the increased urban ambient temperature have been performed around the world (Hassid et al., 2000; Santamouris et al., 2001; Kolokotroni et al., 2012). Santamouris (2014a) and Santamouris (2014b) have calculated the specific energy penalty induced by the UHI for typical buildings in Europe. It is found that on average, the UHI increases the cooling load of buildings by 13%. It is also reported that the calculated increase of the cooling load is more significant in cooling-dominated climates, where the average summer time ambient temperature exceeds 27°C. In parallel, in heating dominated climates, where the average summer ambient temperature is below 23°C, the associated decrease of the heating load is higher than the corresponding increase of the cooling load.

An evaluation of the additional energy penalty caused by the urban overheating (Santamouris, 2014a,b) has shown that the Global Energy Penalty per unit of city space and per degree of UHI intensity is close to 0.74 (±0.67) Kwh/m^2, while the Global Energy Penalty per person is close to 237 (±130) kWh/p.

Link 5: Higher Ambient Temperatures and Heat Waves Increase Peak Electricity Demand and Oblige Utilities to Build Additional Power Plants Operating for Reduced Time Periods. Furthermore, They Create a Requirement for Additional Electricity Transmission Infrastructures. It Results in Increased Cost of Supply Energy

Urban overheating increases the peak electricity load of buildings for cooling purposes. As a result, utility companies must build additional power plants to cover the demand. Santamouris et al. (2015) has analysed the impact of urban overheating on peak electricity demand in various countries. As reported, the peak electricity penalty per degree of temperature increase is between 0.45% and 4.6%. This corresponds to about 21 W of additional power per person and degree of temperature increase. It is also found that the increase in electricity consumption is between 0.5% and 8.5% per degree of temperature rise.

Higher ambient temperatures affect the performance and the capacity of conventional power plants and harm electrical distribution networks (EU, 2013). It is well-known that during the heat waves, nuclear power plants suffered from severe cut backs because of the rise of the water temperature in the rivers used to cool down the plants (Rubbelke and Vogele, 2011). Van Viet et al. (2012) reports that the power capacity of the electricity plants in Europe may decrease between 6.3% and 19% because of the increase of the water temperature and potential water scarcity caused by climate change. This is a highly probable scenario, as IPCC (2008) has foreseen that the frequency of periods of water scarcity and elevated water temperature will increase. The potential water scarcity may affect the production capacity of the hydropower electricity plants by about 25% (EEA, 2012). This may correspond to a capacity reduction of the nuclear and hydropower electricity plants by 6 GW and 12 GW respectively (Rubbelke and Vogele, 2011). A further reduction of about 19 GW is foreseen in the case of serious water scarcity.

The significant uncertainty generated around this issue and the need to maintain the power supply system at very high reliability standards should require additional investments in alternative electricity generation systems. Rubbelke and Vogele (2011) predict that electricity prices in France and

Switzerland may increase up to 30% and 80% respectively because of the above reasons.

Link 6: Climate Change Creates Increased Energy and Environmental Needs for Dwellings Belonging to the Elderly and Vulnerable Population

Climate change has a significant impact on elderly populations. Given that the percentage of the elderly population is continuously increasing, the global influence may be highly significant. There are several physiological, economic, social and geographic risk factors that affect vulnerable populations. High ambient temperatures are a health threat for elderly and vulnerable populations (Gamble et al., 2013) as their climatic sensitivity is much higher than that of the average population (Filiberto et al., 2009).

Recent medical research has shown that the immune system of older adults is highly affected by exposure to high ambient temperatures, air pollutants and toxicants. They frequently result in cardiovascular and respiratory problems, and may cause higher heat-related mortality rates. Especially in the high latitudes, the potential increase of the ambient temperatures and the associated extreme climatic phenomena can disproportionately impact the local older population, as it is not adapted to high ambient temperatures.

In parallel, extreme overheating occurring during heat waves results in a higher energy demand for air conditioning and affects the economic status of vulnerable populations. Additionally, extreme climatic phenomena may result in electricity blackouts and brownouts, and put even the wealthy population at risk. To counter this problem, several efficient adaptation strategies have been proposed for low income and vulnerable populations (Santamouris et al., 2007a). This includes the use of cool and reflecting paints, advanced ventilation systems and efficient solar control devices (Synnefa et al., 2007).

Link 7: UHI Increases the Size of HVAC Systems and Raises Capital Investment Cost

An increase in the peak cooling load of buildings that is induced by higher ambient temperatures results in the installation of HVAC higher cooling capacity systems. Hassid et al. (2000) calculate that the peak electricity demand in urban zones suffering from urban overheating increases by up to 100% compared to other non-overheated urban zones. They also reported that in Athens, Greece, the peak electricity load for cooling a representative office building increases from 13.7 KW to 27.5 KW in urban zones suffering from UHI effects. Additionally, it is found that under higher ambient temperatures the COP of air conditioning may decrease by up to 25% (Santamouris et al., 2001).

Link 8: UHI Increases the Concentration of Ozone in the Atmosphere and Obliges for the Installation of Additional Mechanical Filtration and Ventilation Systems

High ambient temperatures may result in an increased concentration of tropospheric ozone in urban areas. This is mainly because high ambient temperatures act as a catalyst, accelerating photochemical reactions in the atmosphere. In parallel, high urban temperatures modify the turbulent exchange in the atmosphere and may result in a higher concentration of pollutants. Monitoring of the ozone concentration in Athens during a period of heat waves shows that there is a significant increase in the ozone concentration (Stathopoulou et al., 2008). Similar results are also reported in Paris during strong overheating events (Sarrat et al., 2006). Exposure to high concentrations of ozone is found to be associated with increased mortality rates (Chen et al., 2012). Ozone is transferred to the indoor environment through openings, cracks and ventilation systems. Measurements have shown that the indoor to outdoor concentration rate, may vary from 0.09-047 I/O, which means that a very significant fraction of the outdoor ozone concentration is transferred indoors. Given that the byproducts of ozone reactions may include VOCs, carbonyls, and ultrafine particles that can potentially irritate the respiratory system of the occupants (Rai et al., 2013), special ozone removal filters must be installed to protect occupants. Commercially available ozone filters have a relatively high removal efficiency and reduce the exposure of building occupants (Zhao et al., 2005).

Link 9: Higher Ambient Temperatures Decrease the Potential of Passive Cooling Techniques Such as Night Ventilation

The climatic potential of most environmental sinks depends on the levels of the ambient temperature. Higher ambient temperatures affect the performance of sinks and their dissipation capacity, resulting in a more intensive use of air conditioning systems. Techniques to dissipate the excess heat of building are known as "passive cooling dissipation techniques." They include natural and hybrid ventilation techniques and systems, the use of the ground and water as low temperature heat sinks, and also the use of the night sky as a cool sink, known as radiative cooling (Santamouris and Kolokotsa, 2013).

Night cooling is a very efficient technique to decrease the temperature of buildings, provided that ambient temperatures are quite low during the night time and the storage capacity of the building is significant. Night time ventilative cooling may substantially decrease the cooling load of buildings and provide adequate comfort conditions (Santamouris et al., 2010). However, recent climatic data show that the number of warm nights in different parts of the world, such as the UK, Italy, Denmark, and the Balkan and Iberian peninsulas, have increased significantly (Euro4M, 2015).

Increased night time ambient temperatures significantly decrease the dissipative potential of night ventilation techniques and intensify the need for mechanical cooling systems. Geros et al. (1999) have experimentally evaluated the cooling potential of night ventilation techniques under intensive UHI conditions. They found that the cooling potential of night ventilation decreased by 35% in air-conditioned buildings, while under extreme conditions the loss of potential efficacy could be close to 70%. In parallel, they reported that under free floating conditions, the indoor temperature on the following day was about 1.0K higher compared to the reference case, which had a negligible UHI intensity. Kolokotroni et al. (2012) have also reported similar results for offices in London, while Ramponi et al. (2014) draw similar conclusions for buildings located in Amsterdam, Rome and Milan.

Ground cooling techniques are very well known and are extremely well documented (Mihalakakou et al., 1994a). Current ground cooling systems use horizontal or vertical buried heat exchangers to circulate warm indoor air into the ground (Mihalakakou et al., 1994b). UHIs affect the ground temperature and decrease the potential efficacy of ground cooling techniques (Changnon, 2004; Taniguchi, 2006; Lokoshchenko and Korneva, 2015). The potential Increase of the soil temperature may significantly affect the efficiency of ground dissipation technologies and may reduce their cooling capacity.

Synergies and Impacts Between Low Energy Consumption Buildings and Energy Poverty

The relation between the energy consumption of buildings and energy poverty is quite complex. It is evident that once the energy efficiency of buildings improves, the energy consumption should decrease and potentially the energy poverty problem may be amortized. Besides the technological synergies, important social and economic drivers influence and define the relative impacts. Fig. 5.2 below reports the main synergies and links between the energy consumption in buildings and energy poverty, and vice versa. Four specific synergies are identified through which the energy consumption of buildings influences the characteristics of energy poverty. Additionally, two links through which energy poverty affects the energy performance of buildings are identified and analyzed.

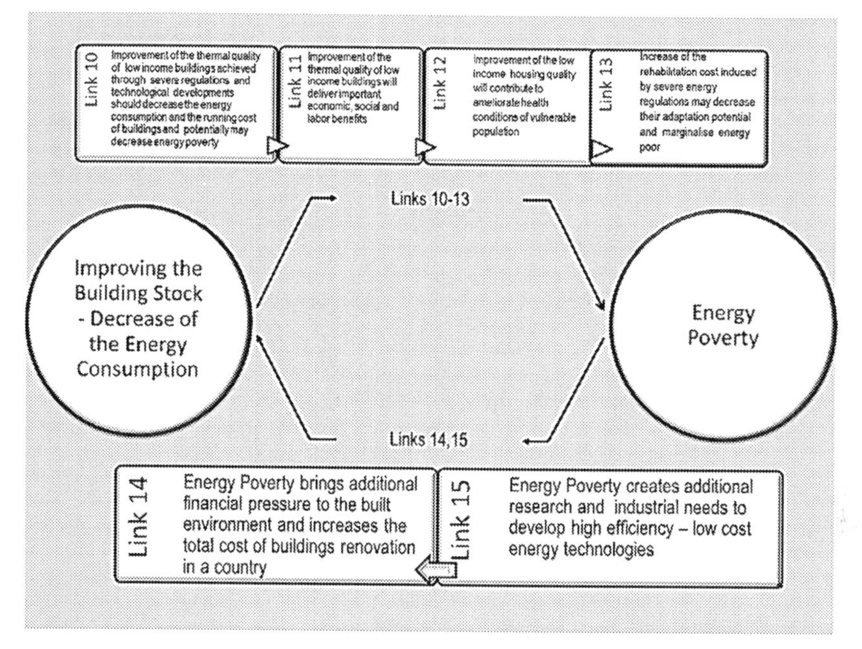

FIGURE 5.2 Causalities that develops between the improved energy performance of buildings and energy poverty. *Adopted from Santamouris, M., 2016. Innovating to zero the building sector in Europe: Minimising the energy consumption, eradication of the energy poverty and mitigating the local climate change, Solar Energy 128, April 2016, Pages 61–94.*

Link 10: Improvement of the Thermal Quality of Buildings Achieved Through Strict Regulations and Technological Developments Should Decrease the Energy Consumption and the Running Cost of Buildings, and may Potentially Decrease Energy Poverty

Analysis of almost all existing data shows that low income families live in poor energy performance houses (Kolokotsa and Santamouris, 2015). It is well known that in the UK, about half of lower income households live in buildings presenting the lowest energy ranking certification. These houses present a very high energy consumption that cannot be met by low income households. Several real-scale projects have shown that when energy efficient techniques are used to retrofit low income houses, their energy consumption decreases considerably, while the indoor environmental quality is substantially improved. Programs for energy efficient retrofitting are designed and implemented in many countries. Most of the projects have succeeded in drastically alleviating energy poverty and improving the economic status of the low income households involved in the projects.

Several planning studies have shown that the implementation of energy efficiency and renewable energy measures in the UK, such as solar water heating, insulation and high efficiency heating systems, may cost GBP£5 billion and can help about 1 million low income households to escape from energy poverty (EAS, 2009). As Preston et al. (2008) report, in the UK, the cost of about 3.9 million measures installed in almost 2 million houses reached 3.6 billion pounds by 2008, while about 70% of the households succeeded in escaping from energy poverty. Additionally, the UK Government is committed to upgrading the energy efficiency of a high number of low income houses by 2030, in accordance with the 2013 Energy Act.

In Greece, energy efficient technologies such as insulation, sufficient heating systems, double windows, solar water heaters, and so on have been installed in about 100,000 low and medium income houses. In Ireland about 60,000 low income buildings have been refurbished by the Warmer Homes Scheme (Irish Department of Communications, Energy and Natural Resources, 2015). Finally, in Romania, it is expected that about 110,000 citizens will benefit from an energy retrofitting program, while greenhouse emissions will be decreased by over 660,000 tonnes of CO_2, eq (UNDP, 2015).

Link 11: Improvement in the Thermal Quality of Low Income Buildings will Deliver Important Economic, Social and Labour Benefits

Improving the energy performance of low income housing may result in important non-energy benefits for the citizens and their communities. According to Skumatz (2010), non-energy benefits may present a value three times higher than the energy profits. According to the same author, the non-energy benefits to low income households may range between EUR€270 and EUR€4500 annually per participant. The main non-energy-related benefits arising from the refurbishment of low income houses are:

- A significant reduction in the greenhouse gas emitted by the energy production plants. In the UK, it is estimated that the possible CO_2 reduction induced by the refurbishment of low income housing stock will be close to 23.6 $MTonCO_2$ (Washan et al., 2014). This corresponds to about 33% of the greenhouse emissions of the transport sector in the country.
- The stimulus of the market will result in new economic opportunities and benefits. Retrofitting low income houses will result in significant economic investments aimed at decreasing the energy consumption of low income buildings. This may result in important economic benefits, such as an increase in national GDP, the creation of additional labor opportunities and a lower energy dependence on external sources. Washan et al (2004) report that in the UK, for each pound invested into improve the energy performance of low income housing, there are almost 3.2 lb returned back through increased GDP and 1.27 lb due to of additional tax revenues. Additionally, it is reported that retrofitting low

income housing may increase the GDP in the UK by 0.6% by 2030 (Washan et al., 2014). Finally, it is estimated that the cost benefit ratio of economic activities relating to the retrofitting of low income housing is very high, at close to 2.27:1.

- Retrofitting low income housing may create additional labor opportunities, especially for the vulnerable population. Washan et al. (2014) reports that because of the refurbishment of low income housing in the UK, about 108,000 new jobs per year may be generated over the period 2020–2030. In parallel, as reported by the Worldwatch Institute (2008), the energy retrofitting of about 200,000 German houses has created 140,000 new jobs.
- Significant increase in the economic value of the retrofitted houses. Rehabilitated houses may present a significantly increased rental and property value.
- Improves the independence of low income households from energy assistance programs. Beneficiaries of retrofitted improved independence of low income houses will no longer be dependent on national assistance programs. This will result in a significant decrease in state expenditure on supporting low income households, while it will help the vulnerable population to improve their social status.
- Positive impact on the economic and social status of deprived areas where low income people use to live. The economic activities undertaken in deprived areas to upgrade the energy performance of low income housing may improve local employment opportunities and financial conditions in the area.

Link 12: Improvement in the Housing Quality will Help to Ameliorate Health Problems Among the Vulnerable Population

It is well accepted that inadequate indoor environmental conditions are associated with significant health problems. Low or high indoor temperatures aggravate serious cardiovascular and respiratory problems, may affect mental health, and may increase mortality rates, especially in vulnerable and aged populations. As reported by the Department of Health (2010), mortality in the colder quarter of houses in the UK is almost three times higher than in the warmest quarter. In parallel, the Chief Medical Officer of the UK (2009) reported that the incidence of respiratory illnesses of children living in low temperature shelters is almost double that of children living in homes with an adequate temperature.

The additional cost required to treat health problems caused by cold or overheated homes seems to be very high. Data from the UK (Marmot Review, 2011) shows that cold homes cost about GBP£1.36 billion pounds in the UK. It is estimated that for each pound spent on the energy rehabilitation of low income houses in the UK, the national health system may save almost 42 pence (Chief Medical Officer, 2009).

Better energy efficiency and more adequate indoor temperature conditions help substantially to improve health conditions in low income households and decrease the cost of the health system. In particular, the implementation of adequate ventilation systems and techniques significantly decreases the concentration of harmful indoor pollutants and contaminants. As shown by a study carried out by Carnegie Mellon University (2005), proper indoor air quality is associated with a seriously reduced incidence of flu and colds (up to 50%). In parallel, as reported by Green and Gilbertson (2008), in low income energy rehabilitated houses the incidence of depression and anxiety was decreased by 50%.

Efficient energy retrofitting of low income housing may significantly improve indoor comfort conditions and reduce excessive indoor moisture. According to Hong et al. (2009), the average indoor temperature in 2500 energy-refurbished low income houses in the UK has resulted in an increase of the average indoor temperature from 17.1°C to 19°C, while almost 79% of the beneficiaries reported that they were living under acceptable thermal comfort conditions.

5Link 13: An Increase in Rehabilitation Cost Induced by Severe Energy Regulations may Decrease Adaptation Potential and Marginalize the Energy Poor

Although the energy rehabilitation of low income buildings contributes greatly to reducing energy poverty, to eradicate energy poverty in a country it must be combined with other social, economic and environmental drivers and policies. In particular, policies regulating the price of energy greatly affect the conditions of energy poverty in a country and may have a serious impact on low income households.

When energy prices increase faster than the income of citizens, they may increase the levels of energy poverty. As reported by the UK House of Commons (2009), an increase in energy prices by 1% results in an additional 40,000 households entering energy poverty. High energy prices may substantially reduce the energy consumption of low income households or even minimize it in a way that basic energy needs are not met. This may result in additional financial assistance and subsidies.

The thermal and construction quality of low income houses is usually not appropriate, and the cost of energy refurbishment may be prohibitive without financial assistance. As reported by Preston et al. (2008), in the UK the cost of the energy refurbishment of houses belonging to households under severe energy poverty is close to GBP£3107, while the average cost of retrofitting a low income house is close to GBP£1229. Because of the very high cost of rehabilitation, almost 29% of energy poor households in the UK, or 730,000 families, may not be included in the national program of housing improvements.

An increase in energy prices has a very significant impact on the budget of low income households. While in 2003 the cost of achieving proper indoor conditions in the UK was close to GBP£248, it increased up to GBP£438 in 2011 due to a substantial increase in energy prices (Palmer and Cooper, 2013).

New, strict building energy regulations aimed at fighting global climate change, such as the new Energy European Directive, may increase the cost of rehabilitation as almost state of the art energy technologies are required. Given that the necessary capital or financial credit are not easily available to low income households, energy-efficient retrofitting may not be attempted, or much less efficiency systems may be implemented, thereby increasing the social and economic disparity between the rich and poor population (Urge-Vorsatz and Tirado Herrero, 2012). This risk is quite high in Europe. According to Urge-Vorsatz et al. (2012), the potential for energy conservation in western Europe for the year 2050 is estimated to be close to 72% compared with 2005 levels, and about 46% of potential energy gains are considered unlikely to be unrealized due to several financial and social problems.

Future societies may present a high energy efficiency for most of the population. Energy poor populations will face a serious risk of marginalization under economic and social terms.

Link 14: Energy Poverty Brings Additional Financial Pressure to the Built Environment and Increases the Total Cost of Building Renovation in a Country

To eradicate the problem of energy poverty, governments must allocate important financial resources to subsidizing the energy cost of low income households and to implement large scale energy efficient retrofitting of their houses. The budget to be spend depends on the quality of the low income building stock in each country, financial subsidies, and the specific energy systems and components to be installed. According to Jenkings (2010), in the UK the cost of refurbishing 500,000 low income houses to achieve a 50% energy reduction varies between GBP£3.9 and GBP£17.5 billion pounds. This corresponds to about GBP£7,000–31,900 per dwelling, according to various technologies considered for retrofitting. Such a budget seems to be very high, however it shouldo be pointed out that between 2000–2008 the British Government has spent almost GBP£20 billion pounds in energy subsidies (UKDECC, 2009). According to Preston et al. (2008), in the UK the average eligible cost spent by the Warmer Front per household is close to GBP£1,825. In Ireland, the so-called Warmer Homes Scheme has undertaken retrofitting activities in about 61,000 low income dwellings, with an average cost close to EUR€1000 per recipient (Irish Department of Communications, Energy and Natural Resources, 2015). Finally, in Greece the maximum eligible retrofitting cost does not exceed EUR€15,000 per retrofitting, out of which low income households receive a state grant of between 15%–70% of the global cost according to household income.

Link 15: Energy Poverty Creates Additional Research and Industrial Needs to Develop High Efficiency-low Cost Energy Technologies

To better face energy poverty, the use of available financial resources must be optimized. The selected and implemented energy technologies for low income houses should present the maximum energy outcome during the lifetime of the building, with the minimum possible financial cost. Several available energy technologies have been assessed regarding their suitability for low income housing retrofitting. For example, Preston et al. (2008) reported that for the UK, the use of photovoltaics is not appropriate until their cost is substantially decreased. Additionally, they point out the need to further improve energy conservation technologies, such as thermal insulation, boilers and lighting.

Serious demand for high energy performance and low financial cost retrofitting of low income housing stock can drive significant technological developments in the field of low cost energy retrofitting. Several advanced and low cost technologies that are able to improve the energy efficiency and global environmental quality of low income housing have been recently proposed (Santamouris et al., 2007b; Kolokotsa and Santamouris, 2015; Santamouris and Kolokotsa, 2015). Recent applications and projects have shown that newly developed, low-cost mitigation and adaptation techniques, such as, reflective coatings for roofs, the use of advanced fans for ventilation, and the use of heat dissipation techniques can increase the survivability and the quality of life of the low income population.

Impacts and Synergies Between Climate Change and Energy Poverty

A higher ambient temperature and more frequent and long lasting extreme events could seriously affect the quality of life of vulnerable population and low income households. Additionally, energy poverty and inadequate indoor and outdoor environmental conditions can affect the performance of mitigation and adaptation policies. The following section discusses three specific links through which energy poverty impacts climate change, and four links that show how climate change influences energy poverty (Fig. 5.3).

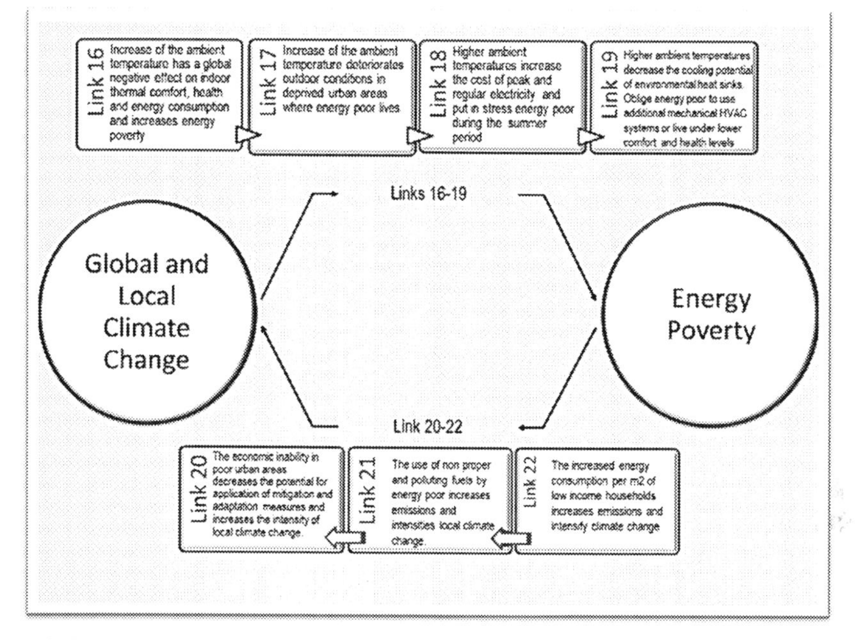

FIGURE 5.3 Causalities that develop between energy poverty and global and local climate change. *Adopted from Santamouris, M., 2016. Innovating to zero the building sector in Europe: Minimising the energy consumption, eradication of the energy poverty and mitigating the local climate change, Solar Energy 128, April 2016, Pages 61–94.*

Link 16: An Increase in the Ambient Temperature has a Global Negative Effect on Indoor Thermal Comfort, Health and Energy Consumption and Increases Energy Poverty

Higher ambient temperatures decrease heating demand, but simultaneously increase the cooling load of buildings to a far greater degree. Several analyses have shown that in cooling-dominated climates the increase in cooling demand is much more significant than the corresponding decrease in the heating load (Santamouris, 2014a, b). A lack of adequate cooling systems, combined with the poor quality of low income houses, result in very high indoor temperatures, especially during heat wave periods, and puts the life of vulnerable populations at risk. Several experimental studies (Mavrogianni et al., 2010; Lomas and Kane, 2013), have shown that during extreme heat events, the indoor temperature in low income households was above the health and comfort thresholds. The energy-poor population is not able to hedge against overheating and presents a much lower ability to adapt. Recent works have demonstrated that climatic hazard is disproportionally higher for the low income population than for other income groups. Experiments carried out during a heat wave in Athens, Greece, revealed that the maximum indoor temperature in low income houses was about 5°C higher than the mean maximum temperature in the rest of the houses, while it persisted for a longer period (Sakka et al., 2012). In parallel, the monitoring of low income houses during a cold spell in Athens, Greece (Santamouris et al., 2014) shows that indoor temperatures were much lower than the recommended indoor temperatures, ranging from 5–9°C. Similar conclusions are also supported by many other similar studies (Kolokotsa and Santamouris, 2015).

Important medical research has been carried out to identify the possible medical problems that extreme weather conditions create for low income and vulnerable populations. It is generally concluded that extreme climatic events seriously affect vulnerable and low income populations living in deprived areas that suffer from overheating problems, people working under very warm conditions, and people suffering from chronic illnesses (Benzie et al., 2011). Data collected in France and Italy during the 2003 heat wave revealed that people living in institutions and vulnerable and low income populations presented the highest percentage of heat related illnesses (Rozzini et al., 2004; Kovats and Ebi, 2006).

Global overheating undoubtedly increases the need for adaptation and mitigation policies and the resources required to implement them. The energy related expenses of people using air conditioners to combat indoor overheating will increase substantially and may drive part of the population under energy poverty levels. Such a risk is significantly higher for the low income population, given the low quality of their houses An assessment of the required additional cost caused by the use of air conditioning in Greece shows that its use disproportionately increases households' electricity expenses. In particular, the use of air conditioning costs on average about EUR€100, or EUR€0.6/m^2 or EUR€12./person, while the corresponding cost for lower income households is EUR€195/household, EUR€1.2/m^2 and EUR€87/person (Santamouris et al., 2007b).

Link 17: An Increase in the Ambient Temperature Deteriorates Outdoor Climatic Conditions in Deprived Urban Areas Where the Energy Poor Live

Low income populations usually live in deprived areas characterized by local overheating problems, the significant generation of anthropogenic heat, high density living, increased atmospheric pollution and a lack of greenery (Gaitani, 2014; Kelly, 2015). As a result, low income people may live under poor outdoor thermal and environmental conditions (Santamouris, 2015). This finding has been reported by many studies aiming to characterize the environmental quality of deprived urban zones (Giannopoulou et al., 2011). Measurements performed in Western Athens show that deprived urban zones presented temperatures that were 2–5°C higher during the summer period than those experienced in the rest of the city. Additionally, the concentration of harmful tropospheric ozone during the summer period was much higher in the low income urban zones than in the rest of the city (Stathopoulou et al., 2008). An evaluation of outdoor thermal comfort conditions in deprived urban areas has shown that very dangerous discomfort conditions occur for the 27.9% and 19.4% in July and August respectively, while comfort conditions may occur for less than 20% of the warm period (Giannopoulou et al., 2013). An assessment of the heat risk performed for the City of London, UK, has resulted in similar conclusions (Arup, 2014).

Link 18: Higher Ambient Temperatures Increase the Cost of Peak and Regular Electricity and put Stress on Energy Poor Populations During the Summer Period

Existing policies aiming at fighting climate change have prompted some countries to introduce carbon taxes, which increase the price of electricity. In particular, Sweden, Finland, Denmark, Norway, Spain, the Netherlands, Italy and Germany have applied carbon levies on electricity (Tiezzi, 2005). The OECD (1996) has raised serious concerns about the distributional problems caused by such taxation. Many researchers and policy makers consider that energy levies have a negative impact on low income households and create additional burdens that cannot be supported (Speck, 1999). For example, Symons et al. (2000) has concluded that energy levies are quite regressive in Germany and France, and also to a certain degree in Spain. Similar conclusions are also drawn by Poltimae and Vork (2009) for Estonia, where additional electricity taxes increase the burden by 40% for the low income group, compared to about 20% for the average population. Callan et al. (2009) has also reported that the levies imposed are regressive in Ireland as well; however, a system of welfare subsidies to the low income population counterbalances the possible negative impacts. As reported by Symons et al. (2000), the additional burden of energy levies to low income households is close to 8%, while it is about 5% for the high income group. In the UK, energy levies have a regressive impact but, as in Ireland, compensation measures are applied for the low income population, although without fully covering their losses (Dresner and Ekins, 2006).

An analysis by Bouzarovski (2015) of the additional electricity taxes imposed in Central and Eastern Europe concludes that they may have a very significant negative impact and may result in increased energy poverty, given that the local population cannot switch away from the use of electricity. Barker and Kohler (1998) analyzed the impact of energy levies on heating fuel in Belgium, Spain, the Netherlands, Ireland, Italy, Luxembourg, Portugal, France, Germany, the UK and Denmark, and they concluded that such taxation is regressive. However, secondary benefits for the low income population may arise from the application of energy levies. As mentioned by Boyd et al. (1995), additional energy taxation may reduce unnecessary energy consumption, decrease pollution and mitigate the magnitude of the

UHI in deprived urban zones where low income populations live. However, the value of the such secondary benefits is not calculated.

Link 19: Higher Ambient Temperatures Decrease the Cooling Potential of Environmental Heat Sinks. They Oblige the Energy Poor to use Additional Mechanical HVAC Systems or Live Under Lower Comfort and Health Levels

As discussed previously in Link 9, an increased ambient temperature decreased significantly the cooling potential of natural cooling techniques based on the use of environmental sinks. The lower cooling potential of passive and hybrid cooling systems and techniques obliges citizens to buy conventional air conditioners to satisfy their thermal comfort requirements. The purchase of air conditioners and the necessary cost of electricity to operate them are serious additional burdens for the low income population and may drive more households towards energy poverty conditions. According to existing statistics, the use of air conditioning by low income populations is much lower at almost half of that on the rest of the society. According to Geros et al. (1999), the expected decrease of the cooling potential of night ventilation techniques may vary between 30% to 70%. In parallel, according to Asimakopoulos et al. (2012), the potential increase of the maximum indoor temperature by 2050 may reach 2.5K, while it is calculated that the increase in cooling degree days may vary between 30% and 108%. It is reasonably concluded that under these specific conditions, economic problems may become more serious, and the corresponding percentage of the population behind the poverty line may increase significantly (Eurostat, 2015).

In parallel, energy poverty affects global and local climate change in the following ways:

Link 20: Economic Disadvantage in Poor Urban Areas Decreases the Potential for the Application of Mitigation and Adaptation Measures, and Increases the Intensity of Local Climate Change

Although it is well accepted that the climatic rehabilitation of outdoor urban areas where vulnerable populations live is one of the best policies to improve local environmental quality, a lack of resources (and in many cases a lack of technical competence) significantly limits the real mitigation potential in deprived urban zones. Assessment of the rehabilitation cost based on real data collected from recent projects shows that it is quite high, and may vary between EUR€300 to EUR€500/m^2 (Fintikakis et al., 2011). To face the problem, resources should be made available by the central government or other authorities who are able to support the expenses. In parallel, the improper maintenance of existing infrastructure can have a significant impact on the real performance of urban mitigation projects. It may intensify local overheating, increase the energy consumption of buildings and put the health of low income and vulnerable populations at risk. Also, it may intensify the local differentiation between rich and poor areas, and decrease the economic value of the specific urban zone.

Link 21: The use of Improper and Polluting Fuels by the Energy Poor Increases Emissions and Intensifies Local Climate Change

To satisfy their energy needs, vulnerable and low income populations mainly use improper and possibly polluting fuels to satisfy their heating needs. In most cases, cheap, uncertified wood, carbon, and other biomass materials are burned in low quality and low efficiency combustion systems, such as wood stoves and so on. The use of polluting fuels generally results in considerable indoor and outdoor pollution problems (Sims et al., 2007). The combustion of improper biomass products increases the emission of greenhouse gases in the atmosphere and intensifies the magnitude of global climate change. In parallel, it creates serious pollution problems because of the increased concentration of suspended particles and smoke. As mentioned by Ecocity (2014), the extensive use of biomass for heating purposes in Athens during the winter period of 2013–15 increased tremendously the concentration of harmful pollutants in the outdoor environment.

According to WHO (2006), almost 5% of European households are using biomass as a heating fuel. However, the use of biomass for heating is much more extensive in several European countries. For example, it is about 23% in Romania, 17% in Bulgaria, 15% in Estonia, 12% in Croatia and 10% in Latvia (WHO, 2006). Additionally, in some non-EU Eastern European countries, biomass seems to be the main fuel for heating and other energy uses: 63% in Moldova, 51% in Bosnia and 50% in Albania. According to UNDP (2004), most of the households that use biomass are low income families lacking sufficient resources, and this 'forces them to adopt dangerous and inefficient strategies' (UNDP, 2004).

Link 22: The Increased Energy Consumption per Square Meter of low Income Households Increases Emissions and Intensifies Climate Change

Several studies have shown that the energy consumption of low income households per unit of house surface is much higher than the corresponding consumption by the high income population. This is due to the low quality of the houses where low income people customarily live, contrary to the rich population, who mostly live in well-insulated and well insulated shelters with low ventilation and solar protection.

Santamouris et al. (2007b) report that in Greece, heating consumption per square meter and per person in low income households are close to 127% higher than the corresponding energy consumption in the high income group. Additionally, the same study has shown that the use of electricity by low income households per person and per unit of house surface is about 67% higher than that of high income group, while the cost of air conditioning when used in low income houses is about 95% higher than that of the average population.

Roberts et al. (2007) has reported that in the UK about the 9% of the households with the lowest income present the highest emissions of greenhouse gases relative to the rest of the population. This is because of the low quality of the global thermal systems of the houses and the desperate effort to keep homes warm. Also in the UK, Roberts (2008) found that about 10% of the poorest households are responsible for about 45% more greenhouse gases emissions than the richest 10% group (Roberts, 2008). It is evident that the additional emissions by the low income population intensifies the magnitude of global and local climate change.

INTEGRATING THE GLOBAL SYNERGIES IN A COMMON FRAME

The previous chapters of this book clearly demonstrated that the high energy consumption of buildings, local and global climate change, and energy poverty are among the major problems in the built environment. This chapter has demonstrated the strong interrelations and synergies between the three sectors. It is evident that policies implemented with the aim of decreasing the energy consumption of the building sector underestimate the impact and the significance of both local and global climate change, as well as the economic, social and technological implications of energy poverty. Failure to consider all problems in an integrated and holistic way may inevitably lead to much higher energy consumption, social problems and future discrepancies.

Innovating the built environment to zero requires a minimization of the energy consumption of the building sector, eradication of energy poverty and an almost complete mitigation of local climate change and urban overheating. This objective may appear to be very ambitious, but such a policy will offer substantial opportunities for future growth, generate a broad set of benefits, and will alleviate the population from the burden of these three specific problems.

REFERENCES

Arup, 2014. Reducing urban heat risk. A study on urban heat risk mapping and visualization. http://publications.arup.com/Publications/R/Reducing_urban_heat_risk.aspx.

Asimakopoulos, D.A., Santamouris, M., Farrou, I., Laskari, M., Saliari, M., Zanis, G., et al., 2012. Modelling the energy demand projection of the building sector in Greece in the 21st century. Energy Build. 49, 488–498.

BPIE, 2011. Buildings performance institute europe: Europe's buildings under the microscope. <http://www.bpie.eu/eu_buildings_under_microscope.html#.VkBlq_nhDWI>.

Barker, T., Kohler, J., 1998. Equity and ecotax reform in the EU: Achieving a 10 per cent reduction in CO2 emissions using excise duties. Fiscal Stud. 19 (4), 375–402.

Benzie, M., Harvey, Alex, Burningham, Kate, Hodgson, Nikki, Siddiqi, Ayesha, 2011. Vulnerability to heat waves and drought case studies of adaptation to climate change in south-west England. Joseph Rountree Foundations, <https://www.jrf.org.uk/report/vulnerability-heatwaves-and-drought-adaptation-climate-change>.

Bohnenstengel, S.I., Hamilton, I., Davies, M., Belcher, S.E., 2014. Impact of anthropogenic heat emissions on London's temperatures. Quart. J. R. Meteorol. Soc 140 (679), 687–698.

Bouzarovski, S., 2015. Social justice and climate change: Addressing energy poverty at the European scale, Available through. <http://www.socialplatform.org/wp-content/uploads/2014/01/Article_energypoverty_Bouzarovski.pdf>.

Boyd, R., Krutilla, K., Viscusi, W. Kip, 1995. Energy taxation as a policy instrument to reduce CO2 emissions: A net benefit analysis. J. Environ. Econ. Manage. 29, 1–24.

Callan, T., Lyons, S., Scott, S., Tol, R.S.J., Verde, S., 2009. The distributional implications of a carbon tax in Ireland. Energy Policy 37, 407–412.

Cellura, M., Guarino, Francesco, Longo, Sonia, Mistretta, Marina, Orioli, Aldo, 2013. The role of the building sector for reducing energy consumption and greenhouse gases: An Italian case study. Renewable Energy 60, 586–597.

Changnon, S.A., 2004. A rare long record of deep soil temperatures defines temporal temperature changes and an urban heat island. Clim. Change 42, 531–538.

Chen, C., Zhao, B., Weschler, C.J., 2012. Assessing the influence of indoor exposure to "outdoor ozone" on the relationship between ozone and short-term mortality in US communities. Environ. Health Perspect. 120 (2), 235–240.

Chief Medical Officer, 2009. Annual report, 2009. <http://www.sthc.co.uk/Documents/CMO_Report_2009.pdf>.

Defaix, P.R., van Sark, W.G.J.H.M., Worrell, E., de Visser, E., 2012. Technical potential for photovoltaics on buildings in the EU-27. Sol. Energy 86, 2644–2653.

Department of Health UK, 2010. Public health white paper. <https://www.gov.uk/government/publications/the-public-health-white-paper-2010>.

Dresner, S., Ekins, P., 2006. Economic instruments to improve UK home energy efficiency without negative social impacts. Fiscal Stud. 27, 47–74.

EPIA, 2014. European photovoltaic industry association: Global market outlook for photovoltaics, 2014–2018. <http://www.cleanenergybusinesscouncil.com/global-market-outlook-for-photovoltaics-2014-2018-epia-2014>.

EURO4M, 2015. 2014 warmest year on record in Europe. <http://cib.knmi.nl/mediawiki/index.php/2014_warmest_year_on_record_in_Europe>.

Ecocity, 2014. Proposal concerning the atmospheric pollution in athens. <https://energia.gr/article.asp?art_id = 80854>, In Greek.

Energy Action Scotland (EAS) and National Energy Action (NEA) (England and Wales), 2009. The cost of affordable warmth, <www.nea.org.uk/assets/Uploads/The-Cost-of-Affordable-WarmthFinal-version2.pdf>.

European Environmental Agency, (EEA), 2012. Climate change, impacts and vulnerability in Europe 2012 An indicator-based report, EEA Report No 12/2012. <http://www.eea.europa.eu/publications/climateimpacts-and-vulnerability-2012>.

European Union: Commission Staff Working Document, 2013. Adapting infrastructure to climate change accompanying the document, communication from the commission to the European parliament, the council, the European economic and social committee and the committee of the regions. An EU strategy on adaptation to climate change, Brussels, 16.4.2013 SWD(2013) 137 final.

Eurostat 2015. Income distribution statistics. <http://ec.europa.eu/eurostat/statistics-explained/index.php/Income_distributionstatistics>.

Filiberto, D., Wethington, E., Pillemer, K., Wells, N., Wysocki, M., Parise, J., 2009. Older people and climate change: Vulnerability and health effects. Generations 33 (4), 19–25.

Fintikakis, N., Gaitani, N., Santamouris, M., Assimakopoulos, M., Assimakopoulos, D.N., Fintikaki, M., et al., 2011. Bioclimatic design of open public spaces in the historic centre of Tirana, Albania. Sust. Cities Soc. 1 (1), 54–62.

Gaitani, N., Santamouris, M., Cartalis, C., Pappas, I., Xyrafi, F., Mastrapostoli, E., et al., 2014. Microclimatic analysis as a prerequisite for sustainable urbanization: Application for an urban regeneration project for a medium size city in the greater urban agglomeration of Athens-Greece. Sust. Cities Soc. 13, 230–236.

Gamble, J., Hurley, Bradford J., Schultz, Peter A., Jaglom, Wendy S., Krishnan, Nisha, Harris, Melinda, 2013. Climate change and older Americans: State of the science. Environ. Health Perspect. 121, 1.

Geros, V., Santamouris, M., Tsangrassoulis, A., Guarracino, G., 1999. Experimental evaluation of night ventilation phenomena. J. Energy Build. 29, 141–154.

Giannopoulou, K., Livada, I., Santamouris, M., Saliari, M., Assimakopoulos, M., Caouris, Y.G., 2011. On the characteristics of the summer urban heat island in Athens, Greece. J. Sust. Cities Soc. 1 (1), 16–28.

Giannopoulou, K., Livada, I., Santamouris, M., Saliari, M., Assimakopoulos, M., Caouris, Y., 2013. The influence of air temperature and humidity on human thermal comfort over the greater Athens area. Sust. Cities Soc. 10, 184–194.

Green, G., Gilbertson, J., 2008. Warm Front Better Health: Health impact evaluation. Centre for Regional, Economic and Social Research, Sheffield Hallam University, <www.google.fr/url?sa = t&source = web&cd = 1&ved = 0CBgQFjAA&url = http%3A%2F%2Fwww.apho.org.uk%2Fresource%2Fview.aspx%3FRID%3D53281&ei = Fp6pTfvNA8qx8QPRoNG4Ag&usg = AFQjCNHdMI2DjMei6aCG9VGTJ3HSdignAA>.

Greening, L.A., Grene, D.L., Difiglio, C., 2000. Energy efficiency and consumption – the rebound effect – a survey. Energy Policy 28, 389–401.

Grepperud, S., Rasmussen, J., 2004. A general equilibrium assessment of rebound effects. Energy Econ. 26, 261–282.

Hamilton, I., Davies, Michael, Steadman, A. Philip, Stone, Andrew, Ridley, Ian, Evans, Stephen, 2009. The significance of the anthropogenic heat emissions of London's buildings: A comparison against captured shortwave solar radiation. Build. Environ. 44, 807–817.

Hanley, N., McGregor, P.G., Swales, J.K., Turner, K.R., 2009. Do increases in energy efficiency, improve environmental quality and sustainability? Ecol. Econ. 68, 692–709.

Hassid, S., Santamouris, M., Papanikolaou, N., Linardi, A., Klitsikas, N., Georgakis, C., et al., 2000. The effect of the Athens heat island on air conditioning load. Energy Build. 32 (2), 131–141.

Hong, S.H., Gilbertson, Jan, Oreszczyn, Tadj, Green, Geoff, Ridley, Ian, the Warm Front Study Group, 2009. A field study of thermal comfort in low-income dwellings in England before and after energy efficient refurbishment. Build. Environ. 44, 1228–1236.

House of Commons, UK Parliament, 2009. Environment, food and rural affairs committee. Energy efficiency and fuel poverty, third report of session, 2008-09.

IPCC (International Panel on Climate Change), 2008. Climate change 2007: Impacts, adaptation and vulnerability. Cambridge University Press, Cambridge.

International Energy Agency (IEA), 2011. CO2 emissions from fuel combustion – highlights. CO2 emissions per kWh from electricity and heat generation. Year 2011. <http://www.iea.org/co2highlights/>.

Irish Department of Communications, Energy and Natural Resources, 2015. Warmer homes a strategy for affordable energy in Ireland, 2015. <http://www.dcenr.gov.ie/en-ie/Pages/home.aspx>.

Jenkings, D.P., 2010. The value of retrofitting carbon-saving measures into fuel poor social housing. Energy Policy 38, 832–839.

Kelly, B., 2015. The process of socio-economic constraint on geographical mobility: England 1991 to 2008. <www.cmist.manchester.ac.uk/medialibrary/archive-publications/working-papers/2013/2013-08-the_process_of_socioeconomic_constraint_on_geographical_mobility.pdf>, 2013.

Kolokotroni, M., Ren, X., Davies, M., Mavrogianni, A., 2012. London's urban heat island: Umpact on current and future energy consumption in office buildings. Energy Build. 47, 302–311.

Kolokotsa, D., Santamouris, M., 2015. Review of the environmental quality and energy consumption studies or low income households in Europe. Sci. Total Environ. 536, 316–330.

Kovats, R.S., Ebi, L.E., 2006. Heatwaves and public health in Europe. Eur. J. Pub. Health 16 (6), 592–599.

Lokoshchenko, M.A., Korneva, I.A., 2015. Underground urban heat island below Moscow city. Urban Climate 13, 1–13.

Lomas, K.J., Kane, T., 2013. Summertime temperatures and thermal comfort in UK homes. Build. Res. Inf. 41, 259–280. Available from: https://doi.org/10.1080/09613218.2013.757886.

Marmot Review, 2011. The health impacts of cold homes and fuel poverty. <http://www.foe.co.uk/sites/default/files/downloads/cold_homes_health.pdf>.

Mavrogianni, A., Davies, M., Wilkinson, P., Pathan, A., 2010. London housing and climate change: Impact on comfort and health — preliminary results of a summer overheating study. Open House Int. 35, 49–59.

Mc Carthy, M., 2009. Report on the CIRCE urban heat island simulations, project No. 036961 – CIRCE, sixth framework programme, European commission, global change and ecosystems, 2009. <www.cru.uea.ac.uk/.../circe/CIRCE>.

Carnegie Mellon, 2005. University center for building performance. As cited in greening America's schools: Costs and benefits. October 2006. G. Kats, capital E. <http://www.usgbc.org/ShowFile.aspx?DocumentID = 2908>.

Mihalakakou, P., Santamouris, M., Asimakopoulos, D., 1994a. On the use of ground for heat dissipation. J. Energy 19 (1), 17–25.

Mihalakakou, P., Santamouris, M., Asimakopoulos, D., 1994b. On the cooling potential of earth to air heat exchangers. J. Energy Convers. Manage. 35 (5), 395–402.

OECD, 1996. Organisation for economic co-operation and development (OECD). implementation strategies for environmental taxes, Paris.

Palmer, J., Ian Cooper, 2013. United kingdom housing energy fact file, 2013, UK department of energy and climate change. <https://www.gov.uk/government/statistics/united-kingdom-housing-energyfact-file-2013>.

Poltimäe, Helen, Võrk, Andres, 2009. Distributional effects of environmental taxes in Estonia. Discuss. Estonian Econ. Policy: Theory Practice Econ. Policy 17.

Preston, I., Moore, R., Guertler, P., 2008. How Much? The Cost of Alleviating Fuel Poverty, Report to the EAGA Partnership Charitable Trust. CSE, Bristol.

Rai, A.C., Guo, B., Lin, C.H., Zhang, J., Pei, J., Chen, Q., 2013. Ozone reaction with clothing and its initiated particle generation in an environmental chamber. Atmos. Environ. 77, 885–892.

Ramponi, R., Gaetani, Isabella, Angelotti, Adriana, 2014. Influence of the urban environment on the effectiveness of natural night-ventilation of an office building. Energy Build. 78, 25–34.

Roberts, S., 2008. Energy, equity and the future of the fuel poor. Energy Policy 36, 4471–4474.

Roberts, S., Vicki White, Ian Preston, Joshua Thumim, 2007. Assessing the social impacts of a supplier obligation a study for Defra, final report.

Rozzini, R., Zanetti, E., Trabucchi, M., 2004. Elevated temperature and nursing home mortality during 2003 European heat wave. J. Am. Med. Dir. Assoc. 5, 138–139.

Rubbelke, D., Stefan Vogele, 2011. Distributional consequences of climate change impacts on the power sector: Who gains and who loses? CEPS working document, No. 349.

Sakka, A., Santamouris, M., Livada, I., Nicol, F., Wilson, M., 2012. On the thermal performance of low income housing during heat waves. Energy Build. 49, 69–77.

Santamouris, M., 2014a. On the energy impact of urban heat island and global warming on buildings. Energy Build. 82 (100–113), 72014.

Santamouris, M., 2014b. Cooling the cities – a review of reflective and green roof mitigation technologies to fight heat island and improve comfort in urban environments. Sol. Energy 103, 682–703.

Santamouris, M., 2015. Regulating the damaged thermostat of the Cities –Status impacts and mitigation strategies. Energy Build. 91, 43–56.

Santamouris, M., April 2016. Innovating to zero the building sector in Europe: Minimising the energy consumption, eradication of the energy poverty and mitigating the local climate change. Solar Energy 128, 61–94.

Santamouris, Mattheos, Kolokotsa, Dionysia, February 2013. Passive cooling dissipation techniques for buildings and other structures: The state of the art review article. Energy Build. 57, 74–94.

Santamouris, M., Kolokotsa, D., 2015. On the impact of urban overheating and extreme climatic conditions on housing energy comfort and environmental quality of vulnerable population in europe. Energy and Buildings 98, 125–133. Available from: https://doi.org/10.1016/j.enbuild.2014.08.050.

Santamouris, M., Papanikolaou, N., Livada, I., Koronakis, I., Georgakis, C., Argiriou, A., et al., 2001. On the impact of urban climate to the energy consumption of buildings. Sol. Energy 70 (3), 201–216.

Santamouris, M., Pavlou, K., Synnefa, A., Niachou, K., Kolokotsa, D., 2007a. Recent progress on passive cooling techniques. Advanced technological developments to improve survivability levels in low –income households. Energy Build. 39, 859–866.

Santamouris, K., Kapsis, D., Korres, I., Livada, C., Pavlou, Assimakopoulos, M.N., 2007b. On the elation between the energy and social characteristics of the residential sector. Energy Build. 39, 893–905.

Santamouris, M., Sfakianaki, A., Pavlou, K., 2010. On the efficiency of night ventilation techniques applied to residential buildings. Energy Build. 42, 1309–1313.

Santamouris, M., Alevizos, S.M., Aslanoglou, L., Mantzios, D., Milonas, P., Sarelli, I., et al., 2014. Freezing the poor—indoor environmental quality in low and very low income households during the winter period in Athens. Energy Build. 70, 61–70.

Santamouris, Matheos, Cartalis, Constantinos, Synnefa, A., 2015. Local urban warming, possible impacts and a resilience plan to climate change for the historical center of Athens. Greece. Sust. Cities Soc. 19, 281–291.

Sarrat, C., Lemonsu, A., Masson, V., Guedalia, D., 2006. Impact of urban heat island on regional atmospheric pollution. Atmos. Environ. 40, 1743–1758.

Sims, R., Schock, R.N., Adegbululgbe, A., Fenhann, J., Konstantinaviciute, I., Moomaw, W., et al., 2007. Energy supply. In: Metz, B., Davidson, O.R., Bosch, P.R., Dave, R., Meyer, L.A. (Eds.), Climate Change 2007: Mitigation. Contribution of Working Group III to the Fourth Assessment Report of the Intergovernmental Panel on Climate Change. Cambridge University Press, Cambridge, United Kingdom and New York, NY, USA.

Skumatz, L., 2010. Non-Energy Benefits: Status, Findings, Next Steps, and Implications for Low Income Program Analyses in California. Skumatz Economic Research Associates, <www.liob.org/docs/LIEE%20Non-Energy%20Benefits%20Revised%20Report.pdf>.

Speck, Stefan, 1999. Energy and carbon taxes and their distributional implications. Energy Policy 27 (11), 659–667.

Stathopoulou, E., Mihalakakou, G., Santamouris, M., Bagiorgas, H.S., 2008. Impact of temperature on tropospheric ozone concentration levels in urban environments. J. Earth Syst. Sci. 117 (3), 227–236.

Symons, E.J., Speck, S. Proops, J.L.R., 2000. The Effects of pollution and energy taxes across the European income distribution. <http://www.researchgate.net/publication/5159193>.

Synnefa, A., Santamouris, M., Akbari, H., 2007. Estimating the effect of using cool coatings on energy loads and thermal comfort in residential buildings in various climatic conditions. Energy Build. 39 (11), 1167–1174.

Taniguchi, M., 2006. Anthropogenic effects on subsurface temperature in 300 Bangkok. Clim. Past 2, 831–846.

Tiezzi, Silvia, 2005. The welfare effects and the distributive impact of carbon taxation on Italian households. Energy Policy 33, 1597.

Turner, K., 2009. Negative rebound and disinvestment effects in response to an improvement in energy efficiency in the UK economy. Energy Econ. 31 (5), 648–666.

UK Government Department of Energy and Climate Change (DECC), 2009. DECC website, 'addressing fuel poverty. <http://www.decc.gov.uk/en/content/cms/what_we_do/consumers/fuel_poverty/fuel_poverty.aspx>.

UNDP, 2004. Stuck in the past - energy, environment and poverty in Serbia and Montenegro. <http://www.stabilitypact.org/energy/041011-bucharest/Stuck_in_the_Past.pdf> (16.06.2011).

United Nations Development Program, 2015. Improving energy efficiency in low-income households and communities in Romania. <http://www.undp.ro/projects.php?project_id = 63>.

Urge-Vorsatz, D., Tirado Herrero, S., 2012. Building synergies between climate change mitigation and energy¨poverty alleviation. Energy Policy. Available from: https://doi.org/10.1016/j.enpol.2011.11.093.

Urge-Vorsatz, D., Eyre, N., Graham, P., Harvey, D., Hertwich, E., Kornevall, C., et al., 2012. Energy end-use: Buildings. The Global Energy Assessment: Toward a more SustainableFuture. IIASA, Laxenburg, Austria and Cambridge University Press, Cambridge, United Kingdom and New York, NY, USA.

Van Vliet, Michelle T.H., Yearsley, John R., Ludwig, Fulco, Vogele, Stefan, Lettenmaier, Dennis P., Kabat, Pavel, 2012. Vulnerability of US and European electricity supply to climate change. Nature Climate Change 2, 676–681. Available from: https://doi.org/10.1038/nclimate1546.

Washan, P., Jon Stenning, Max Goodman, 2014. Building the future: Economic and fiscal impacts of making homes energy efficient, Cambridge Econometrics.

Wigley, K.J., 1997. Assessment of the importance of the rebound effect. Paper presented at the 18th North American conference of the USAEE/IAEE, San Francisco.

World Health Organisation, 2006. Fuel for life: Household energy and health.

Worldwatch Institute, 2008. Green jobs: Towards sustainable work in a low-carbon world.

Zhao, P., Siegel, J.A., Corsi, R.L., 2005. Ozone removal by residential HVAC filters, proceedings: Indoor air 2005.

Chapter 6

Defining Future Targets

THE IDEA OF 'MINIMIZING TO ZERO': ANALYSIS OF THE CONCEPT, ADVANTAGES AND DRAWBACKS

Although the magnitude of the problems in the built environment are well understood, the future targets and objectives set by the international community are still not well defined. Several international organizations and states have set quantified objectives regarding future energy consumption, local climate change and energy poverty, at least for the immediate future. Most of the objectives are defined for each one of the identified problems, without considering an integration approach incorporating all targets and goals under one umbrella and within a common holistic and integrated framework.

A rich dialogue has been developed that aims to identify the capacity of our societies to decrease energy consumption in the building sector, mitigate local and global climate change and to eradicate energy poverty problems. While the technical potential to solve these problems seems to be very high, the economic, financial and social obstacles, as well as the lack of adequate policies, substantially reduces the global capacity to face these problems in a radical way. In the search for adequate strategies and policies for the future, a very interesting approach known as the 'innovation to zero strategy' was proposed by Sarwant Singh (2012). The idea behind the proposed strategy is to develop a zero emission, zero waste and zero defects technological background that will support a society of zero energy consumption and waste in a world of zero crimes and zero diseases. The strategy, when applied in the urban environment, could result in zero energy consumption by the building sector, a complete mitigation of local climate change and a full eradication of energy poverty. It is expected that the achievement of these goals will contribute greatly to decreasing economic disparities and inequalities, promote social equality and enhance standards of health and well-being. It is certain that such an idea appears to be opportunistic, extremely radical or even a utopic. However, it is evident that the concept incorporates some outstanding features that make itattractive and worthy of further analysis and investigation.

Objectives and goals of a quite similar nature have been expressed and adopted in many technological sectors in the developed world, mainly in

Minimizing Energy Consumption, Energy Poverty and Global and Local Climate Change in the Built Environment: Innovating to Zero. DOI: https://doi.org/10.1016/B978-0-12-811417-9.00006-4

Europe and the United States. In particular, the European Directive for the Energy Performance of Buildings (EU, 2010) requires all Member States of the European Union to achieve an almost zero energy consumption by all buildings constructed or renovated after 2020, while all new public buildings must achieve an almost zero energy consumption by 2018. The Directive has a very severe impact on the energy consumption of the building sector in Europe, and it significantly alters the social and economic status quo in Europe. Although it is widely accepted that the Directive will contribute greatly to decreasing the energy consumption of buildings and reduce related greenhouse gas emissions, criticism is raised about the additional cost associated with almost zero energy buildings, the ability of the low income population to respond to the additional financial requirements, and the potential marginalization of vulnerable populations. However, the very first period of the application of the Directive in some European countries shows that when such a radical technological concept is supported by adequate economic and social policies, the potential problems are minimized and the benefits are stronger that the possible drawbacks. Furthermore, the implementation of quite optimistic technological policies helps to increase global standards for the building sector, improve the average quality of constructions, improve the average standards of indoor living and decrease the cost of efficient and advanced energy technologies. The global impact of the implementation of the Directive significantly improves the accessibility of low income households to advanced building energy technologies, while potentially improving the quality of life of the vulnerable population.

Besides the example of the European Directive for the performance of buildings, other, similar policies have been designed and implemented. For example, the European Pathway for Zero Waste (EPOW, 2012) aims to design and implement technological and financial policies to minimize waste in selected European regions. A very successful implementation of the program has already been achieved in the southeastern regions of England, UK.

A third example of a similar nature is Zero Carbon Cities. The objective of Zero Carbon Cities is to minimize their ecological footprint, energy consumption and related greenhouse gas emissions through the use and implementation of advanced energy and environmental technologies as well as through the adoption of green governance practices. Several cities have successfully designed and implemented policies minimizing their ecological footprint, neutralizing the balance of energy consumption and minimizing greenhouse gas emissions. The technologies employed cover most of the economic sectors of the cities, such as buildings, transport, industry, and so on. Although the capital cost of measures implemented appears to be high, the global benefits for the population are also very high, while the amortization period is quite reasonable. Besides the important direct benefits, significant indirect economic and social benefits should be mentioned. For example, the development and implementation of advanced energy and environmental technologies creates a new global market and offers significant economic and technological gains. Moreover, vulnerable populations benefit from an increase in average standards of living. Good examples of Zero Carbon cities include the city of Masdar in the United Arab Emirates, Copenhagen in Denmark, Donton in China, Vaxio in Sweden, and Sydney and Fremantle in Australia. The levels of success and the performance of the policies implemented varies between these cities; however, almost all of them must demonstrate achievements that have resulted in a significant improvement in living standards.

Setting the main energy and environmental problems in the built environment to zero may be translated into the following three specific objectives:

a. Minimization of the energy consumption of the building sector
b. A full mitigation of local climate change and the urban overheating
c. A complete eradication of energy poverty

The satisfaction of the above three objectives seems to be an unequivocal choice. Humanity needs to define an ambitious vision to set ambitious future goals and objectives. It is evident that such a set of goals cannot be satisfied immediately and will be achieved gradually. Intermediate achievements and partial attainments may offer significant benefits for our societies and a considerable progress towards the full achievement of the objectives.

The adoption and achievement of the specific objectives requires planning and following a fully innovative scientific and political agenda full of technological breakthroughs and the implementation of advanced technologies and policies. Such a future plan will require substantial investment in the global building sector, including in cities and individual buildings, which will create widespread economic, scientific and social opportunities for the future and will certainly create major medium and long term benefits for the society. In parallel, it will help to alleviate the intensity and the consequences of the particular problems faced bythe low income and vulnerable citizens.

OBJECTIVES, EXPECTED OUTCOMES, POLICIES AND TOOLS

There are multiple and complex synergies between energy consumption within the building sector, local and global climate change and energy poverty. The following chapters will attempt to investigate the links and the synergies between the three subjects and explore the best path to follow in the future.

We will attempt to address the major issues defining the characteristics of the three subjects and explore a new future strategy. The whole analysis will be based on four fundamental questions concerning possible future objectives, plans, strategies, final developments and implementations:

a. What should be the quantitative and qualitative future targets and objectives for each of the three problems of concern? Targets and objectives should be set, and a clear and well-defined road map must be designed in order to achieve them. The means and the tools used to implement this road map should be judiciously selected, and the necessary financial and social cost should be assessed on a pragmatic and realistic basis. The procedures to achieve the maximum possible social support and acceptability should be analyzed, examined and tested in practice.

b. What will be the major technological, macroeconomic and social forces and trends defining progress and developments in the immediate future that will impact the economy and the sustainable advancement of the societies? How can this impact be quantified, and how can it benefit the majority of the population, not just a small group of privileged and well-connected people?

c. How might the above-mentioned mechanisms, forces and trends change the actual status of the three subjects in question, defining an agenda that is more proactive than reactive? How will it impact problems of energy consumption in buildings, local and global climate change and energy poverty in the developed world? How should this proactive agenda be defined? What kinds of analysis, management, policies and governance should be used? How do we transform these problems in opportunities? How can we increase the added value of technological, economic and social interventions to generate wealth and offer employment opportunities?

In the following chapters, these questions will be discussed for each of the three subjects of concern.

REFERENCES

European Pathway to Zero Waste (EPOW), 2012. Environment Agency, <http://www.environment-agency.gov.uk/aboutus/wfo/epow/123624.aspx>.

European Union, EU, 2010. Directive 2010/31/EU of the European Parliament and of the Council of 19 May 2010 on the energy performance of buildings (recast), 18.6.2010 Official Journal of the European Union L 153/13.

Singh, Sarwant, 2012. New Mega Trends: Implications for our Future Lives. Published by Palgrave Macmillan, London.

Chapter 7

Technological-Economic and Social Measures to Decrease Energy Consumption by the Building Sector

STATE OF THE ART TECHNOLOGIES TO DECREASE THE ENERGY CONSUMPTION OF BUILDINGS AND IMPROVE THEIR ENVIRONMENTAL QUALITY

Although the global energy consumption by the building sector has increased because of various economic and social drivers, the energy consumption of individual buildings has considerably decreased, mainly due to significant technological developments and stricter regulations. More efficient energy conservation technologies and building-integrated renewable systems are being developed with the primary aim of decreasing the winter energy consumption of buildings. As shown, while the average heating consumption of the building stock is close to 220 kWh/m^2/year, buildings constructed recently in line with the passive house standard may present a consumption close to 40 kWh/m^2/year, and in fact current technology permits the construction of near zero energy buildings.

Several factors have contributed to decreasing energy consumption for heating purposes in both commercial and residential buildings, including new industrial products aimed at decreasing heat transmission through the envelope, more efficient ventilation technologies, high performance HVAC systems, and stricter regulations. A comparison of the energy consumption of commercial building stock in the US between 2003 and 2012 shows that heating energy consumption has seriously declined (Fig. 7.1).

As shown in Fig. 7.1, lighting presents a similar pattern of decreasing consumption to heating. New low energy lighting systems for both commercial and residential buildings have resulted in a substantial decrease in energy consumption for lighting purposes.

Contrary to the heating and lighting energy consumption of buildings, the need for cooling has substantially increased in both residential and commercial buildings. This is mainly due to several climatic and economic drivers,

Minimizing Energy Consumption, Energy Poverty and Global and Local Climate Change in the Built Environment: Innovating to Zero. DOI: https://doi.org/10.1016/B978-0-12-811417-9.00007-6

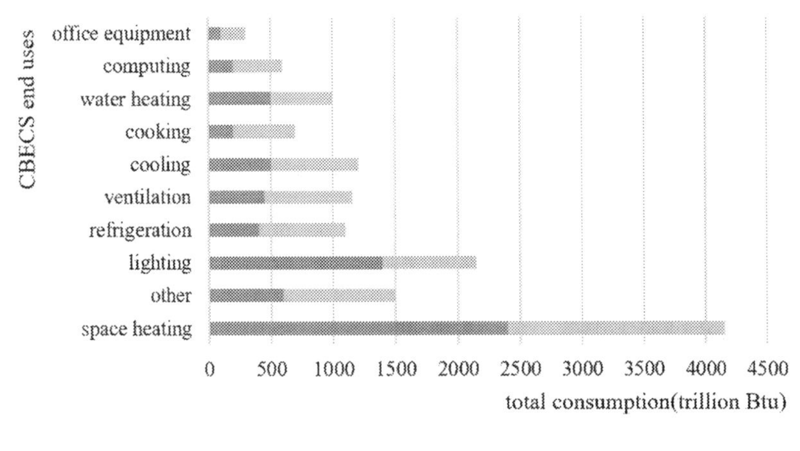

FIGURE 7.1 Energy consumption by commercial buildings in the US in 2003 and 2012. *Adapted from US Energy Information Administrations, 2012. Commercial buildings energy consumption survey. (https://www.eia.gov/consumption/commercial/reports/2012/energyusage/).*

such as an increase in ambient temperature and a substantial decrease in the use of cooling systems, which have made the use of air conditioning more attractive and affordable. Besides these climatic and economic reasons for the increased penetration of air conditioning in the building sector, recent technological developments in the field of heat and solar protection, dissipation and heat amortization technologies, and the development of new, efficient air conditioning systems have resulted in a substantial decrease in the amount of energy necessary to cool a building. Fig. 7.2 presents the evolution of the cooling energy consumption of commercial buildings over the last 20 years in the Mediterranean zone. As shown, while single glazed commercial buildings built mainly in the 1980s presented a cooling energy consumption close to 300 kWh/m^2/year, recent commercial buildings may not consume more than 5 kWh/m^2/year.

To decrease the energy consumption of buildings, four main technological axes of activities must be followed:

a. Activities aiming to increase the efficiency of building energy production systems, such as HVAC and lighting systems to decrease the final energy consumption by the building.
b. Activities aiming to enhance the performance and promote the use of renewable energy systems in buildings in order to replace and substitute the use of conventional energy systems.
c. Activities aiming to optimize the envelope of the building in order to decrease thermal losses and increase solar gains during the winter period, and decrease solar inputs and optimize losses during the warm period of the year.

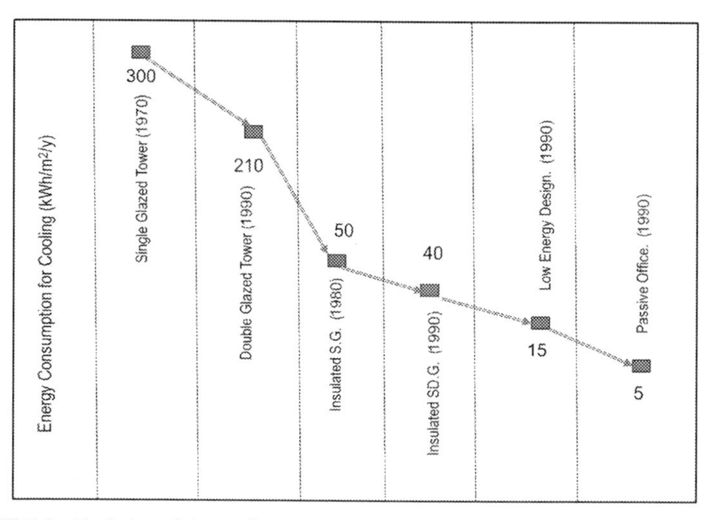

FIGURE 7.2 Evolution of the cooling energy consumption of representative commercial buildings in the Mediterranean zone.

d. Activities aiming to improve the management of the energy and environmental resources and systems in the building through the use of smart and intelligent technologies.

Energy policies should employ integrated approaches based on the use of smart green technologies that offer intelligent and innovative solutions that are able to improve environmental quality in the building sector and decrease energy consumption. In parallel, the economic benefits arising from building-related investments must be seriously considered and optimized so that the maximum percentage of the population and all social groups benefit. Energy-related investments may be the best vehicle to achieve future growth and generate new employment opportunities.

The major problems of the planet generate financial, technological and societal mega-trends that offer additional opportunities for future growth in the building sector. In particular, problems related to the mitigation and adaptation to climate change, new technologies such as nanotechnology, smart metamaterials and advanced building materials, IT and smart green technologies for buildings, new high-performing renewable energy systems, and so on, offer increased opportunities for growth and development, and may contribute significant economic and social benefits for consumers and societies.

The decrease of energy consumption within the building sector is one of the major challenges and priorities of the 21st century. Apart from the important technological benefits, lower energy consumption is associated with significant economic and social benefits. In parallel, the related environmental benefits arising from the decrease of greenhouse gas emissions and increased energy security contribute towards a more sustainable technological development and economic growth. As already mentioned, significant technological developments have been accomplished during the last few years; however, according to the International Energy Agency, almost 65% of the economic potential of energy efficient technologies remains untapped (IEA, 2013). According to BMUB (2014), the volume of the global market for energy efficient products for all sectors of economy was close to EUR€825 billion, representing almost 33% of the global market related to the efficient use of resources and associated environmental technologies. Among them, the share of energy efficient building technologies was close to EUR€126 billion, while the corresponding market for energy efficient appliances was around EUR€183 billion (BMUB, 2014). In parallel, the market for energy efficient cross-sector components represented was worth close to EUR€507 billion (BMUB, 2014). According to the same institution, the expected total volume occupied by energy efficient products may increase to as much as EUR€1.365 trillion (BMUB, 2014), while the market figure associated with energy efficient buildings may reach EUR market 265 billion (BMUB, 2014).

Further economic analyses predict that the market for energy efficient appliances and lighting systems will increase from EUR market 112.5 billion in 2013 to EUR market 200.3 billion in 2025 (BMUB, 2014). In particular, the fast development of highly efficient LED lighting technologies will assist of the lighting sector's market share to achieve an annual growth of close to 6.1% (BMUB, 2014), while by 2020 annual growth rates may exceed 8%. Forecasts for the market development of the lighting sector predict that it may be worth between EUR€73 and EUR€126 billion by 2020, while the corresponding figure for 2030 is close to EUR€160 billion (WRI, 2010).

Building products aiming to improve the thermal performance of the building envelope, such as advanced glazing and insulation materials, represent a sizeable market and significant source of economic growth. In particular, smart glass products for buildings, including selective and chromogenic materials that are able to provide excellent daylight quality and control solar and thermal gains, may achieve an annual market figure of around EUR€635 million by 2020, compared to nearly EUR€76 million in 2015 (Glass, 2015). In parallel, the market for electrochromic glass may reach EUR€2.7 billion by 2019 (Nanomarkets, 2012).

New high performance insulation materials for the building envelope may also undergo significant growth in the near future. According to WRI (2010), the insulation market may reach between EUR€13.6–20 billions by 2020, and EUR€28 billion by 2030. Additionally, according to Grand View Research (2015), the market for smart materials, such as infrared reflective coatings and phase change materials able to amortise the solar and heat gain of buildings and improve their environmental quality, may reach EUR€1.6

billion by 2020, with a growth in market penetration of close to 22.5% per year by 2020.

Modern technology photovoltaic systems for buildings, such as thin film technologies and integrated PV modules in building components like windows, walls and roofing shingles, can substantially reduce the electricity load of buildings and decrease peak electricity demand. While the market for building-integrated technologies represented almost EUR€1.8 billion in 2015, it is expected to increase up to EUR€5.8 billion by 2018 (Nanomarkets, 2012).

Smart grid technologies for buildings and communities incorporating high performance renewable energy systems, together with sophisticated energy management solutions, are considered among the more promising energy and environmental technologies for the building sector. According to Singh (2012), the market for smart grid components may reach EUR€182 billion by 2020, while the 2010 figure was close to EUR€91 billion.

Smart systems and technologies that are able to manage the energy and environmental resources within buildings in an more efficient way may significantly improve the global energy performance of the building sector. Smart technologies based on advanced IT solutions can provide efficient control and management of the available resources, enhance comfort conditions and provide a healthy indoor environment. Smart metering technologies have gained an increasing share of the market and help us to better understand the global performance of our buildings. According to Singh (2012), the market for smart metering technologies for buildings is increasing rapidly. While in 2009 this market was worth close EUR€2.3 billion, it has increased up to EUR€9.1 billion in 2012 and is expected to reach EUR€18.2 billion in 2020. In parallel, the market for advanced green IT technologies for buildings is growing at an annual rate of 7%–10%, and it is estimated that it will reach a global market value of EUR€22.5 – EUR€34.5 billion by 2020 (BMUB, 2014).

The market for advanced heating ventilation and air conditioning systems is experiencing a very rapid annual growth rate of close to 10.5%. In particular, high performance heat pumps that are able to significantly decrease energy consumption by buildings and decrease peak electricity demand may reach close to EUR€162 billion in 2018.

Recent research has succeeded in substantially improving the performance and energy efficiency of HVAC systems. According to Michel et al. (2010) the efficiency of split air conditioning systems in American, European, Asian, and Australian markets has improved by almost 3% per year over the last 15 years. For room air conditioners, new higher efficiency heat exchangers, compressors, expansion valves, and compressors are already being implemented in commercial HVAC systems. In parallel, significant improvements have been achieved in the reduction of crankcase heating and standby load (LBNL, 2013). As reported by LBNL (2013), the development and use of more efficient compressors may improve the efficiency of room air conditioners by 6%–19%, the development of more efficient heat exchangers by 9%–28%, the use of more efficient inverters by 20%–25%, the use of thermostatic expansion valves by 5%–9%, and further reduce the power used for crankcase heating by 10%–11%. Similar efficiency improvements are also planned for commercial air conditioning systems and heat pumps (IEA-ETSAP, 2013). According to Energy Technology Perspectives (2010), the main technological improvements scheduled for commercial heat pumps and other commercial cooling systems rely on the development and use of more efficient compressors and heat exchangers, the implementation of advanced control systems, and the development and use of more efficient fan motors. According to several economic studies, it is estimated that most of the projected technological improvements are cost effective (SEAD, 2013). In parallel, International Energy (2011) estimates that the cost of the energy delivered for cooling purposes may decrease by 50%–65% by 2050.

An important future increase in the energy efficiency of both the commercial and residential HVAC systems is foreseen by many analysts. According to Rong et al. (2007), while the average coefficient of performance of the installed air conditioning stock in the United States is between 2.5, it is expected that by 2050 it my increase up to 3.5 and to 4.5 in 2090. Detailed forecasts concerning the future expected performance of the more common residential and commercial air conditioning systems for up to 2040 are provided by Navigant Consulting, Inc. (2014). The expected performance for the most common air conditioning systems, as predicted by Navigant Consulting, Inc. (2014) is given in Table 7.1 below. As shown, the expected increase in the EER of the residential systems is highly significant, especially for the ground source heat pumps. It is expected that the highest performance of the ground source heat pumps may be close to 46 by 2040, while the performance in 2013 was close to 28. On the contrary, the foreseen increase for the air source heat pumps is significant, but not spectacular. According to the same source residential heat pumps will present a maximum EER of close to 25 in 2040, compared to an EER of 22 in 2013. In parallel, the maximum EER of the residential cooling systems will increase from 11.5 in 2013 to 12.9 in 2040 (Navigant Consulting, Inc., 2014). The future efficiency of commercial heat pumps is expected to increase from 20.6 in 2013 to 26 in 2040. Finally, forecasts for the future performance of commercial chillers show a significant increase in the performance of gas fired absorption chillers and a small increase for centrifugal systems and screw and scroll systems.

Energy labelling schemes contribute significantly to improving the average performance of new air conditioning systems in both residential and commercial sectors, and decrease global energy consumption for cooling. Baillargeon et al. (2011) predicted that in China from 2010 onwards, the EER for all split units of class A with CC <14000, would increase from

TABLE 7.1 Current and Future Performance of Various Residential and Commercial Cooling Technologies

Air Conditioning Type		2013		2020		2030		2040	
		Typical	High	Typical	High	Typical	High	Typical	High
Residential air conditioning	EER	10.8	11.5	10.9	11.9	10.9	12.9	11.1	12.9
Residential air source heat pumps	SEER	14	22	14.5	23	15.5	24	16	25
Residential ground source heat pumps	EER	14.2	28	17.1	36	21	42	24	46
Residential gas heat pumps	COP	0.6		0.7		0.7		0.7	
Commercial Chillers									
Gas fired absorption	COP	1.1		1.2		1.3		1.4	
Gas fired engine driven	COP	1.7		1.8		1.8		1.8	
Centrifugal	COP,(F.L)	6.1	7.8	6.5	8.0	6.8	8.2	7.0	8.4
Reciprocating	COP,(F.L)	2.81	3.52	3.06	3.52	3.06	3.52	3.06	3.52
Screw	COP,(F.L)	2.84	3.46	3.10	3.55	3.20	3.66	3.26	3.74
Scroll	COP,(F.L)	3.02	3.17	3.08	3.23	3.17	3.29	3.23	3.32
Commercial Roof Top									
Air conditioners	EER	11.2	13.9	11.5	13.9	11.5	13.9	11.5	13.9
Gas fired engine driven	COP	1.1		1.1		1.1		1.1	
Heat pumps	EER	11.0	12.0	11.0	12.0	11.0	12.0	11.0	12.0
Ground source heat pumps	EER	17.1	20.6	18.0	22.0	20	24	22	26

Source: *Navigant Consulting, March 2014. Inc, EIA – technology forecast updates – residential and commercial building technologies – reference case.*

3.1–3.3 or higher (Baillargeon et al., 2011), while the efficiency of the worst room air conditioners improved from 2.3 in 2005 to almost 2.9 in 2010. In Europe, a more efficient labelling scheme for air conditioners has been in use since 2013. The new European labelling scheme considers the efficiency of air conditioners based on the seasonal EER value. The minimum Seasonal EER (SEER) required for the highest class of air conditioners, A+++, should exceed 8.5. Finally, in Korea, the minimum energy performance (MEP) of room air conditioners and split systems has increased from 2.86 to 3.37 (Santamouris, 2016a).

ZERO ENERGY BUILDINGS AND ZERO ENERGY SETTLEMENTS: AN EVALUATION OF EXISTING PROGRESS AND FUTURE TRENDS AND PROSPECTS

Zero energy buildings have been at the center of energy-related discussions for at least for the last 15 years. The idea of decreasing the energy needs of buildings as much as possible by using energy conservation technologies together with renewable energy systems is very appealing. The idea has been explored by the research community and the market, and several hundred buildings with an almost zero annual balance have been built and monitored. Zero energy buildings employ a range of procedures to achieve energy targets and cost reduction. The spectrum of existing applications includes buildings in which serious effort has been made to decrease energy requirements ranges from using energy conservation technologies and covering the rest of the load through the use of rational renewable energy systems, to buildings with an excessive use of photovoltaics and an almost negligible energy conservation.

Although the notion of zero energy buildings is insufficiently defined and not fully mature, it has been adopted as future policy by the European Union and partly by the US administration. The forecast document of the EU Directive on Energy Performance of Buildings makes it mandatory for all new buildings be 'almost zero energy buildings' by the end of 2020 (EPBD Recast, 2010). In parallel, the Department of Energy of the United States has set up a strategic goal to achieve 'marketable zero energy homes in 2020 and commercial zero energy buildings in 2025' (US DOE, 2008).

Several definitions are available for zero energy buildings (Torcellini et al., 2006; ECEEE, 2009; ECOFYS, 2013). As mentioned by Sartori et al. (2012), 'there is a very limited agreement on a common definition of zero energy buildings based on scientific analysis'. According to the European Union (EPBD Recast, 2010), 'Near Zero Energy Building means a building that has a very high energy performance as determined in accordance with Annex 1.' According to Annex 1, 'The energy performance of a building shall be determined on the basis of the calculated or actual annual energy that is consumed in order to meet the different needs associated with its typical use and shall reflect the heating energy needs and cooling energy needs (energy needed to avoid overheating) to maintain the envisaged temperature conditions of the building, and domestic hot water needs.' The directive also mentions 'The nearly zero or very low amount of energy required should be covered to a very significant extent by energy from renewable sources, including energy from renewable sources produced on-site or nearby.'

Zero energy buildings (ZEB) can be autonomous and not connected to the grid, or they can be connected to energy infrastructure. According to Sartori et al. (2012), the term ZEB can be used to refer to all zero energy buildings, including autonomous buildings, while for those buildings connected to energy infrastructure the term 'net ZEB' is proposed. The term indicates that there is a balance between the energy supplied by the grid to the building and the energy returned to the grid by the building over a period of one year. A schematic representation of net ZEB is proposed by Sartori et al. (2012) and is shown in Fig. 7.3.

A comprehensive frame and definition of the various types and terms of net Zero Energy Buildings is provided by Tortellini et al. (Table 7.2).

It becomes obvious that the various options regarding the supply of renewable energy is of very high importance. A comprehensive framework of the existing options is provided by Torcellini et al. (2006) (Table 7.3).

There are increasing concerns about the additional cost induced by zero energy buildings. In particular, concerns are raised regarding the additional cost that has to be invested to almost minimize energy cost and the

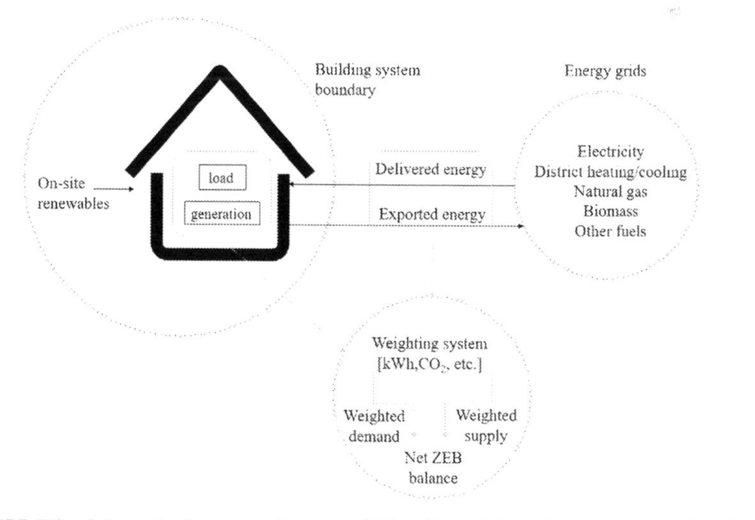

FIGURE 7.3 Schematic Representation a net ZEB. *Adapted from Sartori, I., Napolitano, A., Voss, K., 2012. Net zero energy buildings: A consistent definition framework. Energy and Buildings 48, 220–232.*

TABLE 7.2 Definition and Frame of Net Zero Building

Net Zero Site Energy	A site ZEB produces at least as much energy as it uses in a year, when accounted for at the site.
Net Zero Source Energy	A source ZEB produces at least as much energy as it uses in a year, when accounted for at the source. Source energy refers to the primary energy used to generate and deliver the energy to the site. To calculate a building's total source energy, imported and exported energy is multiplied by the appropriate site-to-source conversion multipliers.
Net Zero Energy Costs	In a cost ZEB, the amount of money the utility pays the building owner for the energy the building exports to the grid is at least equal to the amount the owner pays the utility for the energy services and energy used over the year.
Net Zero Energy Emissions	A net-zero emissions building produces at least as much emissions-free renewable energy as it uses from emissions-producing energy sources.

Source: *Adapted from Torcellini, P., Pless, S., Deru, M., Crawley, D., 2006. Zero energy buildings: A critical look at the definition, national renewable energy laboratory and department of energy, US.*

TABLE 7.3 ZEB Renewable Energy Supply Option Hierarchy

Option Number	ZEB Supply-Side Options	Examples
0	Reduce site energy use through low-energy building technologies	Daylighting, high-efficiency HVAC equipment, natural ventilation, evaporative cooling, etc.
On-Site Supply Options		
1	Use renewable energy sources available within the building's foot print	PV, solar hot water, and wind located on the building.
2	Use renewable energy sources available at the site	PV, solar hot water, low-impact hydro and wind located on the building.
Off-Site Supply Options		
3	Use renewable energy sources available off site to generate energy on site	Biomass, wood pellets, ethanol, or biodiesel that can be imported from off site, or waste streams from on-site processes that can be used on-site to generate electricity and heat.
4	Purchase off-site renewable energy sources	Utility-based wind, PV, emissions credits, or other "green" purchasing options. Hydroelectric is sometimes considered.

Source: *Torcellini, P., Pless, S., Deru, M., Crawley, D., 2006. Zero energy buildings: A critical look at the definition, national renewable energy laboratory and department of energy, US.*

availability of the additional resources necessary, especially for the low income population. Several studies have investigated the additional investments necessary to build and run zero energy buildings. In parallel, several optimization methods have been developed that aim to guide investors on the optimum pathway to design a ZEB building (Kapsalaki et al., 2012).

FUTURE ENERGY CONSUMPTION OF THE BUILDING SECTOR, THE ROLE OF OVERPOPULATION, TECHNOLOGY DEVELOPMENT, ECONOMIC DEVELOPMENT AND CLIMATE CHANGE ON THE FUTURE ENERGY CONSUMPTION OF BUILDINGS

Numerous studies have been carried out to forecast future energy consumption by the building sector. Projections of future energy consumption are necessary to design and plan electricity networks and assess the potential use of conventional and new energy sources. Several models have been developed to predict the building sector's future energy consumption for various parts of the world (Euroheat and Power, 2006; Hadley et al., 2006; Warren et al., 2006; McNeil et al., 2008; Scott et al., 2008; Issac and Van Vuuren, 2009; Siyak, 2009; Mima et al., 2011; Labriet et al., 2013; Zhou et al., 2014; DOE, 2015). Existing models refer to energy consumption by both residential and tertiary buildings.

The proposed models predict the future energy consumption of buildings in specific parts of the globe or for the whole planet, and are based on the forecast of several parameters that impact energy consumption by the building sector, such as the future penetration of heating and cooling systems in the world, increased GDP in various parts of the planet, the impact of global climate change and expected population increase, and estimated technological developments in building energy systems (McNeil et al., 2008; Scott et al., 2008; Isaac and Vuuren, 2009; Mima et al., 2011; Labriet et al., 2013; Zhou et al., 2014). Some less detailed models consider parts of the above parameters, such as increased population or potential temperature increase resulting from climatic change (Hadley et al., 2006; Warren et al., 2006; Sivak, 2009). Most of the predictions focus on the year 2050 or 2100.

While heating energy consumption by the building sector appears to account for the largest portion of total energy consumption today, most studies predict that energy consumption for cooling purposes will increase considerably in the future, and indeed become much more important than heating demand. Fig. 7.4 shows predictions for future heating and cooling demand for the whole world (Isaac and Van Vuuren, 2009). As shown, the expected increase in cooling energy demand is quite impressive, while the heating energy demand is expected to be constant and stabile (Table 7.4).

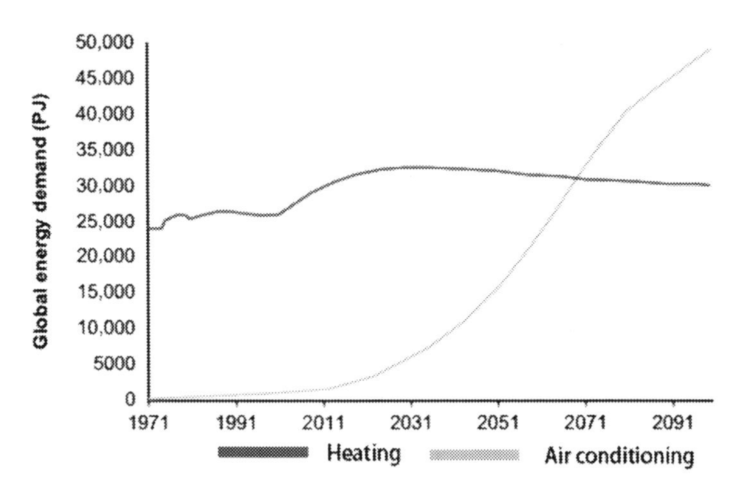

FIGURE 7.4 Modeled global residential energy demand for heating and air conditioning in a reference scenario. *Adapted form Isaac, M., van Vuuren, D.P., 2009. Modeling global residential sector energy demand for heating and air conditioning in the context of climate change. Energy Policy 37, 507–521.*

TABLE 7.4 Percentage Increase of the Cost of Various Types of ZEB Buildings in the District of Columbia, US

	Energy Conservation Measures	Net Zero Energy (Renewables With ECMs)
Office New Construction	1%–6%	5%–10%
Multifamily New Construction	2%–7%	7%–12%
Office Renovation	7%–12%	14%–19%

Source: *From Net Buildings Institute : Net Zero and Living Building Challenge Financial Study: A Cost Comparison Report for Buildings in the District of Columbia, 2013, Assessed through: https://newbuildings.org/wp-content/uploads/2015/11/ZNECostComparisonBuildingsDC1.pdf.*

The drivers that define future heating demand in the major geographic areas of the planet are given in Table 7.5 (Isaac and Van Vuuren, 2009). As shown, the potential increase in the ambient temperature and the world's total population highly influence the energy consumption that will be required for cooling in the future. Fig. 7.5 shows the predicted future residential energy consumption for heating in different parts of the world. Western Europe seems set to experience a very significant decrease in heating energy consumption, mainly because of the intensive energy measures foreseen. On the contrary, China and the USA may experience a very

TABLE 7.5 Change in Regional Drivers for Heating Energy Demand Between 2000–2050 and 2050–2100 (%)

	Canada	USA	Rest of America	Africa	Western Europe	Russia	Rest of Europe	India	China	Pacific OECD	Rest of Asia
2000–2050											
Population	40	39	50	138	3	−22	−7	56	9	−2	76
Floor space	39	39	63	43	11	93	83	240	162	41	101
HDD	−11	−15	−29	−61	−18	−11	−20	−47	−15	−17	−16
Intensity	5	6	−6	0	−12	−25	−15	−25	71	20	−19
Efficiency	9	12	49	81	10	4	14	139	107	17	21
Energy use	66	56	9	−27	−25	−3	1	−12	103	18	100
2050–2100											
Population	−7	7	−5	24	−11	−21	−15	−5	−15	−15	6
Floor space	32	22	45	93	4	26	32	61	28	22	46
HDD	−13	−17	−30	−65	−21	−12	−18	−35	−18	−20	−20
Intensity	5	5	3	−28	−13	−33	−8	−33	42	14	−19
Efficiency	1	−2	11	27	−5	4	4	21	12	−12	0
Energy use	11	18	−9	−53	−34	−43	−18	−44	12	8	1

Source: *Isaac, M., van Vuuren, D.P., 2009. Modeling global residential sector energy demand for heating and air conditioning in the context of climate change. Energy Policy 37, 507–521.*

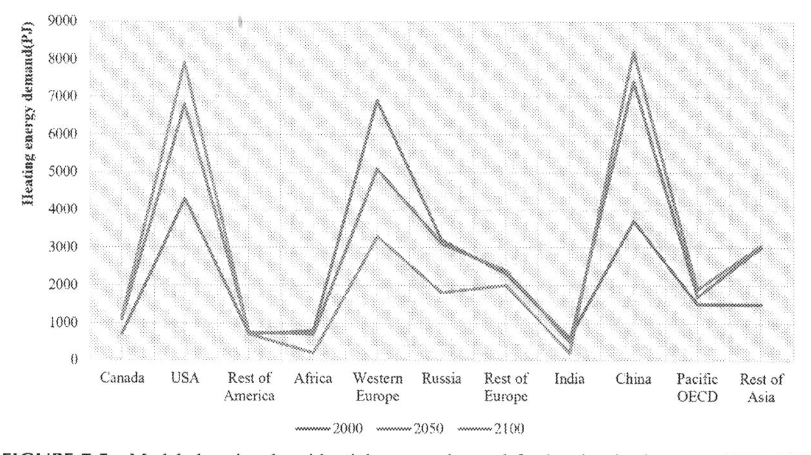

FIGURE 7.5 Modeled regional residential energy demand for heating in the years 2000, 2050 and 2100 (reference scenario). *Adapted from Isaac, M., van Vuuren, D.P., 2009. Modeling global residential sector energy demand for heating and air conditioning in the context of climate change. Energy Policy 37, 507–521.*

significant increase in heating demand, mainly due to population growth in China and a corresponding increase in the total space used.

As per cooling energy consumption, Fig. 7.6 shows predictions from several published models regarding residential cooling energy demand for the major geographical areas of the planet for 2030, 2050 and 2100. All models used considered the full set of inputs rather than a reduced number to evaluate the final cooling energy consumption. As shown, there are significant discrepancies between the predictions given by the models considered. This is mainly due to uncertainty regarding the input data and the future evolution of the market.

It is clear that the expected increase in residential energy demand for cooling will be highly significant. For 2030, the highest cooling demand is still predicted to take place in the USA and Europe. However, from 2050 onwards, China and India are predicted to consume very high amounts of energy for cooling, mainly because of the very rapid penetration of air conditioning, the tremendous increase in the local population, and a significant increase in the ambient temperature as a result of global climate change.

As predicted by McNeil and Letschert (2008) and Issac and van Vuuren (2009), in India the residential energy consumption for cooling will be 110 TWh in 2030 and will increase up to 780 TWh in 2050 and 4700 TWh in 2100. In China, forecasts show than in 2050 residential energy consumption for cooling will reach 1250 TWh and will increase up to 1610 in 2100. Finally, for the rest of the Asian countries, residential energy demand may increase up to 4100 TWh by 2100 (Isaac and van Vuuren, 2009; Mima et al., 2011).

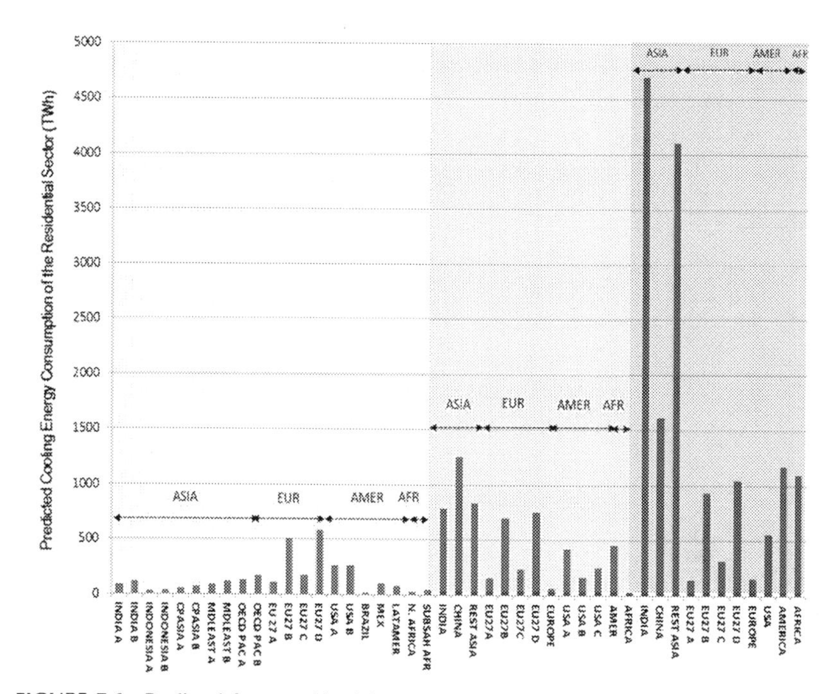

FIGURE 7.6 Predicted future residential cooling energy consumption by the various existing models. The light gray zone (left part of the figure) is for 2030, the dark gray (middle part of the figure) is for 2050, and the black (right part) is for 2100. *Adapted from Santamouris, M., 2016b. Innovating to zero the building sector in Europe: Minimising the energy consumption, eradication of the energy poverty and mitigating the local climate change. Solar Energy 128, 61–94.*

For Europe, predictions for future residential cooling energy consumption vary considerably across the three existing studies (Issac and van Vuuren, 2009; Mima et al., 2011). This is because of the variability in their assumptions regarding the future penetration of air conditioning and the magnitude of global climate change in Europe. The forecasts for 2050 range between 60 TWh and 750 TWh, while for 2100 they range between 150–1050 TWh. As concerns the commercial sector in Europe, predictions offered by Mima et al. (2011) estimate that the cooling energy demand in 2030 will fall between 930 and 1050 TWh, for 2050 between 1050 and 1500 Twh, and for 2100 it may range between 1400 to 2850 TWh.

As it concerns future energy consumption for residential cooling in the United States, three forecast models are proposed (Scott et al., 2008; Isaac and van Vuuren, 2009; DOE, 2015). It is estimated that for 2030 cooling consumption may be close to 260 Twh, and that it will increase up to 420 in 2050 and 555 in 2100. The corresponding consumption for the commercial sector will range from 270–563 TWh (DOE, 2015). All other countries on the American continent may present a consumption close to 450 TWh in 2050 and 1170 TWh in 2100 (Isaac and van Vuuren, 2009).

In Africa a significant increase in residential cooling demand is foreseen. Isaac and van Vuuren (2009) predict that it may increase from 30 TWh in 2050 to about 1100 TWh in 2100.

A method to determine future energy consumption for cooling purposes in commercial buildings, Ect(t), has been proposed by Santamouris (2016b). According to the proposed methodology, the future energy consumption of commercial buildings may be determined as a function of the future total floor area, FA, their specific energy consumption per square meter, EPM(t), the impact of the global and local climate change, f(CDD), the potential future increase of air conditioning in commercial buildings, f(ACavlblty), the potential increase of the performance of air conditioners, f(ACeff), and the future energy performance of commercial buildings, f(BldgPerf):

$$\text{Ect}(t) = \text{FA} \times \text{EPM}(t) \times f(\text{CDD}) \times f(\text{ACavlblty}) \times f(\text{ACeff}) \times f(\text{BldgPerf}) \tag{7.1}$$

Three scenarios have been developed to predict future cooling consumption by commercial buildings for 2050: High, average and low development scenarios. The basic assumptions for each scenario are given in Santamouris (2016a). The results of the model for all three scenarios are given in Fig. 7.7. As shown, the predicted future energy consumption of the commercial buildings is 1.13 PWh, 1.55 PWh and 2.03 PWh for the low, average, and high development scenarios respectively. Predictions correspond to an increase of close to 200%, 275% and 360% for the three scenarios.

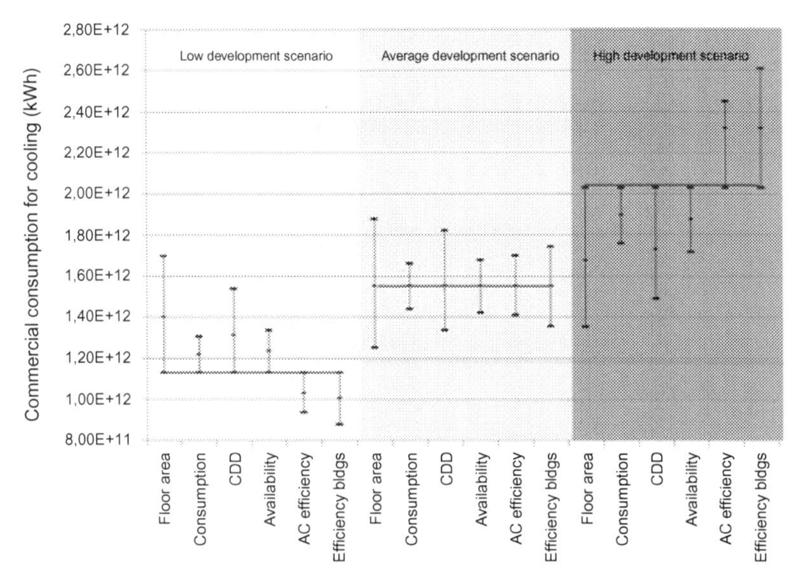

FIGURE 7.7 Predicted cooling energy consumption of commercial buildings globally for 2050. Results of the low, average, and high development scenarios and variability as a function of the main parameters and drivers. *Adapted from Santamouris, M., 2016a. Cooling of buildings. Past, Present and Future, Energy and Buildings, 128, 617–638, 2016.*

In light of the considerable uncertainty of the input data considered, a sensitivity analysis has been performed to identify the levels of uncertainty. It is estimated that cooling energy consumption by commercial buildings in 2050 may vary from 0.9 PWh to 2.61 Pwh, with a more probable prediction close 1.6 PWh.

Future cooling energy consumption by the residential sector for 2015 is also assessed by Santamouris (2016a). Calculations have been performed in a similar way as for commercial buildings. In particular, the annual cooling energy consumption of the residential sector, Ect(r), may be calculated by the following expression, (7.2):

$$\begin{aligned}\mathrm{Ect}(r) = {} & \mathrm{POP} \times \mathrm{FAP} \times \mathrm{EPM}(r) \times f(\mathrm{CDD}) \times f(\mathrm{ACavlblty}) \\ & \times f(\mathrm{ACeff}) \times f(\mathrm{BldgPerf})\end{aligned} \tag{7.2}$$

Where POP is the world population, FAP is the residential floor area per person, EPM(r) is the current average cooling consumption per square meter in the residential sector, f(CDD) is a function that considers the impact of global climate change, while all other parameters are as in Eq. 7.1. Similar to the case of commercial buildings, three scenarios (low, average, and high development) have been generated for 2050. The assumptions considered for each scenario are given in Santamouris (2016a). The values calculated for cooling energy demand by the residential sector in 2050 are given in Fig. 7.8. As shown, the predicted residential cooling energy consumption is

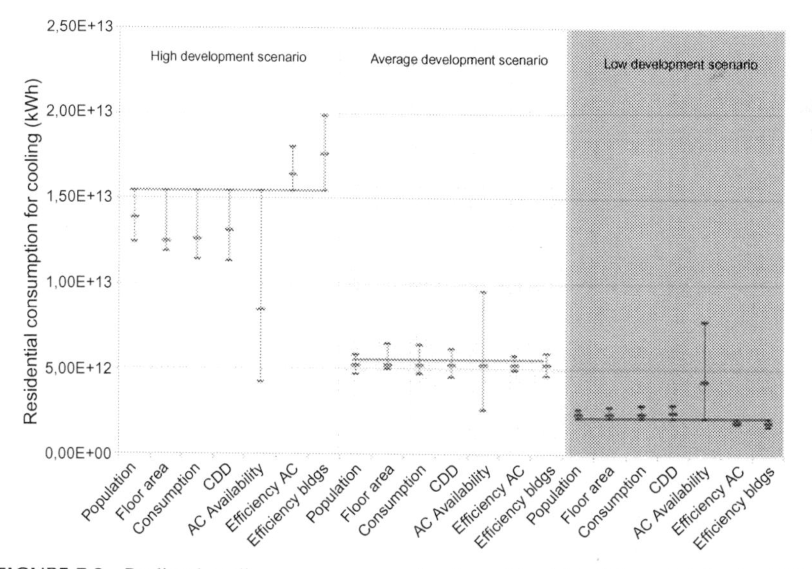

FIGURE 7.8 Predicted cooling energy consumption by residential buildings globally for 2050. Results of the low, average, and high development scenarios and variability as a function of the main parameters and drivers. *Adapted from Santamouris, M., 2016a. Cooling of buildings. Past, Present and Future, Energy and Buildings, 128, 617–638, 2016.*

calculated as being close to 2.15 PWh for the low development scenario, 5.27 PWh per year for the average scenario and 15.4 PWh for the high development scenario. These figures correspond to an increase in residential energy consumption by 2050 of close to 320%, 750% and 2270% compared to actual levels.

As in the case of commercial buildings, a sensitivity analysis has been performed to define the uncertainty levels. The results of the sensitivity analysis are shown in Fig. 7.8. The estimated potential variability is close to 360%. In conclusion, residential cooling energy consumption in 2050 is predicted to vary between 1.67 PWh/year and 19.8 PWh/year, with an average predicted value of close to 5.3 PWh/year.

POLICY AND REGULATORY ISSUES TO CONTROL AND MINIMIZE ENERGY CONSUMPTION IN THE BUILDING SECTOR

Analysis of the Potential Impact of the Minimization of Energy Consumption in the Building Sector on Labor, Economy, Environment and Global Development

Minimization of the energy consumption within the building sector should be based on a very drastic decrease in energy consumption by the existing building stock and a radical reduction of the energy needs of future buildings. Given the size of the existing building stock and the significant energy consumption it represents, most of the effort required is related to the energy retrofitting of existing buildings. Improving the energy performance of existing buildings will require a sizeable budget to be invested in the near future. Building-related investments are labor intensive and may generate new jobs, while the impact on the economy and society may be significant. Building-related investments generate intensive economic growth and substantial opportunities for global development. Parallel to economic and financial growth, energy related investments will result in a significant generation of innovative knowledge and will boost energy-related scientific developments, and even technological breakthroughs.

In the following, the order of magnitude of the necessary investments to minimize energy consumption by the building sector in Europe until 2050 will be investigated. In parallel, the related environmental, economic and social benefits will be analysed.

Several International institutions, including the European Union, have published numerous scenarios for the future energy consumption of the building sector (BPIE, 2011; GEA, 2012; European Commission, 2014). The existing studies forecast the future energy consumption up to 2030 or 2050, and propose appropriate technological solutions to achieve the targets set. Given that these scenarios use different initial and boundary conditions, it is

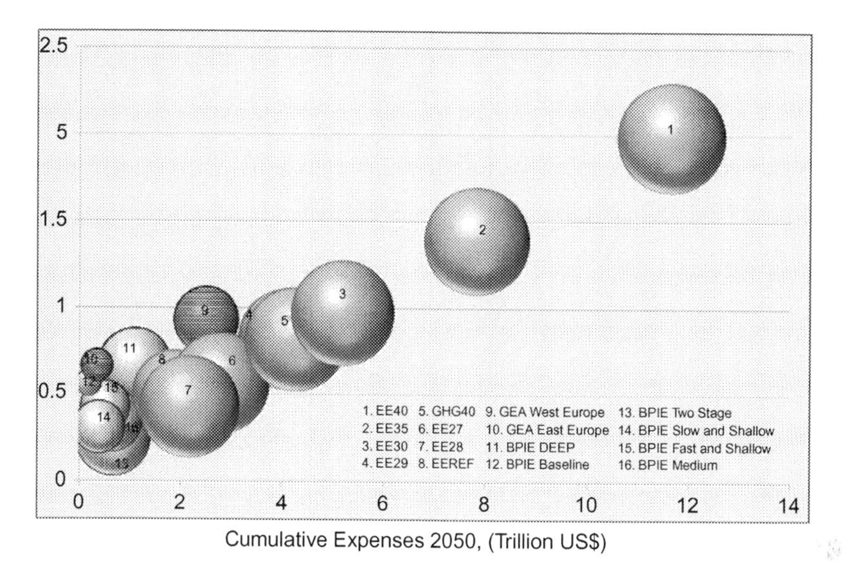

FIGURE 7.9 Cumulative expenses and required cost per PWh of energy conservation predicted by 16 scenarios for 2050. *Adapted from Santamouris, M., 2016a. Cooling of buildings. Past, Present and Future, Energy and Buildings, 128, 617–638, 2016.*

not feasible to compare their results. However, almost all of scenarios estimate the potential reduction in energy consumption by the building sector, together with the required investments to achieve the specific energy objectives. In total, 17 different scenarios have been published for energy consumption by the building sector up to 2050. Fig. 7.9. presents the characteristics of the existing energy conservation scenarios in Europe for 2050 (BPIE, 2011; GEA, 2012; European Commission, 2014).

Fig. 7.9 shows the estimates for the cumulative investments needed to renovate European building stock by 2050 as a function of the cost per PWh of energy conservation. The cost per conserved PWh of energy varies between EUR€0.227 and EUR€1.79 trillion, with an average value of near to EUR€0.66 trillion. The significant variability in cost is mainly due to the different mixture of energy technologies considered by each study and scenario. In parallel, the investment cost per PWh is found to increase as a function of the total expected energy conservation. This is mainly due to the higher cost of some of the technologies that may be employed to achieve a very high energy conservation figure. In other words, the incremental cost of the energy conservation technologies increases once more sophisticated technologies are employed to minimize energy consumption by the building sector.

The energy conservation targets for 2050 that were considered by the different studies vary from 40%–63% compared to 2007 consumption levels (European Commission, 2014). The calculated cost per PWh of energy conservation is found to increase almost exponentially with the percentage of

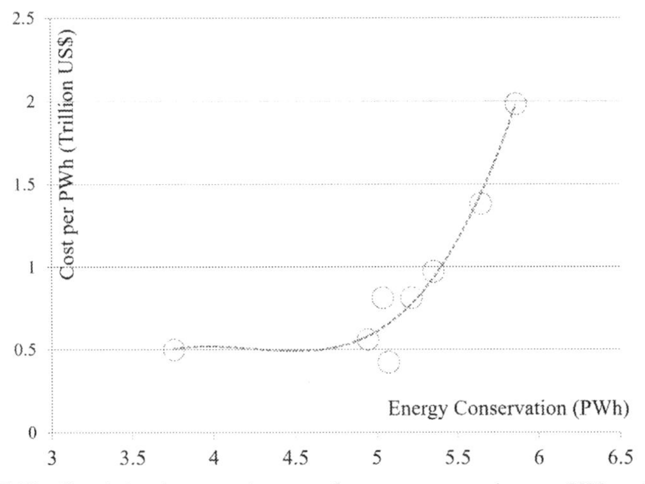

FIGURE 7.10 Correlation between the cost of energy conservation per PWh and the total energy conservation in PWh.

considered energy conservation (Fig. 7.10). It varies between EUR€0.39 trillion per PWh to EUR€1,8 trillion per PWh for the scenario with the highest energy consumption, EE40. For this quite advanced energy conservation scenario, EE40, the final energy consumption by the building sector is predicted to reach 3.4 PWh in 2050, compared to 9.3 PWh in 2007, while the calculated related cumulative expenses are close to 10.6 trillion euros. Based on the assumptions of the EE40 scenario, it can be estimated that in order for the potential minimization of energy consumption within the building sector will decrease to 1 TWH by 2050, additional investments should be between EUR€4.3–5.45 trillion, and the total cumulative investments should be between EUR€15–23.7 trillion.

A substantial decrease in energy consumption by the building sector is associated with significant environmental benefits. Based on the assumptions of the EE40 scenario of the European Commission (2014), carbon dioxide emissions by the residential building sector will decrease by 90% in 2050 compared to 2007 levels, while the corresponding decrease for the commercial sector is estimated at close to 87.7%. It is well known energy conservation technologies are labour intensive. Thus, besides significant energy, financial and environmental benefits, the minimization of energy consumption by the building sector will create a substantial number of additional jobs in the construction sector and the global economy.

The impact of energy conservation technologies on labour it is well recognized by many international institutions. The IPCC (2008) reports that the employment of energy efficiency measures may create new business opportunities, while the indirect effects arising from the spending of conserved energy money on other economic activities may significantly boost employment and generate important new labour opportunities. The Energy

Efficiency Industrial Forum (EEIF) (2012) compiled information from different sources on the impact of energy efficient measures on employment. They report that the additional full time jobs created per million euros invested in energy efficient measures varies between 6.4 and 39, with an average of close to 19 new jobs per million euros invested in Europe. For the USA and Canada, existing estimations suggest that 6.8 and 57.5 new jobs will be generated per million euros invested on energy efficiency measures, with an average near 20.6 jobs per million invested.

The European Trade Association (2007) reports that the intensive implementation of energy efficiency measures in Europe aiming to reduce greenhouse gas emissions in the residential sector may generate almost 1.4 million additional full time jobs annually. It is estimated that for each million euros invested in energy efficiency measures, there are almost 18.8 new jobs created in the global economy. Additionally, the multiplication factor may vary between 2 and 6.25 additional jobs generated per million euros invested. Thus, in total, the additional new jobs that are indirectly generated may range from 146,000−456,250 per year.

To estimate the annual generation of new jobs caused by additional investment in building-related energy efficiency measures, the assumptions considered by the European Union (2014) and the EEIF (2011) are considered. Calculations have been performed for low, average, and high multiplication factors, and in particular 2.6 and 19 jobs per million euros invested. According to the previously presented analysis, the lower level of investments necessary to decrease energy consumption by the building sector down to 1 TWH in 2050 is close to EUR€14.5 trillion. Thus, the number of new full time jobs that are expected to be generated annually up to 2050 may vary from 674,000−6,407,000, as a function of the selected multiplication factor.

REFERENCES

Pierre Baillargeon, Normand Michaud, Luc Tossou, Paul Waide, Cooling: BENCHMARKING study part 1: Mapping component report, June 2011, in partnership with the collaborative labeling and appliance standards program (CLASP).

BPIE, 2011. Buildings performance institute Europe: Europe's buildings under the microscope. <http://www.bpie.eu/eu_buildings_under_microscope.html#.VkBlq_nhDWI>.

Claus Händel, 2011. Ventilation with heat recovery is a necessity in "nearly zero" energy buildings. REHVA Journal.

DOE/EIA, 2015. Annual energy outlook 2015. Energy information administration, D.o.E.

ECEEE, 2009. Net zero energy buildings: Definitions, issues and experience, European council for an energy efficient economy, EU.

ECOFYS, 2013. Towards nearly zero energy buildings definition of common principles under the EPBD.

EPBD Recast, directive 2010/31/EU of the European parliament and of the council of 19 May 2010 on the energy performance of buildings (recast), official Journal of the European Union, (2010) 18/06/2010.

Energy Efficiency Industrial Forum, 2012. A survey of the employment effects of investment in energy efficiency of buildings, Brussels, Belgium. <http://www.euroace.org/EuroACEActions/PolicyOverview/PublicationsReports.aspx>.

Energy Technology Perspectives 2010–Scenarios & strategies to 2050, International energy agency, Paris, 2010.

Euroheat and Power, The European cold market, final report of the EcoHeatCool project, European Union, 2006.

European Commission, 2014. Commission staff working document impact assessment accompanying the document communication from the commission to the European parliament and the council energy efficiency and its contribution to energy security and the 2030. Framework for climate and energy policy, COM (2014) 520 final}{SWD(2014) 256 final.

European Trade Union Confederation, 2007. Climate change and employment. <https://www.etuc.org/climate-change-andemployment>.

GEA, 2012. Global Energy Assessment – Toward a Sustainable Future. Cambridge University Press, Cambridge UK and New York, NY, USA, and the International Institute for Applied Systems Analysis, Luxemburg, Austria. <http://www.iiasa.ac.at/web/home/research/Flagship-Projects/Global-Energy-Assessment/GEA-Summary-web.pdf>.

Glass, 2015. Dynamic glass market poised for growth as new production capacity comes on line. http://glassmagazine.com/article/commercial/smart-future-1210320.

Grand View Research, 2015. Global advanced phase change material (APCM) market by application. <http://www.grandviewresearch.com/press-release/global-advanced-phase-change-materialmarket>.

Hadley, Stanton W., Erickson, s III, David, J., Hernandez, Jose Luis, Broniak, Christine T., Blasing, T.J., 2006. Responses of energy use to climate change: A climate modeling study. Geophys. Res. Lett. 33, L17703. Available from: https://doi.org/10.1029/2006GL026652.

IEA-ETSAP and IRENA© Technology brief E12 – January 2013: Heat pumps, technology brief.

IPCC (International Panel on Climate Change), 2008. Climate Change 2007: Impacts, Adaptation and Vulnerability. Cambridge University Press, Cambridge.

International Energy Agency: Technology roadmap energy-efficient buildings: Heating and cooling equipment, 2011.

International Energy Agency, 2013. World energy outlook 2013. <http://www.worldenergyoutlook.org/weo2013/>.

Isaac, Morna, van Vuuren, Detlef P., 2009. Modeling global residential sector energy demand for heating and air conditioning in the context of climate change. Energy Policy 37, 507–521.

Kapsalaki, M., Leal, V., Santamouris, M., 2012. A methodology for economic efficient design of net zero energy buildings. Energy Build. 55, 765–778.

Labriet, Maryse, Joshi, Santosh R., Vielle, Marc, June 2013. Amit Kanudia in Collaboration with Phil Holden, Neil Edwards: Impacts of Climate Change on Heating and Cooling: a Worldwide Estimate from Energy and Macro-economic Perspectives. International Energy Workshop, Paris, France.

Lawrence Berkeley National Laboratory: Super-efficient equipment and appliance deployment (SEAD) initiative: Lessons from the technical analysis of room air conditioners, April 2013.

A.J. Marszal, J.S. Bourrelle, E. Musall, P. Heiselberg, A. Gustavsen, K. Voss, Net zero energy buildings – calculation methodologies versus national building codes, in: EuroSun Conference, Graz, Austria, 2010.

MBUB, Federal Ministry for the Environment, Nature Conservation, Building and Nuclear Safety (BMUB), 2014. Greentech made in Germany 4.0 – environmental technology Atlas for Germany. http://www.greentech-made-in-germany.de/en/.

McNeil, Michael A., Letschert, Virginie E., 2008. Future Air Conditioning Energy Consumption in Developing Countries and what can be done about it: The Potential of Efficiency in the Residential Sector. Lawrence Berkeley National Laboratory. Available at: <http://escholarship.org/uc/item/64f9r6wr>.

Anette Michel, Eric Bush, Jürg Nipkow, Conrad U. Brunner, Hu Bo, 2010. Energy efficient room air conditioners – best available technology (BAT), Available through: <http://www.topten.eu/uploads/File/023-Anette-Michel final paper-S.pdf>.

Mima, Silvana, Criqui, Patrick, Watkiss, Paul, 2011. The Impacts and Economic Costs of Climate Change on energy in the European Union: Summary of Sector Results from the ClimateCost Project. Funded by the European Community's Seventh Framework Programme. Technical Policy Briefing Note Series. Stockholm Environment Institute, Oxford.

Nanomarkets: Smart Window Markets, 2012. <http://nanomarkets.net/Downloads/ES/Nano-515%20ES.pdf>.

Navigant Consulting, Inc, EIA – technology forecast updates – residential and commercial building technologies – reference case, March 2014.

Rong, F., Clarke, L., Smith, S., 2007. Climate Change and the Long-Term Evolution of the U.S. Buildings Sector. Pacific Northwest National Laboratory.

Santamouris, M., 2016a. Cooling of buildings. Past, Present and Future, Energy and Buildings 128, 617–638.

Santamouris, M., 2016b. Innovating to zero the building sector in Europe: Minimising the energy consumption, eradication of the energy poverty and mitigating the local climate change. Solar Energy 128, 61–94.

Sartori, I., Napolitano, Assunta, Voss, Karsten, 2012. Net zero energy buildings: A consistent definition framework. Energy and Buildings 48, 220–232.

Scott, Michael J., Dirks, James A., Cort, Katherine A., 2008. The value of energy efficiency programs for US residential and commercial buildings in a warmer world. Mitig. Adapt. Strategies Global Change 13, 307–339.

SEAD, Technical Analysis Working Group, 2013. Cooling the Planet: Opportunities for Deployment of Superefficient Room Air Conditioners, Environmental Energy Technologies Division International Energy Studies Group. Lawrence Berkeley National Laboratory.

Singh, Sarwant, 2012. New Mega Trends: Implications for our Future Lives. Published by Palgrave Macmillan, London.

Sivak, Michael, 2009. Potential energy demand for cooling in the 50 largest metropolitan areas of the world: Implications for developing countries. Energy Policy 37, 1382–1384.

The Net Buildings Institute : Net zero and living building challenge financial study: A cost comparison report for buildings in the district of Columbia, 2013, assessed through: <https://newbuildings.org/wp-content/uploads/2015/11/ZNECostComparisonBuildingsDC1.pdf>.

P. Torcellini, S. Pless, M. Deru, D. Crawley, 2006. Zero energy buildings: A critical look at the definition, national renewable energy laboratory and department of energy, US.

US DOE, Building technologies program, planned program activities for 2008–2012, department of energy, US, <http://www1.eere.energy.gov/buildings/mypp.html>, 2008 (downloaded 01/07/2010).

US Energy Information Administrations, 2012. Commercial buildings energy consumption survey.

Rachel Warren, Nigel Arnell, Robert Nicholls, Peter Levy, Jeff Price, 2006. Understanding the regional impacts of climate change research report prepared for the stern review on the economics of climate change, working paper 90, Tyndall centre for climate change research.

World Resources Institute, 2010. Green investment horizons: Effects of policy on the market for building energy efficiency technologies. <http://www.wri.org/sites/default/files/green_investment_horizons.pdf>.

Zhou, Yuyu, Clarke, Leon, Eom, Jiyong, Kyle, Page, Patel, Pralit, Kim, Son H., et al., 2014. Modeling the effect of climate change on U.S. state-level buildings energy demands in an integrated assessment framework. Appl. Energy 113, 1077–1088.

Chapter 8

Mitigating the Local Climatic Change and Fighting Urban Vulnerability

INTRODUCTION

Urban Heat Islands (UHIs) and local climate change significantly increase the temperature of cities and cause serious health, comfort, energy, economic and social problems. A full discussion and analysis of the magnitude and the characteristics of urban overheating is provided in Chapter 3. Specific mitigation technologies and techniques have been developed recently to face the serious problem of present and future urban overheating. These aim to decrease the strength of heat sources in the urban environment and increase the potential of cool sinks to drop the temperature. In particular, mitigation technologies and policies are designed to decrease the absorption of incoming solar radiation by the city structure and decrease the release of anthropogenic heat. They also aim to increase losses of solar radiation through latent heat processes such as evaporation and evapotranspiration, the increased emission of infrared radiation, and the dissipation of excess urban heat to cool ambient sinks such as the ground and the sky. In addition, air flow and ventilation processes in the urban environment could also contribute to improving the ambient microclimate, provided that the advected or circulated air is of a lower temperature than the temperature of the urban ambient air.

Based on the above, mitigation techniques to decrease the absorption of solar radiation aim to either increase the solar reflectivity of the materials used in the urban fabric or to provide shading and solar control in urban spaces. An increase of the latent losses may be achieved through evapotranspiration processes, using additional urban greenery placed either in open urban spaces or in the envelope of the buildings. In parallel, the use of water in pools, ponds, fountains, sprays, and so on helps to decrease the ambient temperature through evaporation processes. An increase of infrared thermal losses in the urban environment may be achieved using materials presenting a high thermal emissivity. Dissipation techniques aim to decrease the ambient temperature through coupling the ambient air with an atmospheric sink presenting a low temperature where the excess ambient heat may be

Minimizing Energy Consumption, Energy Poverty and Global and Local Climate Change in the Built Environment: Innovating to Zero. DOI: https://doi.org/10.1016/B978-0-12-811417-9.00008-8

dissipated, such as the ground or the sky. Earth to air heat exchangers are used for this purpose, while radiative cooling techniques may benefit from the lower night time surface temperature of metallic radiators. Finally, several conventive cooling techniques, such as wind towers and solar chimneys, are proposed and tested to enhance the air flow in cities.

This chapter discusses and analyzes the main mitigation technologies used to decrease the ambient temperature in cities. Existing knowledge on each of the main mitigation technologies is presented, and the knowledge acquired from the known large scale applications is reported.

DESCRIPTION AND PERFORMANCE OF MITIGATION TECHNOLOGIES TO FIGHT LOCAL CLIMATE CHANGE

General

Four major clusters of mitigation technologies that are able to counterbalance the impact of UHI and local climate change are identified in Fig. 8.1: reflective and shading technologies, greenery technologies, heat dissipation technologies, and techniques aiming to decrease the anthropogenic heat in the urban environment.

All of the above technologies are well developed and large scale applications have shown that their mitigation potential is quite significant. Results from about 210 large scale mitigation projects show that the use of these technologies can reduce the average temperature of the urban ambient air by

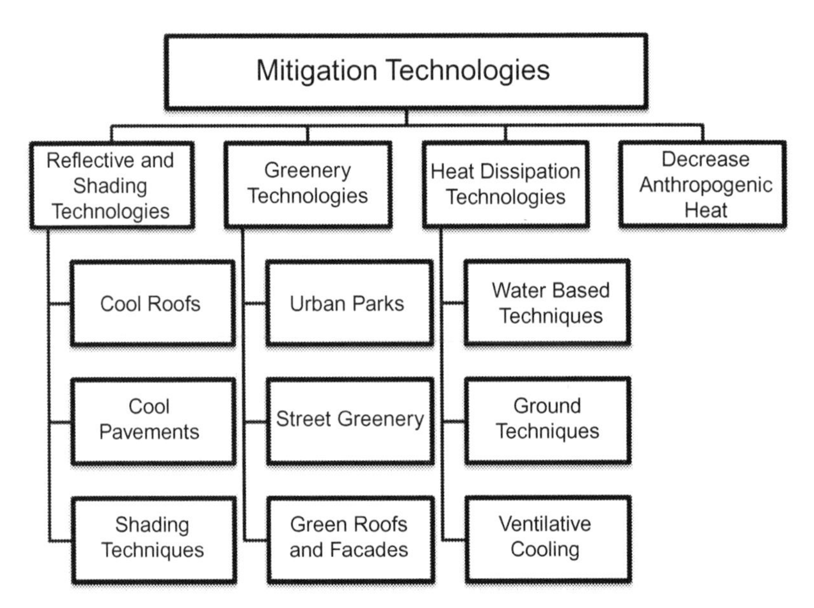

FIGURE 8.1 The main mitigation technologies to counterbalance the impact of local climate change in cities.

up to 2.5°C, while the local temperature decrease can be as high as 4–5°C. In the following chapters, the main developments in the four major mitigation technologies are presented and discussed.

Reflective Technologies – Increasing the Albedo of Cities

Materials absorb solar radiation and increase their temperature. The excess heat stored in their mass is released into the ambient environment in the form of sensible heat and increases their temperature. In parallel, materials emit infrared radiation, which is partly absorbed by humans, affecting their comfort conditions. To reduce the surface temperature of the materials in the urban environment, the absorbed solar radiation must decrease, either by applying solar control and shading techniques or by increasing their reflectivity. The use of reflective materials, especially white external coatings applied to building facades and roofs, is a well-known technique in warm climates. Intensive research carried out in recent years has resulted in the development of innovative and well-performing reflective materials. These materials present a very high reflectivity in the solar spectrum, together with a high emissivity value. These are commonly known as 'cool materials' and may take the form of a coating, paint, membrane, tile or pavement. Actually, there are thousands of industrial cool materials available on the market. Most of them are certified regarding their solar reflectivity and emissivity values, as well as regarding the optical durability of the materials, through the cool roof councils of the USA and Europe.

White Materials

White materials present a high solar reflectivity. However, their mitigation potential depends highly on their emissivity value. Thousands of white reflective materials, presenting a quite high solar reflectivity, are commercially available. Experimental testing of several types of white and silver colored materials (Table 8.1) presenting various values of solar reflectivities and emissivities has shown that the surface temperature of the materials may vary significantly.

The spectral reflectivity of the tested materials is given in Figs. 8.2A–C.

The specific infrared emissivity of the tested materials is given in Table 8.2.

Figs. 8.3A–B report the range of the measured surface temperature of all the tested materials during the day and night periods.

As shown, white materials presenting both a high solar reflectivity and a high thermal emissivity may present surface temperatures that are up to 10–15°C lower than white or silver colored materials that have a quite similar reflectivity but a much lower emissivity value. In particular, during the

TABLE 8.1 Description of the Tested White and Silver Colored Materials

Sample	Sample Description	Applications	Sample Colour
Sl	Aluminum pigmented acrylic coating	B	Silver gray
S2	Acrylic, ceramic coating	B	White
S3	Acrylic, elastomeric coating	B	White
S4	Acrylic, elastomeric coating	B	White
S5	Alkyd, chlorine rubber coating	UB, B	White
S6	Aluminum pigmented, alkyd coating	B	Silver gray
S7	Emulsion paint	–	Black
S8	Acryl-polymer emulsion paint	B	White
S9	Acrylic latex	UB, B	White
S10	Aluminum pigmented coating	B	Silver
S11	Acrylic insulating paint	B	White
S12	Aluminum pigmented acrylic coating	B, UB	Silver
S13	Epoxy polyamide coating	UB	White
S14	Acrylic paint	B	White
S15	Uncoated tile (reference)	B, UB	White
S16	Acrylic elastomeric coating	B	White

B: Buildings UB: Urban environment.
Source: Synnefa, M., Santamouris, I. August 2006. Livada: A study of the thermal performance of reflective coatings for the urban environment, Solar Energy 80 (8), 968–981.

night time when solar reflectivity is not playing any role, the value of the thermal emissivity defines the surface temperature of the materials (Fig. 8.4).

Colored Infrared Reflective Materials

It is well known that almost half of the energy content of solar radiation that reaches the ground surface is included in the visible part of the solar radiation, while the rest is in infrared (Fig. 8.5).

Conventional dark colored materials used in the built environment present a low solar reflectivity in both the visible and infrared part of solar radiation. This results in high surface temperatures and a considerable release of sensible heat into the atmosphere. Cool colored materials present an increased reflectivity in the infrared part of solar radiation. High reflectivity in the IR increases the broadband reflectivity of the materials and decreases their surface temperature, while their color is kept unchanged.

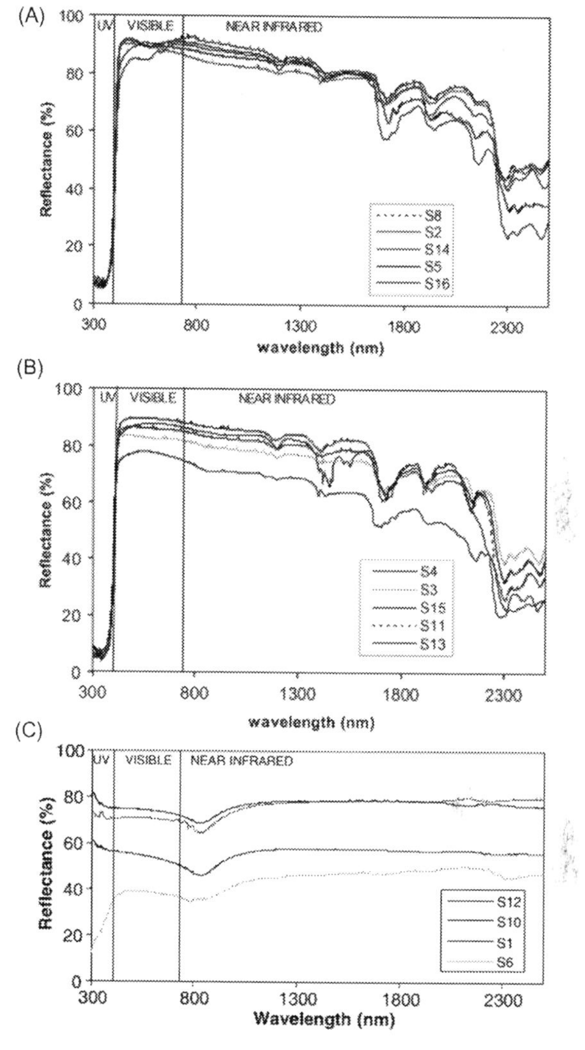

FIGURE 8.2 (A–B) Spectral reflectivity of the tested white materials. Source: Synnefa et al. (2006). (C) Spectral reflectance of the silver coatings. *Adapted from Synnefa, M., Santamouris, I. August 2006. Livada: A study of the thermal performance of reflective coatings for the urban environment, Solar Energy 80 (8), 968–981.*

An increase in the solar reflectivity of the infrared part can be achieved by using colored pigments that present a high reflectivity in the IR spectrum. A full list of the existing IR reflective pigments is provided by Levinson et al. (2005). Using such pigments, several cool colored materials have been designed and developed by Synnefa et al. (2007). The specific spectral reflectivity of the developed materials and the conventional materials of the same color are given in Fig. 8.6. As shown, the reflectivity of the developed

TABLE 8.2 Infrared Emissivity of the Tested Materials. Source: Synnefa et al. (2006)

Infrared Emittance Values of the 16 Samples			
Samples	Infrared Emittance (Error = ± 0.01)	Samples	Infrared Emittance (Error = ± 0.01)
S1	0.68	S9	0.89
S2	0.92	S10	0.49
S3	0.93	S11	0.89
S4	0.93	S12	0.35
S5	0.91	S13	0.9
S6	0.71	S14	0.89
S7	0.91	S15	0.89
S8	0.91	S16	0.91

materials in the IR spectra is considerably higher compared to conventional materials of the same color.

A picture of the developed and the conventional materials of the same colour is given in Fig. 8.7.

The spectral and broadband solar reflectance of IR reflecting materials and conventional materials of the same color is measured using appropriate spectrophotometers (Table 8.3). As shown, the developed IR reflecting materials present a much higher reflectivity than the corresponding conventional ones.

All developed materials, as well as the corresponding conventional ones, were exposed to solar radiation during the whole summer period and their surface temperature was measured. It is observed that the surface temperature of IR reflecting materials was up to 12°C lower than that of the conventional materials of similar color (Fig. 8.8).

That there is a very clear relation between the solar reflectance and the maximum surface temperature (Fig. 8.9).

The measured maximum surface temperature difference between the IR reflecting materials and the corresponding conventional materials is given in Fig. 8.10. As already mentioned, the IR reflecting colored materials present a substantially lower surface temperature than the conventional materials of the same color.

Infrared Reflective Materials Doped with Phase Change Materials

To further decrease the surface temperature of materials in the urban environment, construction materials with high thermal capacity must be used.

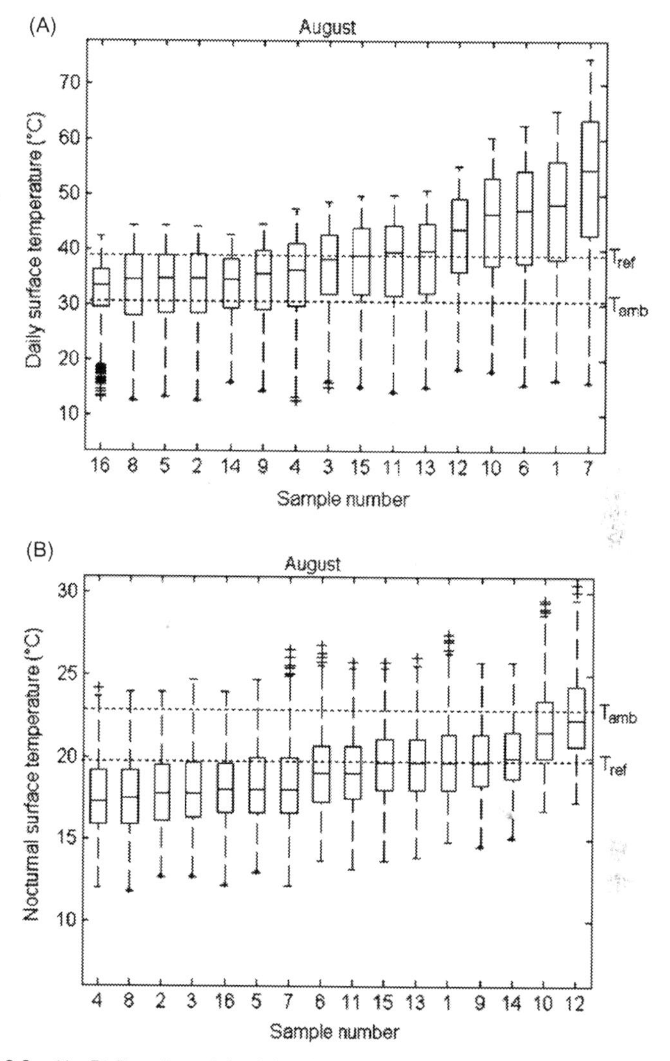

FIGURE 8.3 (A–B) Boxplots of the daily (08:00–19:00) and nocturnal (22:00–05:00) surface temperatures during the experimental period. *Adapted from Synnefa, M., Santamouris, I. August 2006. Livada: A study of the thermal performance of reflective coatings for the urban environment, Solar Energy 80 (8), 968–981.*

However, existing paving and roofing materials present a very high thermal capacity and it seems quite impossible to further increase it. To enhance the storage capacity of the construction materials used in the urban environment, composites presenting a high latent heat may be used. Phase change materials are added into the mass of the paving or roofing materials and can store additional heat under the form of latent heat. Usually, phase change materials

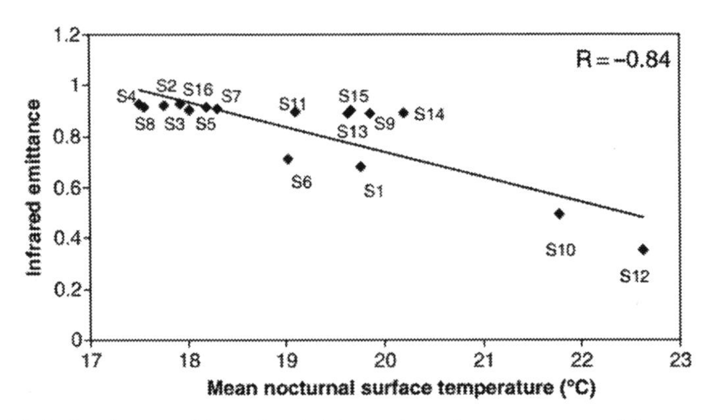

FIGURE 8.4 The impact of thermal emissivity on the surface temperature of the materials during the night period. *Adapted from Synnefa, M., Santamouris, I. August 2006. Livada: A study of the thermal performance of reflective coatings for the urban environment, Solar Energy 80 (8), 968–981.*

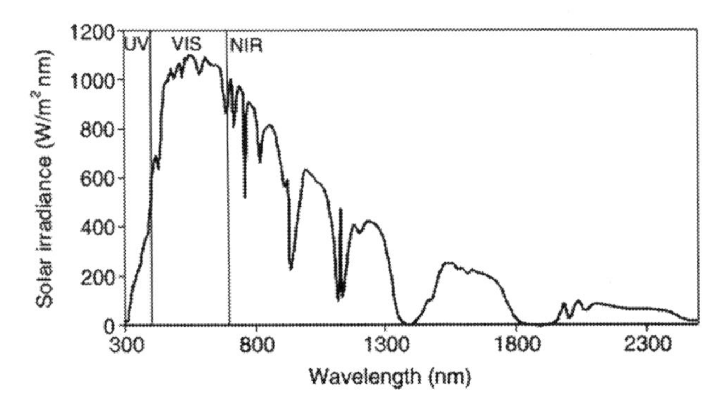

FIGURE 8.5 The solar spectrum (ASTM Standard G159-91). *Adapted from Synnefa, A., Santamouris, M., Apostolakis, K. 2007. On the development, optical properties and thermal performance of cool colored coatings for the urban environment. Solar Energy 81, 488–497.*

are paraffins or various salts that change phase at a temperature close to the ambient one These materials can melt or solidify at a specific temperature while they absorb or release latent heat respectively.

To decrease the surface temperature of concrete paving tiles, nano-phase materials have been added in the mass of IR reflective coatings. Thirty-six coatings of six colors have been developed and tested (Figs. 8.11 and 8.12). The PCM coatings presented various concentrations of phase change materials and different melting temperatures. The PCM doped coatings were tested in comparison with simple IR and conventional coatings of the same color (Karlessi et al. 2011).

Organic paraffin PCMs were used for the development of PCM-doped IR reflective colored coatings. All pigments were microencapsulated with an

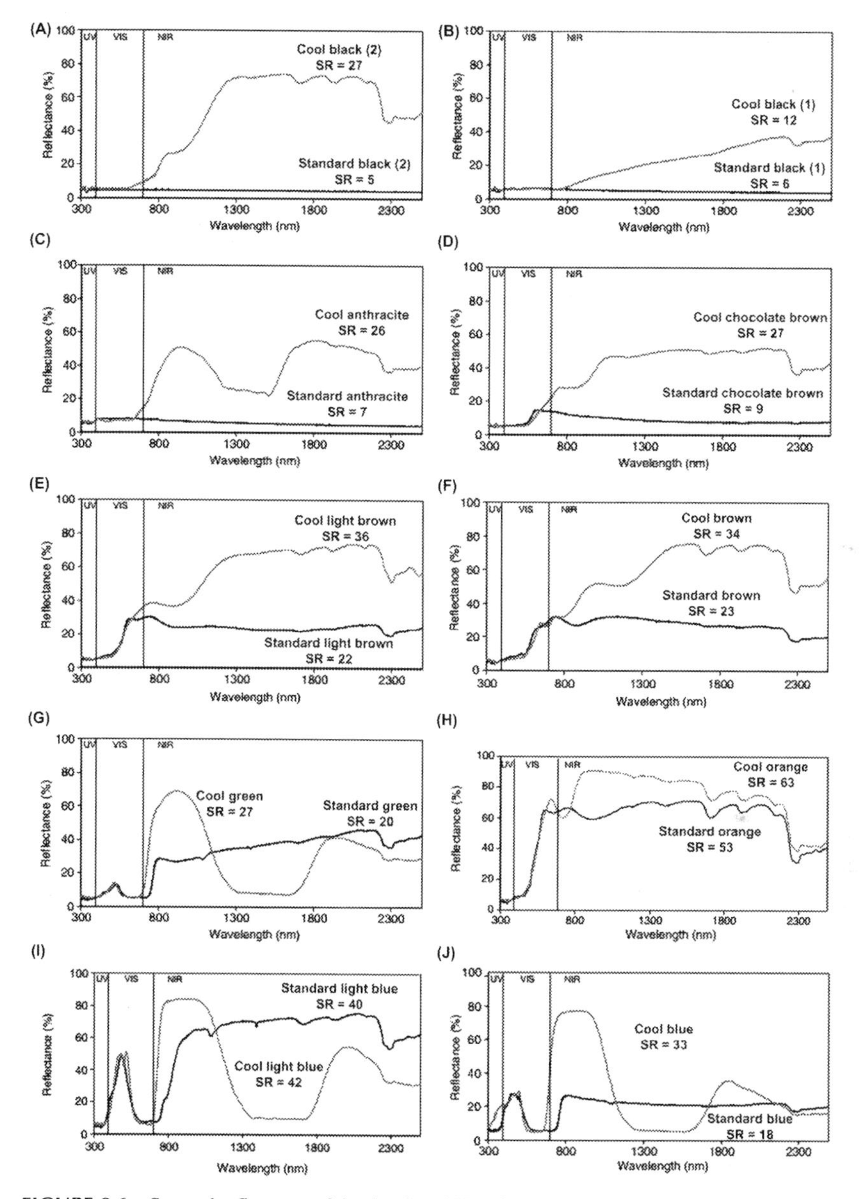

FIGURE 8.6 Spectral reflectance of the developed IR reflective materials. *Adapted from Synnefa, A., Santamouris, M., Apostolakis, K. 2007. On the development, optical properties and thermal performance of cool colored coatings for the urban environment. Solar Energy 81, 488–497.*

average particle size of 17–20 mm. The latent heat of fusion (solid to liquid) of the selected PCM is around 180 J/g and their specific gravity is 0.9. The volume expansion of the phase change material by itself can be as much as 10%. The coatings developed were acryl-based like those used on building

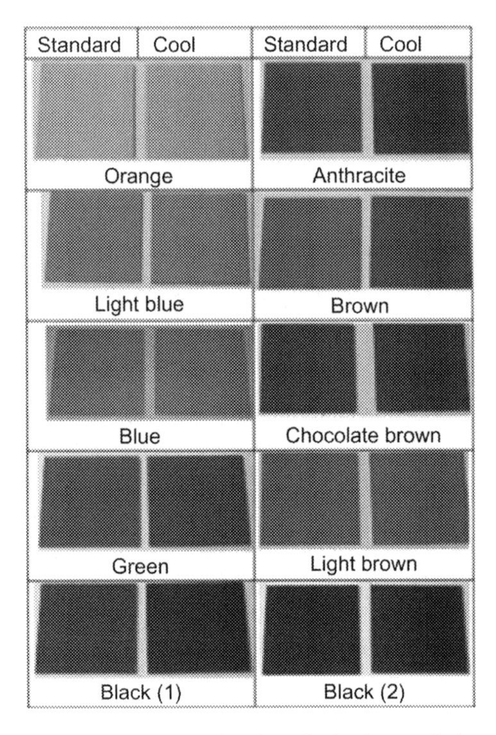

FIGURE 8.7 The developed and tested cool and standard color-matched coatings. *Adapted from Synnefa, A., Santamouris, M., Apostolakis, K. 2007. On the development, optical properties and thermal performance of cool colored coatings for the urban environment. Solar Energy 81, 488–497.*

TABLE 8.3 Measured Broadband Solar Reflectivity of the Developed IR Reflecting Materials and the Corresponding Conventional Ones. Source: Synnefa et al. (2007)

Color	Solar Reflectance(%)		%Increase $SR_{(cool-stand)}$
	Cool	Standard	
Orange	63	53	19
Light blue	42	40	5
Blue	33	18	83
Green	27	20	35
Black(1)	12	6	100
Anthracite	26	7	271
Brown	34	23	48
Chocolate brown	27	9	200
Light brown	36	22	64
Black(2)	27	5	440

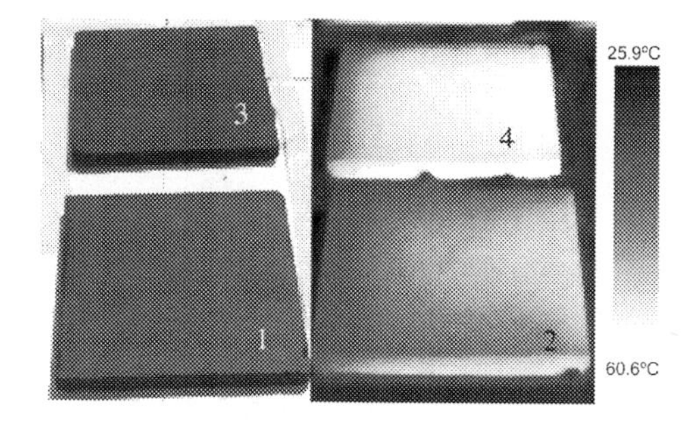

FIGURE 8.8 Measured surface temperature of an IR reflective and a conventional black colour tile. *Adapted from Synnefa, A., Dandou, A., Santamouris, M., Tombrou, M. 2008. On the use of cool materials as a heat island mitigation strategy. J. Appl. Meteorol. Climatol. 47, 2846–2856.*

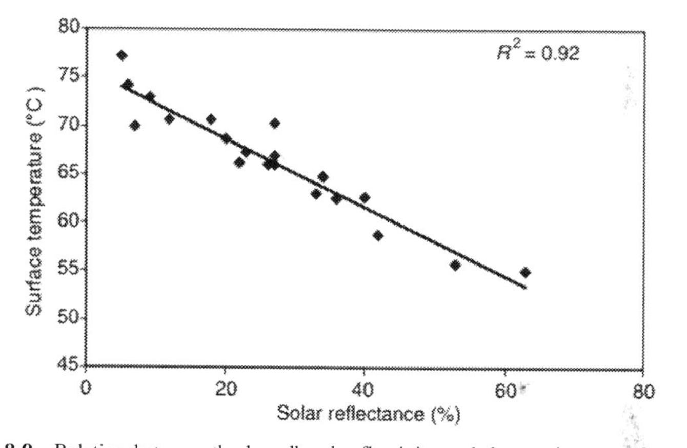

FIGURE 8.9 Relation between the broadband reflectivity and the maximum surface temperature of the developed IR reflective colored materials. *Adapted from Synnefa, A., Santamouris, M., Apostolakis, K. 2007. On the development, optical properties and thermal performance of cool colored coatings for the urban environment. Solar Energy 81, 488–497.*

envelopes and other urban structures. The different PCMs with melting temperatures equal to 18, 24 and 28°C at different concentrations (20% w/w, 30% w/w) were used. Coatings of six different colors were tested: blue, green, brown, black, grey and golden brown (Figs. 8.11 and 8.12). The developed coatings were applied on white concrete tiles placed on an unshaded horizontal platform that was insulated from below (Fig. 8.12). The size of the tiles was 33 × 33 cm. The surface temperature of the developed tiles was measured and recorded on a 24 hour basis during the experimental period. Measurements were performed during the summer period in Athens, Greece.

Figs. 8.13 and 8.14 present the maximum and minimum temperatures differences for common-PCM and cool-PCM coatings respectively and the

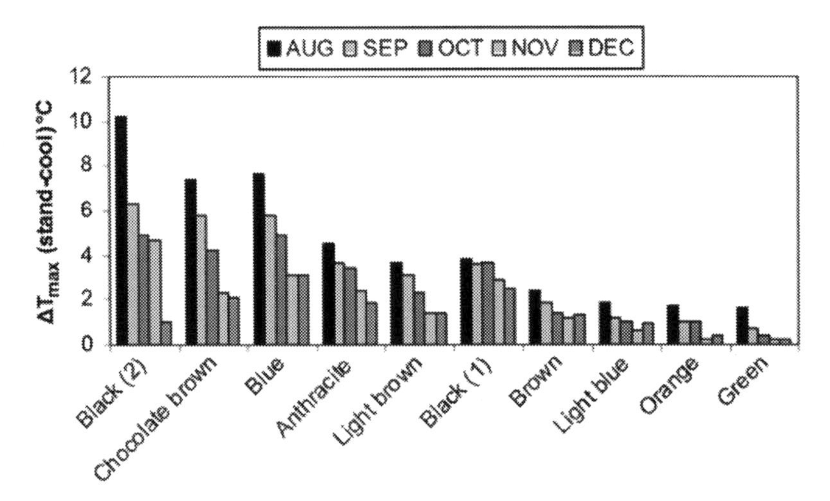

FIGURE 8.10 Maximum measured surface temperature difference between IR reflecting materials and the corresponding conventional materials. *Adapted from Synnefa, A., Santamouris, M., Apostolakis, K. 2007. On the development, optical properties and thermal performance of cool colored coatings for the urban environment. Solar Energy 81, 488–497.*

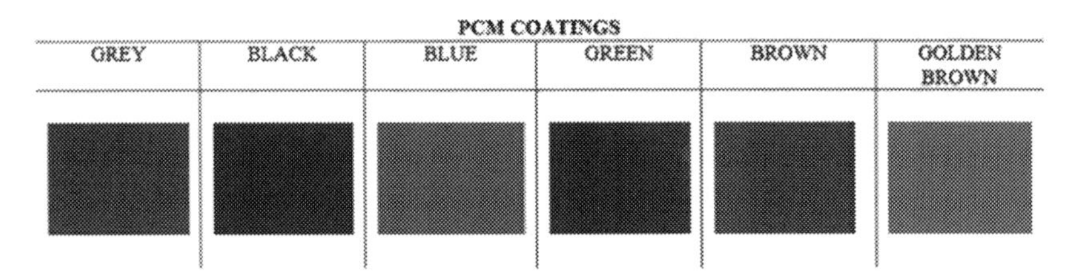

FIGURE 8.11 Color appearance of the PCM coatings. *Adapted from Karlessi, T., Santamouris, M., Synnefa, A., Assimakopoulos, D., Didaskalopoulos, P., Apostolakis, K.March 2011. Development and testing of PCM doped cool colored coatings to mitigate urban heat island and cool buildings. Building Environ 46 (3), 570–576.*

percentages of temperature decrease. Measurements show that the PCM coatings present lower temperatures at a range of 2.9–8.3°C compared to the common coatings. The maximum temperature difference is noticed for the samples of the black color coatings doped with PCM melting at 28°C and with a concentration of PCM pigments equal to 20% and 30% w/w. The minimum temperature drop was observed for the for golden brown PCM coating melting at 24°C, and with a concentration of PCM coating close to 30°C. The temperature difference against the simple IR reflective colored coatings was between 0.6–2.6°C. The highest temperature difference is observed for the PCM doped black coating melting at 28°C with a concentration of 20°%.

FIGURE 8.12 The tested PCM and other tiles. *Adapted from Karlessi, T., Santamouris, M., Synnefa, A., Assimakopoulos, D., Didaskalopoulos, P., Apostolakis, K.March 2011. Development and testing of PCM doped cool colored coatings to mitigate urban heat island and cool buildings. Building Environ 46 (3), 570–576.*

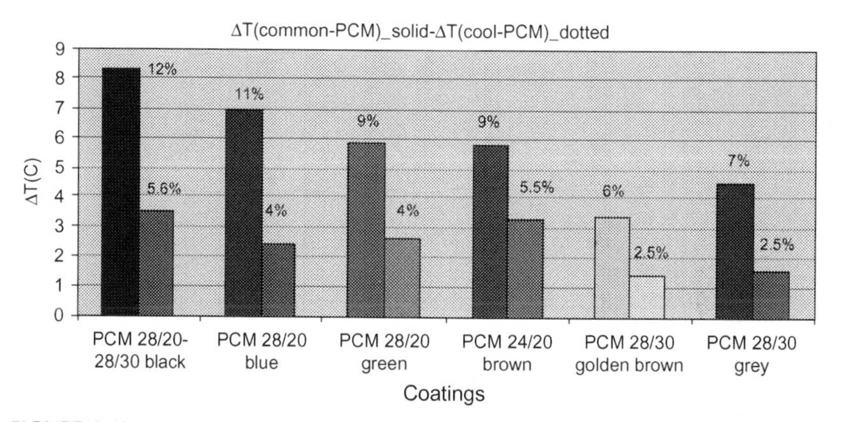

FIGURE 8.13 Maximum temperature differences between PCM, common and cool coatings. *Adapted from Karlessi, T., Santamouris, M., Synnefa, A., Assimakopoulos, D., Didaskalopoulos, P., Apostolakis, K. March 2011. Development and testing of PCM doped cool colored coatings to mitigate urban heat island and cool buildings. Building Environ 46 (3), 570–576.*

The whole experimental procedure has shown that the integration of phase change nanomaterials in cool coatings can help to further decrease their surface temperature by up to 3–4°C.

Color Changing Reflective Technologies – Thermochromic Coatings for the Urban Built Environment

Color changing materials are quite well known technologies. In particular, thermochromic technologies were developed many decades ago, mainly for indoor

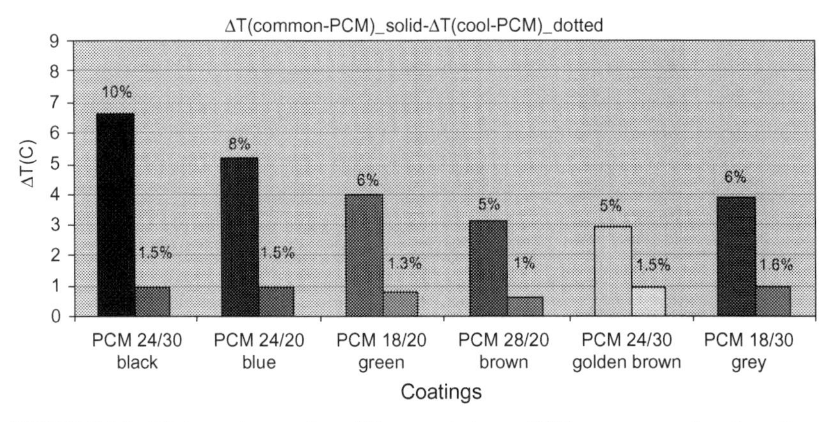

FIGURE 8.14 Minimum temperature differences between PCM, common and cool coatings. *Adapted from Karlessi, T., Santamouris, M., Synnefa, A., Assimakopoulos, D., Didaskalopoulos, P., Apostolakis, K. March 2011. Development and testing of PCM doped cool colored coatings to mitigate urban heat island and cool buildings. Building Environ 46 (3), 570–576.*

use. Thermochromic coatings change their color as a function of the surface temperature. They can be dark during cool periods and thus absorb solar radiation, and then change to a light color during the warm season. The transition temperature over which color may change varies as a function of the characteristics of the materials used to prepare the thermochromic coatings.

Thermochromic coatings can be an excellent solution for urban sites located in both heating- and cooling-dominated climates. Thermochromic materials may keep the surface temperature of cities quite warm during the winter period and cool during the summer time.

Thermochromic coatings based on organic water thermochromic pigments of powder and slurry form were developed and tested by Karlessi et al. (2009). All pigments had a transition temperature of close to 30°C, and were colored in their cold state and translucent in their warm state. Pigments had an average size of 5μm and were microencapsulated. A binder system that is non-absorbing in the infrared was used to develop the thermochromic coatings. Two groups of thermochromic coatings were developed in order to test the optical performance of the thermochromic pigments without the interference of any other type of pigments and in parallel to avoid transparency of the coating in the warm state. The first coating comprised the thermochromic pigments and the binder, and the second one the thermochromic pigments, the binder and titanium dioxide (TiO2) (Fig. 8.15).

The coatings were applied on white concrete tiles placed on an unshaded horizontal platform insulated from below. The size of the tiles was 33 × 33 cm (Figs. 8.16 and 8.17). Measurements were performed during the summer period, August and September. The thermochromic coatings were evaluated against IR reflective cool and conventional coatings of the same color.

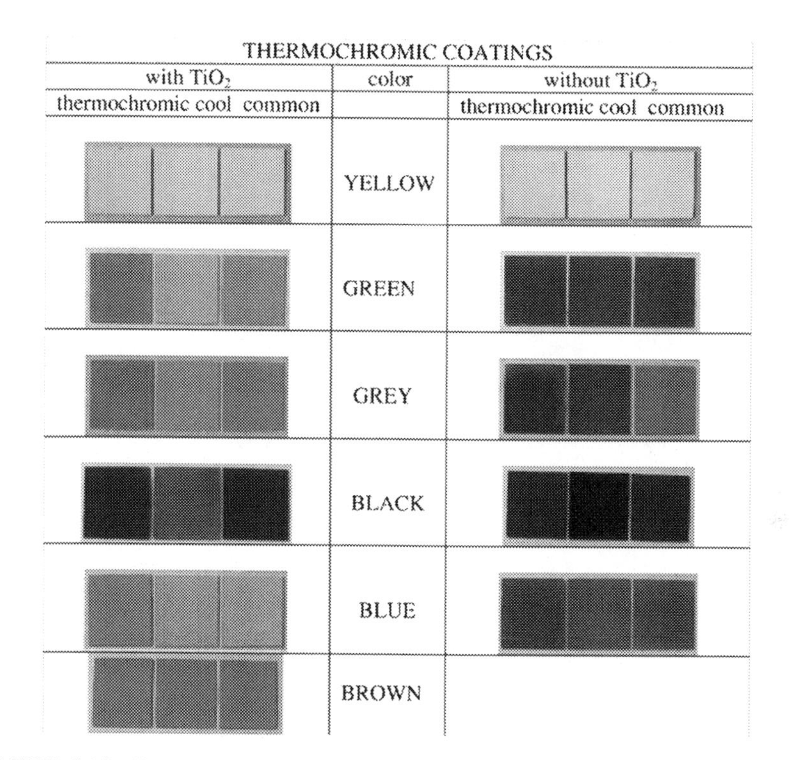

FIGURE 8.15 The developed and tested thermochromic, cool and common color-matched coatings. *Adapted from Karlessi, T., Santamouris, M., Apostolakis, K., Synnefa, A., Livada, I. April 2009. Development and testing of thermochromic coatings for buildings and urban structures. Solar Energy 83 (4), 538–551.*

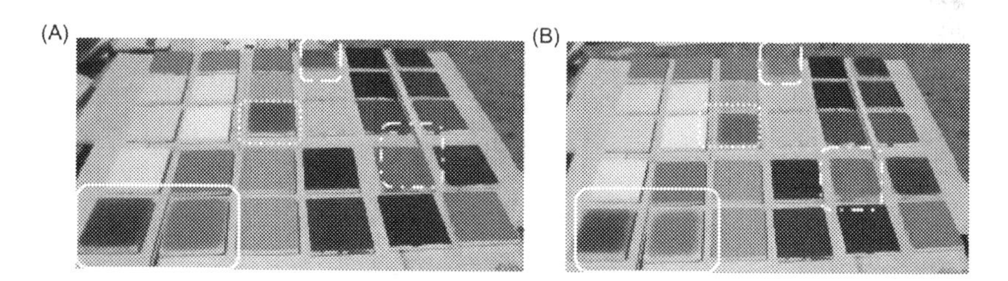

FIGURE 8.16 Color-changing phase of the thermochromic coatings. Colored state below 30°C (A) start becoming colorless above 30°C (B). *Adapted from Karlessi, T., Santamouris, M., Apostolakis, K., Synnefa, A., Livada, I. April 2009. Development and testing of thermochromic coatings for buildings and urban structures. Solar Energy 83 (4), 538–551.*

The monitoring shows that the mean daily surface temperatures of the thermochromic coatings in August range from 31–38.4°C, for the cool coatings from 34.4–45.2°C, and from 36.4–48.5°C for the conventional coatings. In September, the corresponding range varies from 26.1–31.6°C for

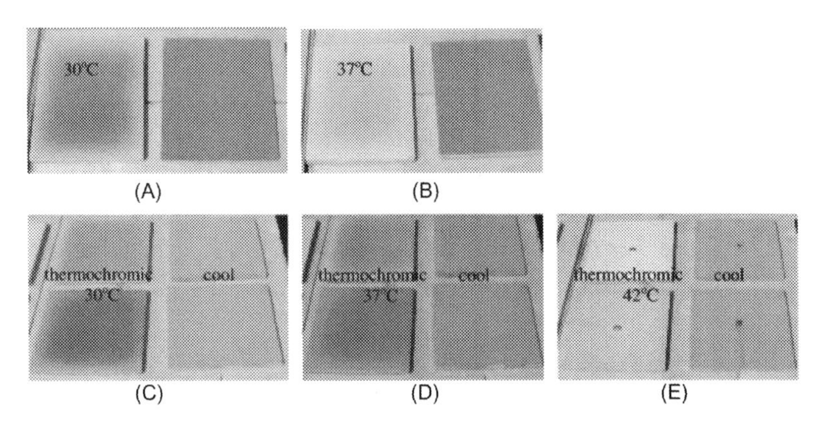

FIGURE 8.17 Transition phase of thermochromic green (A, B), and blue (C, D, E) coatings. Thermochromic coatings are on the left side, becoming white as the temperature rises above 30°C. On the right side, color-matched cool coatings are presented. *Adapted from Karlessi, T., Santamouris, M., Apostolakis, K., Synnefa, A., Livada, I. April 2009. Development and testing of thermochromic coatings for buildings and urban structures. Solar Energy 83 (4), 538–551.*

thermochromic coatings, from 28.1–39.2°C for the IR reflective coatings, and from 29.8–42.3°C for the conventional coatings (Fig. 8.18).

In conclusion, thermochromic coatings demonstrate 10–15°C lower mean max daily temperatures than cool coatings, and 18–20°C lower than common coatings (Fig. 8.19).

Spectral measurements of the reflectivity of the thermochromic coating at both the colored and colorless phase is found to be substantially higher than that that of the IR reflective and conventional coatings of the same color (Fig. 8.20).

A detailed aging study was performed to investigate the potential color degradation and impact of solar radiation on the thermochromic effect of the samples. Rflectance values increase for the colored phase of thermochromic coatings, while values decrease for the colorless phase (Fig. 8.21).

Thermochromic coatings therefore present a very high mitigation potential. However, aging problems may seriously limit their performance. To improve the life span of thermochromic coatings designed for urban mitigation purposes, several techniques have been proposed. Technologies to decrease the absorption of UV solar radiation by the coatings were tested by Karlessi and Santamouris (2013). The measurements of reflectance and color prove that the use of a UV filter did not significantly improve the dark phase of thermochromic effect, while during the white phase SR was reduced by 5% and SRvis was reduced by 9.7% compared with the uncovered sample. Figs. 8.22–8.24 show the spectral transmittance of a non-protected coating during dark and white phases, as well as that of a sample covered with a UV filter during the dark phase. The use of optical filters to cover the coatings was found to be much more effective. Covering the coatings with a red filter,

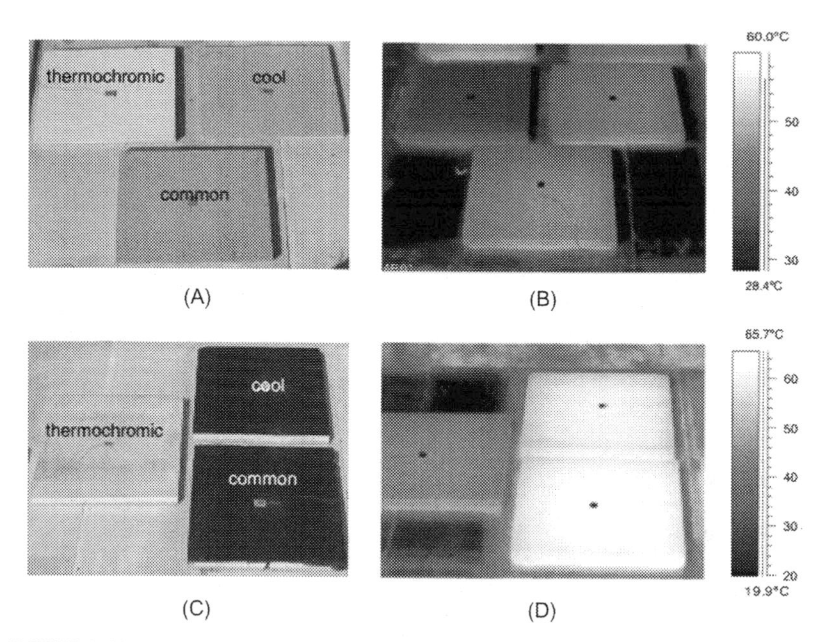

FIGURE 8.18 Temperature differences of thermochromic, cool and common coatings: visible (A, C) and infrared (B, D) images of blue coatings with TiO2 and black coatings without TiO2, respectively. *Adapted from Karlessi, T., Santamouris, M., Apostolakis, K., Synnefa, A., Livada, I. April 2009. Development and testing of thermochromic coatings for buildings and urban structures. Solar Energy 83 (4), 538–551.*

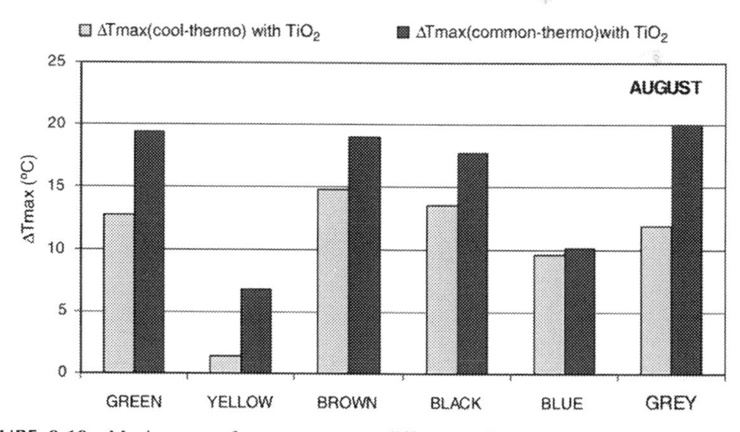

FIGURE 8.19 Maximum surface temperature difference (DNmax) between thermochromic, cool and common coating with TiO2 in August. *Adapted from Karlessi, T., Santamouris, M., Apostolakis, K., Synnefa, A., Livada, I. April 2009. Development and testing of thermochromic coatings for buildings and urban structures. Solar Energy 83 (4), 538–551.*

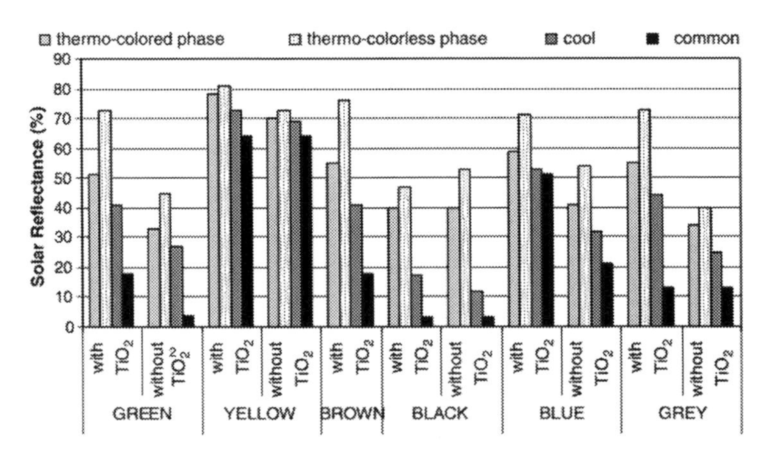

FIGURE 8.20 Solar reflectance (%) of the studied coatings. *Adapted from Karlessi, T., Santamouris, M., Apostolakis, K., Synnefa, A., Livada, I. April 2009. Development and testing of thermochromic coatings for buildings and urban structures. Solar Energy 83 (4), 538–551.*

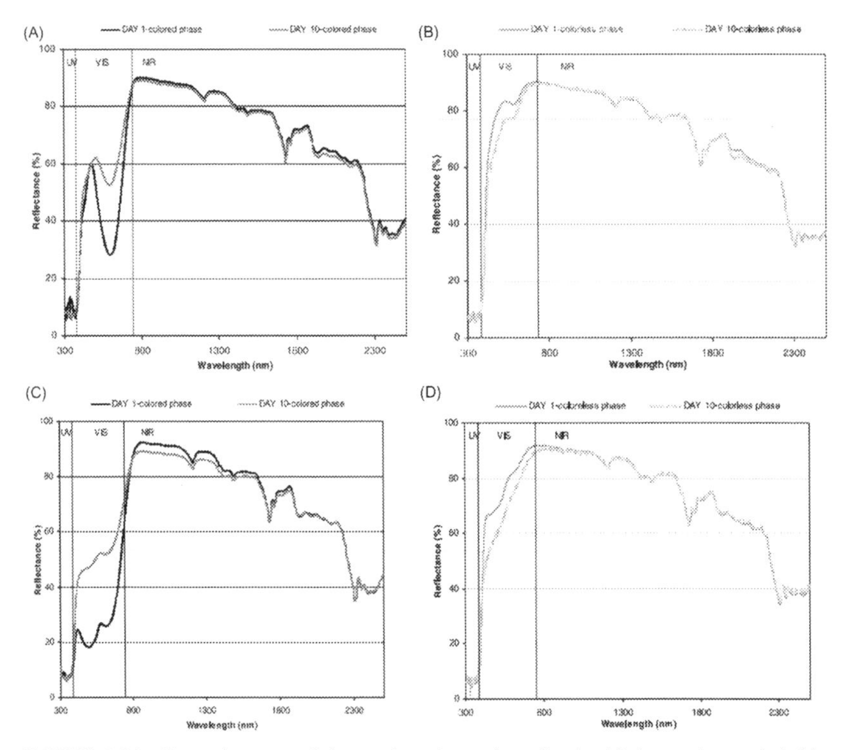

FIGURE 8.21 Spectral curves of thermochromic coatings for the 10-day testing period: blue with TiO2 (A and B) and brown with TiO2 (C and D). No significant change in the reflectance in the near infrared was observed. Source: *Karlesi et al. (2009).*

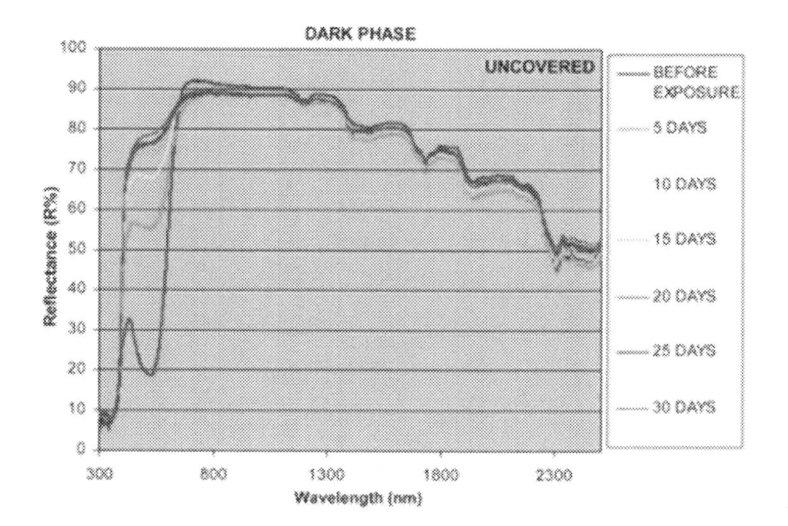

FIGURE 8.22 Spectral reflectance of the uncovered sample during the dark phase. *Adapted from Karlessi, T., Santamouris, M. May 12, 2013. Improving the performance of thermochromic coatings with the use of UV and optical filters tested under accelerated aging conditions. J Low Carbon Technologies, 2015. First published online May 12, 2013 doi:10.1093/ijlct/ctt027.*

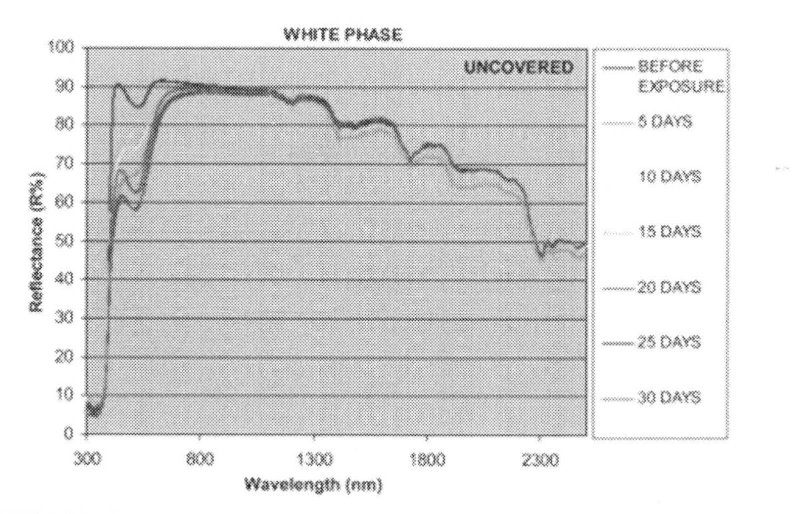

FIGURE 8.23 Spectral reflectance of the sample covered with a UV filter during the white phase. *Adapted from Karlessi, T., Santamouris, M. May 12, 2013. Improving the performance of thermochromic coatings with the use of UV and optical filters tested under accelerated aging conditions. J Low Carbon Technologies, 2015. First published online May 12, 2013 doi:10.1093/ijlct/ctt027.*

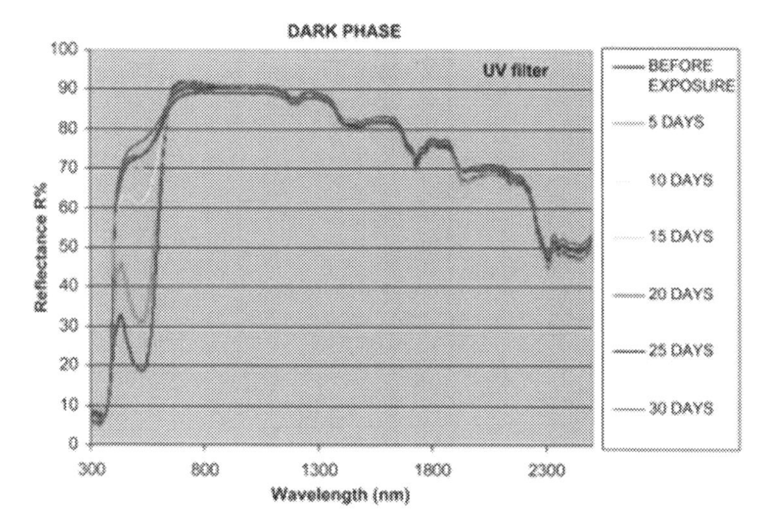

FIGURE 8.24 Spectral reflectance of the sample covered with a UV filter during the dark phase. *Adapted from Karlessi, T., Santamouris, M. May 12, 2013. Improving the performance of thermochromic coatings with the use of UV and optical filters tested under accelerated aging conditions. J Low Carbon Technologies, 2015. First published online May 12, 2013 doi:10.1093/ijlct/ctt027.*

which cuts off wavelengths below 600 nm, is found to most efficiently protect the reversible color change of the thermochromic coating as the solar reflectance at the dark phase remains unaffected during the whole experimental period (Karlessi and Santamouris, 2013) (Figs. 8.25–8.27).

Fluorescent Cooling for Urban Materials

The use of IR reflective color coatings contributes significantly to decreasing the temperature of the materials used in the urban environment. A higher efficiency can be achieved when the active phenomenon of fluorescence is used. In this case, the absorbed photons will cause fluorescence and produce heat. Berdahl et al. (2016) used Al2O3:Cr, a material that presents an important emission in the red spectrum, 694 μm, and the near infrared, 700–800 μm, as a coating on a white surface (Fig. 8.28). As shown, the effect of fluorescence considerably increased the effective solar reflectance and contributed to substantially lowering the surface temperature of the materials.

Greenery Technologies

Urban greenery significantly improves the climate of cities and mitigates the UHI. An increase in urban greenery helps to decrease the surface and ambient temperatures in cities and mitigate the UHI effect. As reported by

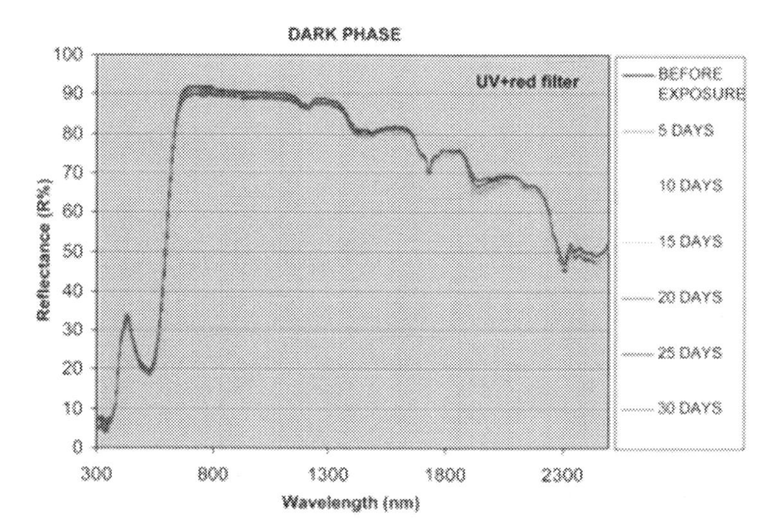

FIGURE 8.25 Spectral reflectance of the sample covered with UV + red filters during the dark phase. *Adapted from Karlessi, T., Santamouris, M. May 12, 2013. Improving the performance of thermochromic coatings with the use of UV and optical filters tested under accelerated aging conditions. J Low Carbon Technologies, 2015. First published online May 12, 2013 doi:10.1093/ijlct/ctt027.*

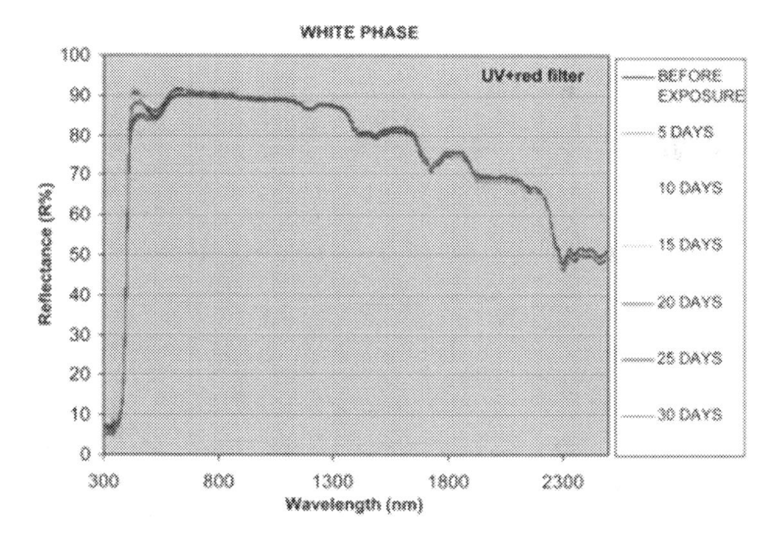

FIGURE 8.26 Spectral reflectance of the sample covered with UV + red filters during the white phase. *Adapted from Karlessi, T., Santamouris, M. May 12, 2013. Improving the performance of thermochromic coatings with the use of UV and optical filters tested under accelerated aging conditions. J Low Carbon Technologies, 2015. First published online May 12, 2013 doi:10.1093/ijlct/ctt027.*

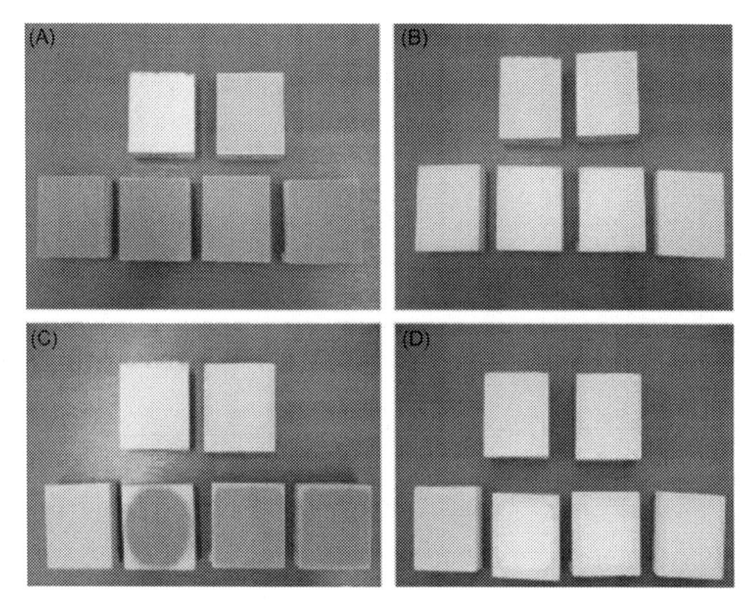

FIGURE 8.27 (A) Samples on the 5th day of exposure during the dark phase, (B) samples on the fifth day of exposure during the white phase, (C) samples on the 30th day of exposure during the dark phase, (D) samples on the 30th day of exposure during the white phase. *Adapted from Karlessi, T., Santamouris, M. May 12, 2013. Improving the performance of thermochromic coatings with the use of UV and optical filters tested under accelerated aging conditions. J Low Carbon Technologies, 2015. First published online May 12, 2013 doi:10.1093/ijlct/ctt027.*

Gill et al. (2007), a 10% increase in urban greenery in Manchester, UK, could amortize the predicted increase of the ambient temperature by 4 K over the next 80 years. Greenery contributes to decreasing ambient temperature mainly through evapotranspiration processes, while trees provide shading and control the air flow and heat exchange. The main types of urban greenery are parks, street trees, and building-integrated greenery such as planted roofs. Urban parks help to decrease ambient temperatures not just inside their boundaries but also in the adjacent urban areas. In parallel, urban parks help considerably to filter pollutants, mask noise, provide relaxation to visitors and prevent erosion. According to the American Forestry Association, the economic value of an urban tree is close to $57,000 for a 50 years mature specimen (Santamouris, 2001). This value includes the benefits of air conditioning, soil protection, air pollution mitigation and the provision of wildlife habitats. Among the many benefits arising from the existence of urban parks, the provision of comfort to visitors, the reduction of pollution levels, improvements in health, and an increase in property values are the most important ones (Hull, 1992).

The following section analyzes the main characteristics defining the mitigation potential of urban parks and building integrated greenery (planted roofs).

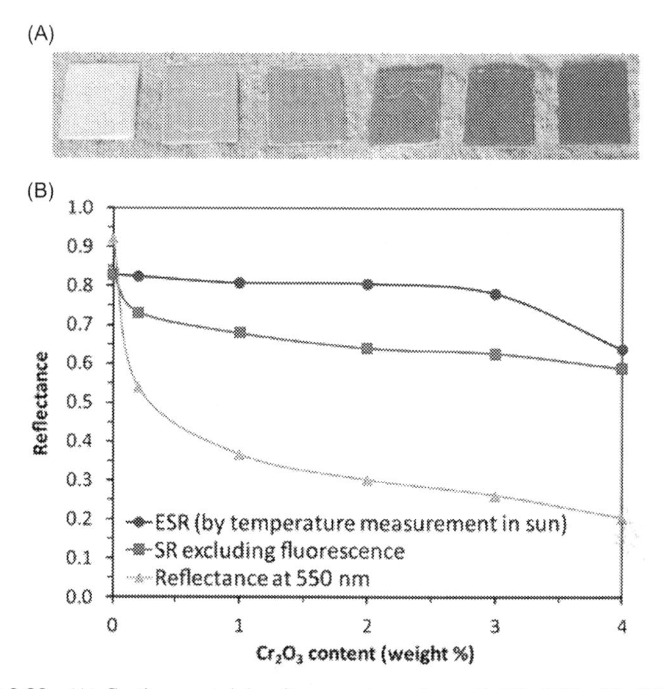

FIGURE 8.28 (A) Coatings containing fluorescent powders with 0%, 0.2%, 1%, 2%, 3% and 4% concentration. The higher the chromium concentration, the darker the coating. (B) Effective solar reflectance (ESR), solar reflectance (SR) and visual brightness: reflectance at 550 nm. *Adapted from Berdahl, P., Chen, S.S., Destaillats, H., Kirchstetter, T.W., Levinson, R.M., Zalich, M.A. 2016. Fluorescent cooling of objects exposed to sunlight – The ruby example. Sol. Energy Mater. Sol. Cells 157, 312–317.*

Urban Parks

The mitigation potential of urban trees is usually defined as a function of the ambient temperature difference between the park area and the rest of the urban zones. Parks develop lower temperatures during the day and night because of the increased evapotranspiration and solar shading during the day and the increased radiative cooling and the lower convective heat released during the night period (Oke et al., 1991; Spronken-Smith and Oke, 1998).

The mitigation potential of urban parks depends on several main parameters: (1) the size and the characteristics of the park, (2) the water frequency, sky obstruction and the type of plants in the park, (3) the thermal balance of the urban zones surrounding the park, and (4) the local weather conditions in the area under study (Skoulika et al., 2014).

Several studies have been performed to identify the relation between the park size and its mitigation potential. Most of the studies conclude that the larger the size of the park the higher its mitigation potential (Upmanis et al., 1998). Nevertheless there are many experimental studies that demonstrate the significance of even small and medium size parks (Thorsson et al., 2007).

Many research projects have been carried out to identify the cooling rate in and around parks during both the day and night periods. Bowler et al. (2010) reviewed seven relevant studies and found that the average temperature difference between parks and urban areas during the night period was close to −1.15 K. In parallel, other studies have identified that the cooling rate of the surrounding urban areas varies between 1 C/h to 2.9 C/h, (Spronken-Smith and Oke, 1998; Upmanis et al., 1998). Skoulika et al. (2014) reviewed a high number of articles reporting the magnitude of the night time cool island intensity of urban parks (Table 8.4). It is found

TABLE 8.4 Existing Studies Where the Nocturnal nocturna Cooling island index in parks is measured

No.	Reference	City/Country	Nocturnal CII (K)
1	Eliasson (1996)	Gotemborg. Sweden	4
2	Upmanis et al. (1998)	Gotemborg. Sweden	5.9
3	Bacci et al. (2003)	Aorence, Italy	0.5–2.4
4	Ca et al. (1998)	Around Tokyo	1.5
5	Bencheikh and Ameur (2012)	Algeria	10
6	Lee et al. (2009)	Seoul, Korea	2–6
7	Zoulia et al. (2009)	Athens, Greece	6
8	Jauregui (1990)	Mexico City, Mexico	4
9	Sundersingh (1991)	Madras, India	4
10	Gaffin et al. (2008)	New York USA	3
11	Lewis et al. (1971)	Washington, USA	3
12	Chow et al. (2010)	Tempe, USA	3
13	Spronken-Smith and Oke (1998)	Sacramento, USA	2.3–3.3
14	Spronken-Smith and Oke (1998)	Vancouve Canada	1.5–2.5
15	Jusuf et al. (2007)	Singapore	0.5–1
16	Potchter et al. (2006)	Tel Aviv, Israel	0.5–1
17	Cohen et al. (2012)	Tel Aviv, Israel	2.7

Source: Skoulika, F., Santamouris, M., Boemi, N. Kolokotsa, D. 2014. On the thermal characteristics and the mitigation potential of a medium size Urban Park in Athens, Greece. Landscape and Urban Planning 123, 73–86.

that in all cases, during the night time parks present a much lower temperature than the surrounding urban zones. The magnitude of the CII may vary between 0.5–10 K as a function of the characteristics of the park and the urban area.

Several studies show that parks during the day time period are on average 0.94 K cooler than the reference urban areas (Bowler et al., 2010). The existing studies and the reported results are given in details in Table 8.5. The reported daytime park CII is found to vary between −0.3 K and −7 K.

Green Roofs

Green or planted roofs are greenery systems that are fully or party covered by vegetation developed over a waterproofing membrane or soil. There are two main types of planted roofs: (1) extensive planted roofs, which are light structures covered by a thin layer of vegetation, and (2) intensive planted roofs, which are heavier structures and can support small trees and shrubs. Planted roofs present important advantages such as decreased energy consumption, increased UHI mitigation potential, management of storm water, increased durability of roofing materials, noise reduction, better air quality, and creating space for urban wildlife (Parizotto and Lamberts, 2011).

While the potential of planted roofs to decrease the energy consumption of buildings is very well documented, the potential contribution of planted roofs to mitigate the UHIs is still under investigation (Castleton et al., 2010). Green roofs provide cooling through latent heat processes, and may operate in a more efficient way in dry climates where the evaporative potential is substantially higher than in humid climates. The mitigation potential of the planted roofs depends mainly on the local climate, the type and the characteristics of the plants, and the frequency of watering the roof (Santamouris, 2014).

The potential decrease in the ambient temperature caused by green roofs depends mainly on the magnitude of the released latent heat. Several studies have been performed to determine the latent heat release by planted roofs under different climatic conditions. Table 8.6 summarizes the results of the existing studies (Santamouris, 2014). The proposed values of the released latent heat vary significantly as a function of the experimental boundary conditions and of the water content in the roof.

As already mentioned, very few studies of the heat island mitigation potential of green roofs are available. Most of the studies are based on mesoscale modeling and have been carried out for planted roofs of extensive type. Simulations have been performed for New York, Chicago, Hong Kong and Tokyo, while important knowledge has been provided by an experimental study in Singapore. The main characteristics of the existing mitigation studies are summarized in Table 8.7. It is concluded that when green roofs are installed in high or even medium rise buildings, their mitigation potential is greatly reduced and is almost negligible. The calculated drop of the ambient

TABLE 8.5 Existing Studies Where the Daytime Cooling Island Index in Parks is Measured and the Main Results

No.	Reference	Type of Climate	City/Country	Day Time Cool Island Index (K)
1	Oliveira et al.(2011)	Mediterranean	Lisbon, Portugal	Median cool island intensity was −1.6 K, while the maximum temperature difference between the park and the reference urban area was 6.9 K
2	Alcoforado (1996)	Mediterranean	Lisbon, Portugal	Cool island intensity varied between −3.9 K and −5.7 K
3	Andrade and Vieira (2007)	Mediterranean	Lisbon, Portugal	Cool island intensity was −4.1 K
4	Gomez, Gaja, and Reig (1998)	Mediterranean	Valencia, Spain	Cool island intensity was −2.5 K
5	Zoulia et al. (2009)	Mediterranean	Athens, Greece	Cool island intensity was −1.0 K
6	Bacci et al. (2003)	Mediterranean	Florence, Italy	Cool island intensity was between −1.0 K and −3.0 K
7	Potchter et al. (2006)	Mediterranean	Tel Aviv, Israel	Cool island intensity varied between −0.7 K and −2.5 K
8	Spronken-Smith and Oke(1998)	Mediterranean	Sacramento, USA	Cool island intensity varied between −1.0 K and −2.5 K
9	Sani (1990)	Tropical/subtropical	Kuala Lumpur, Malaysia	Cool island intensity was between − 4 K and − 5 K
10	Barradas (1991)	Tropical/subtropical	Mexico city, Mexico	Cool island intensity varied between −0.9 K and −4.6 K
11	Jauregui (1990)	Tropical/subtropical	Mexico city, Mexico	The temperature reduction in the park was low. 0.5 K, while occasionally the park presented higher temperatures than the reference urban station

12	Padmanabhamurty(1990)	Tropical/subtropical	Delhi, India	Cool island intensity was −2.5 K
13	Chang et al.(2007)	Tropical/subtropical	Taipei, Taipei	Cool island intensity was −0.8 K
14	Jonsson (2004)	Tropical/subtropical	Botswana	Cool island intensity was −2.0 K
15	Lee et al. (2009)	Tropical/subtropical	Seoul, Korea	Cool island intensity varied between −0.5 K and −2.5 K
16	Lu et al. (2012)	Tropical/subtropical	Chqongqing, China	Cool island intensity was − 2.5 K. In some cases the temperature in the parks was to about 0.3 K higher than in the reference urban station
17	Sugawara, et al. (2006)	Tropical/subtropical	Tokyo. Japan	Cool island intensity was −1.1 K
18	Thorsson et al. (2007)	Tropical/subtropical	Around Tokyo. Japan	Cool island intensity was −1.1 K
19	Ca et al. (1998)	Tropical/subtropical	Around Tokyo. Japan	Cool island intensity was −2.0 K
20	Hamada and Ohta (2010)	Tropical/subtropical	Nagoya. Japan	Cool island intensity varied between −1.9 K and −0.3 K
21	Spronken-Smith and Oke (1998)	Oceanic	Vancouver. Canada	Cool island intensity varied between −1 K and −2.5 K
22	Kjelgren and Clark (1992)	Oceanic	Seattle, US	Cool island intensity varied between −0.5 K and −1.5 K
23	Watkins et al.(2002)	Oceanic	London, UK	Cool island intensity varied between −0.6 K and −1.1 K
24	Lahme and Bruse (2003)	Oceanic	Essen. Germany	Cool island intensity was −1.0 K
25	Bencheikh and Ameur (2012)	Hot desert	Algeria	The average CII was −4.5 K while for—some periods the park presented higher temperatures than the reference urban station
26	Jansson et al. (2006)	Continental	Stockholm, Sweden	Cool island intensity varied between −0.5 K and −2.0 K

Source: Skoulika, F., Santamouris, M., Boemi, N. Kolokotsa, D. 2014. On the thermal characteristics and the mitigation potential of a medium size Urban Park in Athens, Greece. Landscape and Urban Planning 123, 73–86.

TABLE 8.6 Latent Heat Release by Green Roofs According to Various Studies

Reference	Characteristics of Green Roof	Peak Solar Radiation Intensity (W/m^2)	Latent Heat (W/m^2)
Takebayashi and Moriyama (2007)	Lawn	900	300–400
Hodo-Abalo et al. (2012)	For Bi = 8.61		
	LAI = 2	520	250
	LAI = 3	520	300
	LAI-4	520	350
	LAI = 5	520	450
	LAI = 6	520	500
	LAI = 7	520	560
Lazzarin et al. (2005)	I. I <LAI< 1.5		
	Dry Roof	900	110
	Wet Roof	900	230
Feng et al. (2010)	Extensive, LAI = 4.6	900	600
Rezaei (2005) and Berghage et al. (2007)	Extensive		350
(10)	Grass	1000	100
	Shrubs	1000	150
Alexandri and Jones (2007)	Grass	500–800	26–593
Jim and He (2010)	Turf Grass	800	250
	Ground cover herb	800	280
	Shrubs	800	400

Source: Adapted from Santamouris, M. 2014. Cooling the cities – A review of reflective and green roof mitigation technologies to fight heat Island and improve comfort in Urban Environments. Solar Energy 103, 682–703.

temperature varies as a function of local characteristics and the density of the planted roofs. Although some studies have concluded that the potential temperature drop of the urban ambient temperature may exceed 2 K, it is widely accepted that the cooling potential of green roofs is below 1 K.

TABLE 8.7 Characteristics of Existing Studies on the Mitigation Potential of Green Roofs

Reference	City	Type of Research	Types of Green Roof	Results
Smith and Roeber (2011)	Chicago, US	Simulation using the Weather Research and Forecasting Model	Extensive Type	Urban temperatures during 19:00–23:00 were 2–3 K cooler compared to the temperatures simulated without the use of cool roofs.
Savio et al. (2006)	New York, US	Simulation using MM5	Extensive Type	Peak temperatures at 2 m height decrease 0.37–0.86 K, while daily average temperature decrease between 0.3 and 0.55 K
Chen et al. (2009)	Tokyo, Japan	Simulation using the CSCRC model	Extensive Type	Almost negligible impact because of the high of the buildings where green roofs are installed
Ng et al. (2012)	Hong Kong, China	Simulation using the Envi-met tool	Extensive Type	Almost negligible impact because of the high of the buildings where green roofs are installed

Source: Adapted from Santamouris, M. 2014. Cooling the cities – A review of reflective and green roof mitigation technologies to fight heat Island and improve comfort in Urban Environments. Solar Energy 103, 682–703.

Heat Dissipation Techniques

Heat dissipation mitigation techniques are based on the rejection of excess urban heat into an atmospheric sink of a much lower temperature than the ambient one. There are four main atmospheric heat sinks that may be used for mitigation purposes (Santamouris and Assimakopoulos, 1997): The ambient air, the ground, the water and the sky. When the temperature of the ambient air is lower in the suburban or rural areas of cities, or even in coastal zones, proper channeling techniques may be designed and implemented to transfer cool air in the overheated areas of the city. The soil temperature at a certain depth presents a constant temperature that may be much lower than the ambient one. The warm air of the city may be transferred through earth to air heat exchangers to the ground, where the excess heat is rejected and then the cool air is redistributed in the urban environment. Water significantly decreases the ambient temperature through evaporation processes. Evaporation requires latent heat that it is extracted from the ambient air, reducing its temperature. Finally, during the night period the emitted thermal radiation from a metal collector may be higher than the thermal radiation absorbed and the possible convective losses. In this case, the thermal balance of the collector may be negative and its surface temperature may be lower than the ambient one.

Dissipation mitigation techniques are very well developed and are used to cool down individual buildings and agricultural greenhouses (Santamouris et al., 1995; Santamouris et al., 1996). However, only the water-based techniques are widely used to mitigate the UHI. In the following section, existing knowledge about the mitigation potential of water-based and ground technologies is presented.

Water-Based Mitigation Technologies

Water-based evaporative cooling technologies have been used for many centuries to provide comfort in cities. Cooling is achieved because of the extraction of the evaporation latent heat from the ambient air. As mentioned by Dominnguez and de la Flor (2016), the evaporation of 1 kg of water can decrease the temperature of 2000 M^3 of water by 1°**C**. In many cases, the surface temperature of the water may be lower than the ambient temperature, and thus cooling can be provided through convective processes.

The cooling potential of natural water bodies such as lakes and rivers is very well investigated (Xu et al 2009; Sun and Chen, 2012; Sun et al., 2012). Their cooling potential depends on many parameters, such as the characteristics of the water body and the wetland's proximity to the urban environment. Most of the existing studies agree that urban wetlands contribute highly to mitigating urban heat and improving comfort in cities, and may decrease the urban ambient temperature by one or two degrees (Manteghi et al., 2015).

Several artificial water-based cooling technologies and techniques are proposed and used in the urban environment to provide cooling (Kleerekoper et al., 2012). Most of the techniques are passive systems, such as pools, ponds and fountains, while active or hybrid evaporative systems such as sprinklers, wind towers, and water curtains are developed and implemented in cities (Dominnguez and de la Flor, 2016).

The performance of several water-based mitigation projects is presented in the following section.

Ground Mitigation Technologies

As already mentioned, ground cooling systems are used extensively to decrease the temperature inside buildings and agricultural premises. The

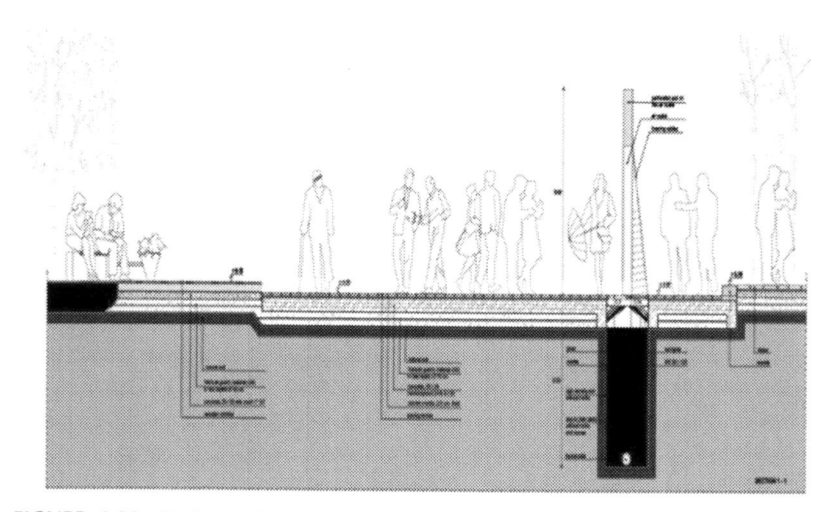

FIGURE 8.29 Earth to air heat exchangers. *Adapted from Fintikakis, N., Gaitani, N., Santamouris, M., Assimakopoulos, M., Assimakopoulos, D.N., Fintikaki, M., et al. February 2011. Bioclimatic design of pen public spaces in the historic centre of Tirana. Albania Original Research Article Sustainable Cities and Society, 1 (1), 54–62.*

most well-known earth cooling system is the so called 'earth to air heat exchanger' (Fig. 8.29).

The system consists of horizontal metallic, plastic or ceramic tubes placed at a depth of 2–3 meters underground. The warm ambient air is circulated through the pipes using fans. Because of the important temperature difference between the ground and the ambient air, heat is transferred to the soil, while the temperature of the circulated ambient air is significantly reduced. The air at the exit of the exchangers is distributed to the ambient environment of the city to decrease the urban temperature and provide comfort to pedestrians.

The mitigation efficiency of earth to air heat exchangers depends on many factors, such as the temperature difference between the ground and the ambient air, the geometric and thermal characteristics of the exchangers, and the speed and mass of the circulated air (Mihalakakou et al., 1994a,b).

Several accurate numerical models are available to calculate the thermal performance and the mitigation potential of earth to air heat exchangers (Mihalakakou et al., 2014a). However, the system is not considered by most of the urban computational tools.

In the following section, several examples of cooling mitigation projects that have been designed and implemented around the world are presented.

PRESENTATION AND EVALUATION OF EXISTING LARGE-SCALE MITIGATION PROJECTS GLOBALLY

Existing and proposed mitigation technologies and techniques present a very high potential for decreasing the ambient temperature of cities. The specific performance of the available technologies when applied to cities is assessed, either through detailed measurements carried out within large scale real projects or through numerical simulations. The collected experimental and numerical data corresponding to specific boundary and climatic conditions of cities must always be interpreted in a comparative and critical way, and always within the frame of their validity. A comparative analysis of the existing mitigation projects employing reflective technologies was attempted by Santamouris (2014), and relations between the possible increase of the urban albedo and the corresponding decrease of the average and peak ambient temperatures were proposed. A detailed analysis of about 220 mitigation projects involving most of the existing technologies and their combinations was carried out. Their analysis examines several passive and active mitigation technologies and systems, as well as their combination simulated implementation under various climatic conditions. When urban cooling is achieved using mechanical means, the system is considered to be an active one. In the following, the results of the whole analysis are presented and discussed in detail.

Performance of Evaporative Mitigation Technologies

The mitigation performance of seventeen large scale projects has been comparatively analyzed. Formation is collected through the following sources: Nishimura et al. (1998), O'Malley et al. (2015), Chadzidimitriou et al. (2013), Tumini (2014), Taleghani et al. (2014a), Soutullo Castro et al. (2012), Velazquez et al. (1992), Dominnguez and de la Flor (2016), Theeuwes et al. (2013), Martins et al. (2016), Amor et al. (2015). The principal characteristics of the considered projects are reported in Table 8.8.

Among the 17 projects reported, five of them deal with the mitigation performance of pools, ponds and open water bodies, three refer to the performance of evaporative wind towers, four are on water sprinklers, two on fountains, and finally, four case studies report results for the performance of various combinations of water-based mitigation technologies, such as sprinklers, pools, ponds, fountains, and water curtains.

The mitigation potential of the water-based systems is highly affected by the geometric characteristics of the system and the urban area. In parallel, the climatic conditions, particularly the humidity content of the atmosphere, wind speed and turbulence, and solar radiation levels determine the potential for evaporation.

Pools and ponds are the more common evaporative systems used in the urban environment. Heat transfer and cooling of the ambient temperature is achieved through evaporative and convective procedures when the water temperature is lower than the ambient one. The evaporative capacity of pools and ponds depends highly on the water content in the atmosphere and the moisture gradient between the water surface and the ambient air. The cooling

TABLE 8.8 Characteristics of the Considered Large Scale Evaporative Mitigation Projects

No	Location of Project	Theoretical or Experimental Study	Simulation Tool	Area of Project (m^2)	Area of Water Bodies (m^2)	Characteristics of the Location	Maximum Design Temperature, (C)	Maximum Temperature Decrease (C)	Mean Temperature Decrease (C)	Reference
Mitigation System : Pools and Ponds										
1	Seagrave London, UK	Theoretical	Envi- Met	7150	700	Low Density Buildings + Open Spaces	25,6	0,4	0,1	O'Malley et al. (2015)
2	Salonica, Greece	Theoretical	Envi- Met	110000	550	Highly Density Buildings with Open Spaces	40,9	7,1	1,9	Chadzidimitriou et al. (2013)
3	Pavones, Fontarron and Horcajo Madrid, Spain	Theoretical	Envi- Met			High Density Buildings with Open Spaces	29	2,6	1,5	Tumini (2014)
4	Portland, USA	Theoretical	Envi- Met			Urban Zone	27	1,1	0,3	Taleghani et al. (2014a)
Mitigation System : Evaporative Towers										
5	Vallecas, Madrid, Spain	Experimental				Open Space	30	3,5	1,0	Soutullo Castro et al. (2012)
6	Seville, Spain	Experimental				Open Space	42	5	2	Velazquez et al. (1992)
7	Seville, Spain	Experimental				Open Space	42	7	4	Alvares and De la Flor (2016).
8	Salonica, Greece	Theoretical	Envi-Met	11000		Highly Density Buildings with Open Space	40,9	7,1		Chadzidimitriou et al. (2013)
9	Beirut, Lebanon	Theoretical	Manual Calculations			Open Space	35	6	4	Nunes et al. (2016)

(Continued)

TABLE 8.8 (Continued)

No	Location of Project	Theoretical or Experimental Study	Simulation Tool	Area of Project (m^2)	Area of Water Bodies (m^2)	Characteristics of the Location	Maximum Design Temperature, (C)	Maximum Temperature Decrease (C)	Mean Temperature Decrease (C)	Reference	
10	Seville, EXPO 92	Experimental				Open Space	42	10	3	(Alvares and De la Flor (2016).	
12	Seville, Mairena del Alcor	Experimental				Open Space	30	7	4	(Alvares and De la Flor (2016).	
Mitigation System: Fountains											
11	Setif, Algeria	Experimental		5600	214	Open Space	35	0,9	0,9	Amor et al. (2015)	
12	Osaka	Experimental				Park	35	4	1	Nishimura et al. (1998)	
Mitigation System : Other or/and Combination of the above											
	Location of Project	System	Theoretical or Experimental Study	Simulation Tool	Area of Project (m^2)	Area of Water Bodies (m_2)	Characteristics of the Location	Maximum Design Temperature, (C)	Maximum Temperature Decrease (C)	Mean Temperature Decrease (C)	Reference
13	Toulouse, France :	Pools + Ponds + fountains	Theoretical	Envi- Met			Low Density Buildings + Open Space	29	6	2	Martins et al. (2016)
14	European Fictitious City	Open Water Body	Theoretical	WRF			Urban Zone	25	1,7	0,5	Theeuwes et al. (2013)
15	Salonica, Greece	Fountains + Water Courtains + sprinklers	Theoretical	Envi-Met	11000	550	Low DensityBuildings + Open Spaces	40,9	7,1	1,9	Chadzidimitriou et al. (2013)
16	Setif, Algeria	Fountain + Ponds	Theoretical	Envi-Met	5600	770	Open Space	35	1	1	Amor et al. (2015)
17	Osaka	Falling Water	Experimental				Park	35	4	1	Nishimura et al. (1998)

potential of the ponds and pools is more significant in the leeward space zone of the water system, given that cooler air is transferred from the wind. The specific performance of the four studies given in Table 8.8, based on the use of pools and ponds, shows that the decrease of the ambient temperature in the surrounding area varies on average between 0.1 K and 1.9 K, while the calculated maximum temperature drop is between 0.4 K and 7.1 K. The average maximum temperature drop is found to be close to 2.8 K.

Fountains are passive evaporative systems that are extensively used in the urban environment to decrease the ambient temperature and provide thermal comfort. The evaporative capacity of fountains is determined by the temperature and radius of the initial and final water drop (Dominnguez and de la Flor, 2016). Measurements from two evaporative applications using fountains is given in Table 8.8. In both cases, the average ambient temperature drop was about 1 K, while the maximum temperature drop was 0.9 K and 4 K respectively.

Evaporative cooling towers are usually used to decrease the air temperature of very high volumes of air (Fig. 8.30). In cooling towers, water is sprayed in the top of the tower while air is circulated to the tower by mechanical or natural means. The circulating air is cooled because of evaporation. It then flows to the lower part of the tower and to the ambient environment. Evaporative towers have been studied extensively by Ford et al. (2010) and additional information may be found in the relevant articles. Three projects involving the use of evaporative cooling towers are reported

FIGURE 8.30 An urban evaporative tower in Expo '92 in Seville, Spain. *Adapted from https://www.mimoa.eu/projects/Spain/Seville/Avenue%20of%20Europe%20Expo%20%2792/.*

in Table 8.8. As shown, the average ambient temperature drop varies between 1 K and 4 K. In parallel, the maximum temperature drop is found to vary between 3.5 K and 7 K. The specific performance data refer to very dry urban zones where the evaporative potential is quite high. In less dry climates, it can be considered that evaporative cooling towers may present a substantially lower mitigation performance.

Urban misting systems based on the use of water sprinklers may supply the air with evaporating water droplets that help to decrease the ambient temperature. The evaporative performance of the misting systems depends on the characteristics of the nozzles, the size of the droplets and the wind speed and humidity content in the atmosphere. The size of the droplets depends highly on the pressure applied to the nozzles (Nunes et al., 2016), while use of non-proper nozzles may significantly reduce the efficiency of the system (Yamada and Yoon, 2008; Yoon, 2008). Four case studies employing evaporative misting systems are reported in Table 8.8. As shown, the average temperature drop in the ambient air varies between 3 K and 4 K, while the maximum decrease is between 6 K and 10 K. It is evident that the mitigation potential of misting systems is quite high. However, the temperature drop is reduced to the space around the misting system, while the performance of the system is very low in non-dry climates.

Analysis of the above data may result in the conclusion that the use of outdoor evaporative systems can reduce the average ambient temperature by up to 2°C, while the maximum temperature decrease may be as high as 7°C at a very local level. The range of the average and maximum temperature drop for all the examined systems is given in Fig. 8.31. The mitigation potential of the water-based evaporative systems is on average close to 1.9 K, while the average maximum expected temperature drop is close to 4.5 K.

Further analysis of the existing performance information indicates that there is a significant correlation between the average and the maximum ambient temperature drop, and the maximum design temperature in the urban zone (Fig. 8.32). It is evident that the mitigation potential of water-based evaporative cooling systems and techniques in urban areas improves with increasing ambient temperatures. This is because high ambient temperatures help to increase the saturation capacity of the air and the evaporative potential of the water systems.

Mitigation Potential of Urban Greenery Systems

About sixty eight studies aiming to identify the mitigation potential of urban greenery systems have been identified and analyzed. The complete list and the characteristics of each case study is given in Table 8.9. Among them, twenty-nine articles analyze the mitigation potential of trees and hedges in cities, about twenty-one papers discuss the mitigation of green roofs, seven

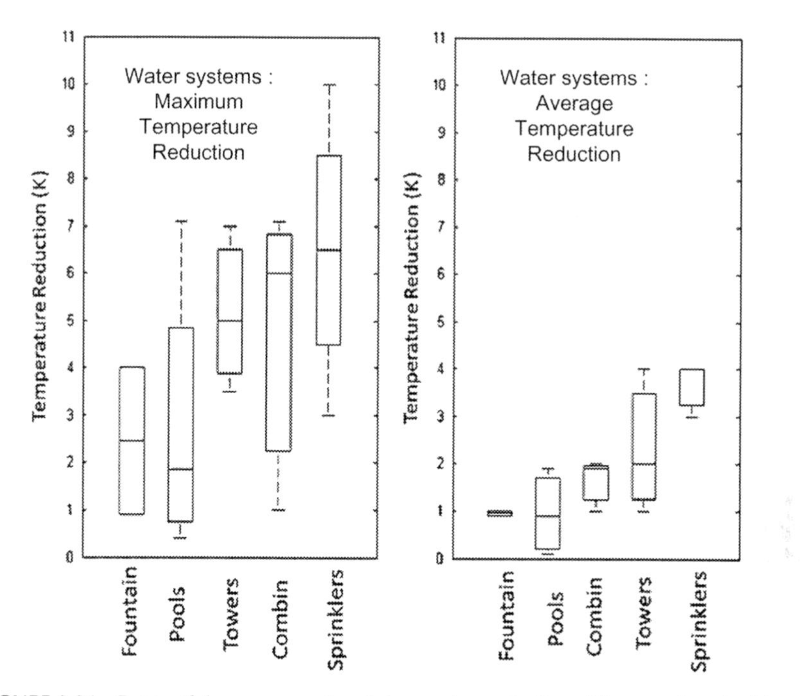

FIGURE 8.31 Range of the average and peak temperature reduction of the water-based mitigation systems and technologies. *Adapted from Santamouris et al. (2017).*

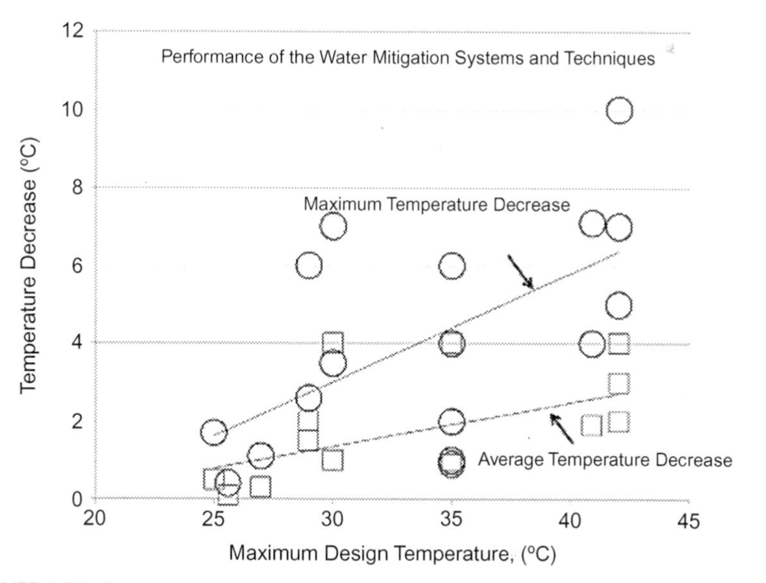

FIGURE 8.32 Decrease of the peak and average ambient temperature for water-based mitigation systems and technologies as a function of the peak ambient temperature. *Adapted from Santamouris et al. (2017).*

TABLE 8.9 Characteristics and Reported Performance of Urban Greenery Projects

a/a	Location	Experimental /Theoretical	Simulation Tool	Maximum Design Temperature (C)	Maximum Decrease of the Ambient Temperature (C)	Average Decrease of the Ambient Temperature (C)	Reference
Trees and Hedges							
1	Paphos, Cyprus	Theoretical	Envi-met	36	0.1	0.1	Pisello et al. (2016)
2	Rimini, Italy	Theoretical	Envi-met	34	1.5	1.0	Pisello et al. (2016)
3	York, UK	Theoretical	Envi-met	26	0.2	0.1	Pisello et al. (2016)
4	Grenoble, France	Theoretical	Envi-met	34	0.1	0.1	Pisello et al. (2016)
5	Athens, Greece	Theoretical	CFD	39	3	1.5	GRBES (2016)
6	Nea Smirni, Greece	Theoretical	Envi-met	37	6	3.5	Polyzos and Istapolou (2014)
7	Singapore	Theoretical	Envi-met	30	3	1.5	Kardinal Jusuf et al. (2006)
8	Toronto, Canada	Theoretical	Envi-met	33	0.6	0.4	Wang et al. (2016)
10	Hania, Greece	Theoretical	Envi-met	26	1	0.6	Tsilini et al. (2014)

11	Ho Chi Minh City, Vietnam	Theoretical	Envi-met	28	0.5	0.3	Chau Huynh and Ronald Eckert (2012)
12	Setif, Algeria	Theoretical	Envi-met	35	1.3	0.3	Amor et al. (2015)
13	Vienna, Austria	Theoretical	Envi-met	33.7	0.7	0.5	Maleki and Mahdavi (2016)
14	Campinas, Brazil	Theoretical	Envi-met	33	5.2	3	Alchapar et al. (2016)
15	Mendoza, Argentina	Theoretical	Envi-met	36	1.9	0.6	Alchapar et al. (2016)
16	Belgrade, Serbia	Theoretical	Envi-met	25	2	0.8	Djukic (2015)
17	Hong Kong	Theoretical	Envi-met	32	1.8	1	Ng et al. (2012)
18	Phoenix, USA	Theoretical	Envi-met	33	2.4	1.1	Middel and Chhetri (2014)
19	Dubai, UAE	Theoretical	Envi-met	48	7	2.5	Rajabi and Abu-Hijleh (2014)
20	Peristeri, Greece	Theoretical	Envi-met	35.7	2	0.6	Ferrante (2016)
21	Hong Kong			30	1.2		Santamouris and Kolokotsa (2016)
22	Mexico City, Mexico			35	3.5		Santamouris and Kolokotsa (2016)
23	Los Angeles, USA	Theoretical	MM5	33	3	1.2	Taha (2008a)

(*Continued*)

TABLE 8.9 (Continued)

a/a	Location	Experimental /Theoretical	Simulation Tool	Maximum Design Temperature (C)	Maximum Decrease of the Ambient Temperature (C)	Average Decrease of the Ambient Temperature (C)	Reference
24	Pomona, USA	Theoretical	MM5	34	1	0.3	Taha (2008a)
25	San Fernando, USA	Theoretical	MM5	34	1	0.35	Taha (2008a)
26	Toulouse, France	Theoretical	Envi-met	29	0.5	0.2	Martins et al. (2016)
27	Manchester, UK	Theoretical	Envi-met	26	1.5	0.6	Skelhorn et al. (2014)
28	Portland, USA	Theoretical	Envi-met	27	1.6	0.6	Taleghani et al. (2014a)
29	Portland, USA	Experimental		27	4.7	2.8	Taleghani et al. (2014b)
30	Tokyo, Japan	Theoretical	WRF	35	0.0	0.0	Huang et al (2009)
Green Roofs							
31	Ho Chi Minh City, Vietnam	Theoretical	Envi-met	28	0.2	0.1	Chau Huynh and Ronald Eckert (2012)
32	Toronto, Canada	Theoretical	MC2	30	0.5	0.3	Bass et al. (2002)

33	Toronto, Canada	Theoretical	MC2	30	1.5	0.9	Bass et al. (2002)
34	California, USA	Theoretical	WRF	30	1.0	0.2	Georgescu et al. (2014)
35	Arizona, USA	Theoretical	WRF	35	0.8	0.1	Georgescu et al. (2014)
36	Texas, USA	Theoretical	WRF	34	1.4	0.3	Georgescu et al. (2014)
37	Mid Atlantic, USA	Theoretical	WRF	30	1.5	1.0	Georgescu et al. (2014)
38	Chicago,	Theoretical	WRF	25	1.8	1.0	Georgescu et al. (2014)
39	Tokyo, Japan	Theoretical	CFD	35	0.0	0.0	Huang et al. (2009)
40	Tokyo, Japan	Theoretical	CFD	34	1.0	0.3	Chen et al. (2009)
41	Tokyo, Japan	Theoretical	CFD	34	0.1	0.1	Chen et al. (2009)
42	Chicago, USA	Theoretical	WRF	26	3.0	2.0	Smith and Roeber (2011)
43	New York, USA	Theoretical	MM5	28	0.86	0.5	Savio et al. (2006)
44	Tokyo, Japan	Theoretical	CSCRC	36	0.0	0.0	Chen et al. (2009)
45	Hong Kong	Theoretical	Envi-met	32	0.0	0.0	Ng et al. (2012)
46	Chicago, USA	Theoretical	WRF	28	0.6	0.35	Sharma et al. (2016)

(*Continued*)

TABLE 8.9 (Continued)

a/a	Location	Experimental /Theoretical	Simulation Tool	Maximum Design Temperature (C)	Maximum Decrease of the Ambient Temperature (C)	Average Decrease of the Ambient Temperature (C)	Reference
47	Melbourne, Australia	Theoretical	Envi-met	25	1.4		Bruce and Skinner (1999)
48	Vienna, Austria	Theoretical	Envi-met	33.5	0.3	0.2	Maleki and Mahdavi (2016)
49	Changwon City, Korea	Theoretical	Envi-met	31	0.1	0.1	Song and Park (2015)
50	Cambera, Austalia	Theoretical	LUMPS	30	0.4		Mitchell et al. (2008)
Grass							
51	Belgrade, Serbia	Theoretical	Envi - Met	25	1.4	0.5	Djukic (2015)
52	Manchester, UK	Theoretical	Envi - Met	26	0.2	0.05	Skelhorn et al. (2014)
53	Kobe, Japan	Experimental		38	0.1	0.1	Takebayashi and Moriyama (2009)
54	Dubai, UAE	Theoretical	Envi - Met	48	3.0	1.0	Rajabi and Abu-Hijleh (2014)
55	Acquila, Italy	Theoretical	Envi - Met	39	1.2	0.5	Ambrosini et al. (2014)
56	Veneto, Italy	Theoretical	Envi - Met	34	1.0	0.3	Campostrini (2013)
57	Hong Kong	Theoretical	Envi - Met	32	0.9	0.4	Ng et al. (2012)

Trees and Green Roofs							
58	Vienna, Austria	Theoretical	Envi - Met	33.5	1.0	0.6	Maleki and Mahdavi (2016)
59	Melbourne, Australia	Theoretical	Envi - Met	25	2.4		Bruce and Skinner (1999)
60	Dubai, UAE	Theoretical	Envi - Met	48	7.0	2.5	Rajabi and Abu-Hijleh (2014)
61	Atlanta, USA	Theoretical	WRF	32		0.38	Stone et al. (2014)
62	Philadelphia, USA	Theoretical	WRF	30		0.2	Stone et al. (2014)
63	Phoenix, USA	Theoretical	WRF	36		0.1	Stone et al. (2014)
Trees and Grass							
64	London, UK	Theoretical	Envi-Met	25	0.9	0.4	O'Malley et al. (2015)
65	Phoenix, USA	Theoretical	LUMPS	35	1		Gober et al. (2010)
66	Cambera, Australia	Theoretical	LUMPS	30	0.8		Mitchell et al. (2008)
67	Dubai, UAE	Theoretical	Envi - Met	48	7.0	2.8	Rajabi and Abu-Hijleh (2014)
Grass and Green Roofs							
68	Dubai, UAE	Theoretical	Envi - Met	48	3.6	1.4	Rajabi and Abu-Hijleh (2014)

Adapted from Santamouris et al. (2017).

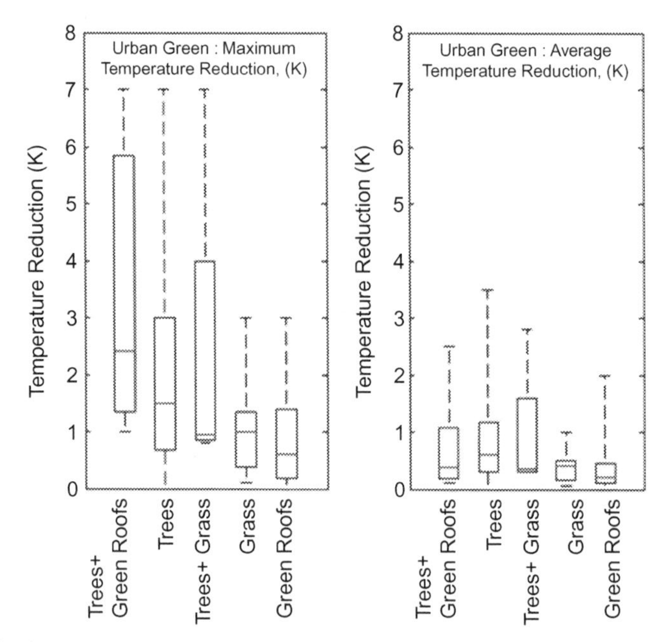

FIGURE 8.33 Range of the average and peak temperature reduction of urban greenery mitigation techniques. *Adapted from Santamouris et al (2017).*

studies analyze the cooling effect of grass, and eleven works discuss the cooling impact of various combinations of urban greenery.

The specific cooling performance of all the considered projects regarding their maximum and average temperature drop is given in Fig. 8.33.

The whole analysis shows that the mitigation potential of urban greenery systems depends on the characteristics and the density of the considered greenery, the thermal properties of the city and the prevailing climatic conditions. The main conclusions drawn are the following:

- Urban trees and hedges may cause a maximum peak ambient temperature drop ranging between 0.1 K and 7 K, with an average drop of close to 1.5 K. Urban scenarios based on a quite extreme and almost unrealistic assumption of urban greenery seem to present a very high temperature drop. Under realistic considerations, the maximum temperature mitigation potential may not exceed 4 K, and the average temperature decrease may vary between 0 K and 3.5 K, with an average value close to 0.6 K.
- When the addition of ground greenery is not possible and important space limitations apply at the city level, then the use of planted roofs seems to be an excellent alternative (Karachaliou et al., 2016). The maximum temperature decrease of the peak ambient temperature is found to vary between 0 K and 3 K, with an average value of around 0.6 K. In

parallel, the average temperature drop is found to vary between 0 K and 2 K, with an average of around 0.2 K. The whole analysis shows that the mitigation potential of planted roofs is impacted by the irrigation levels, the LAI value of the vegetation, and the characteristics of the city (Kolokotsa et al., 2013). Reports from applications where green roofs are applied on high rise buildings show that the temperature drop at the street level is almost negligible. It must be pointed out that mitigation potential of green roofs is significantly lower than that of urban trees.

- The use of grass helps to decrease ambient temperatures and mitigate the UHI considerably. It is found that the use of grass decreases the peak ambient temperature between 0.1 K and 3 K, with an average value close to 1 K. The average decrease of the ambient temperature ranges between 0.1 K and 1 K, with a mean value close to 0.4 K.
- The highest mitigation potential is achieved through the combined use of urban trees and green roofs. However, it should be noted that the number of projects in the sample is quite small.

Projects Using Reflecting Materials to Mitigate UHI

Almost 75 projects involving cool roofs, cool pavements and a combination of the two have been analyzed by. The characteristics of all the considered case studies are presented below in Tables 8.10 and 8.11.

Among the various projects, 15 case studies report the mitigation performance of cool pavements, 24 projects analyze the mitigation potential of cool roofs, and the remainder report on the performance of projects based on a combined use of cool roofs and pavements. The range of the average and the maximum temperature drop at ground level for each of the three considered technologies is given in Fig. 8.34.

When cool roof technologies are used, the average peak temperature drop achieved is close to 1 K. It increases up to 1.3 K for the cool pavements and 1.4 K for the combination of cool roofs and pavements.

At the same time, the absolute maximum temperature decrease for cool roofs is 2.3 K, 2.5 K for cool pavements and 3.4 K for their combination. The calculated median of the average temperature decrease varies between 0.3 K and 0.4 K for all considered technologies, while the maximum average decrease is close to 1 K. Based on the above results, the following conclusions may be drawn:

- Comparison between the various systems shows that cool pavements present a higher mitigation potential than cool roof technologies. This is because pavements are located at the ground level and may decrease the temperature of the air at this level, while cool roofs are installed at a certain height from the ground and mainly cool the air above the roof. Their mitigation capacity depends on the ability of the cool air to reach

TABLE 8.10 Characteristics and Reported Performance For All Projects Using Reflective Materials

No	Location	Increase of the Albedo in the Study Area, (0–1)	Design Temperature (C)	Maximum Temperature Reduction (C)	Average Temperature Reduction, (C)	Experimental /Simulation	Simulation Tool Used	Reference
Cool Roofs								
1	Rimini, Italy	0.05–0.15	35.5	0.9	0.21	Simulation	Envi-Met	Pisello et al. (2016)
2	Toronto, Canada	0.05–0.15	33	0.7	0.17	Simulation	Envi-Met	Wang et al. (2016)
3	Ho Chi Minh City, Vietnam	0.03–0.07	28	0.1	0.1	Simulation	Envi-Met	Chau Huynh and Ronald Eckert (2012)
4	Vienna, Austria	0.2–0.3	33.5	0.5	0.35	Simulation	Envi-Met	Maleki and Mahdavi (2016)
5	Colombo, Sri Lanka	0.15–0.25	30	1.2	0.5	Simulation	Envi-Met	Emmanuel et al. (2007)
6	Athens, Greece	0.3–0.4	35.3	2.2	1	Simulation	MM5	Synnefa et al. (2008)
7	Athens, Greece	0.1–0.2	35.3	1.5	0.43	Simulation	MM5	Synnefa et al. (2008)
8	Changwon City, Korea	0.03–0.08	31	0.2	0.1	Simulation	Envi-Met	Song and Park (2015)

9	Phoenix, USA	0.05–0.15	33	0.6	0.3	Simulation	Envi-Met	Middel and Chhetri (2014)
10	Phoenix, USA	0.15–0.25	33	1.5	0.6	Simulation	WRF	Georgescu et al. (2012)
11	Aquila, Italy	0.2–0.3	39	1.8	0.5	Simulation	Envi-Met	Ambrosini et al. (2014)
12	New York, USA	0.05–0.15	31	0.31	0.25	Simulation	MM5	Lynn et al (2009)
13	Phoenix, USA	0.1–0.2	35	1		Simulation	LUMPS	Gober et al. (2009)
14	Toulouse, France	0.10–0.15	29	0.9	0.4	Simulation	Envi-Met	Martins et al. (2016)
15	California, USA	0.15–0.25	30	1.85	0.7	Simulation	WRF	Georgescu et al. (2014)
16	Arizona, USA	0.15–0.25	35	1	0.5	Simulation	WRF	Georgescu et al. (2014)
17	Texas, USA	0.2–0.3	34	2	0.5	Simulation	WRF	Georgescu et al. (2014)
18	Florida, USA	0.05–0.15	33	0.7	0.3	Simulation	WRF	Georgescu et al. (2014)
19	Mid Atlantic, USA	0.25–0.35	30	2.3	0.7	Simulation	WRF	Georgescu et al. (2014)
20	Chicago	0.25–0.35	25	2.2	0.6	Simulation	WRF	Georgescu et al. (2014)
21	Tokyo, Japan	0.10–0.15	35	0.4	0.1	Simulation	CFD	Huang et al. (2009)

(Continued)

TABLE 8.10 (Continued)

No	Location	Increase of the Albedo in the Study Area, (0–1)	Design Temperature (C)	Maximum Temperature Reduction (C)	Average Temperature Reduction, (C)	Experimental /Simulation	Simulation Tool Used	Reference
Cool Roofs								
22	Atlanta, USA	0.10–0.20	32		0.18	Simulation	WRF	Stone et al. (2014)
23	Philadelphia	0.10–0.20	30		0.16	Simulation	WRF	Stone et al. (2014)
24	Phoenix	0.10–0.20	36		0.195	Simulation	WRF	Stone et al. (2014)
Pavements								
25	Western Athens, Greece	0.15–0.25	36	2.5	1	Simulation	CFD	GRBES (2016)
26	Historical Center, Athens, Greece	0.15–0.20	37	1.8	0.45	Simulation and experimental	CFD	GRBES (2016)
27	Aegaleo, Athens, Greece	0.15–0.25	35	1.3	0.4	Simulation and experimental	Envi-Met	Kyriakodis et al. (2016)
28	Athens, Greece	0.15–0.25	29	1	0.42	Simulation and experimental	CFD	Georgakis et al. (2014)
29	Toronto, Canada	0.03–0.08	33	0.5	0.12	Simulation	Envi-Met	Wang et al. (2016)

30	London, UK	0.05–0.15	25	0.5	0.2	Simulation	Envi-Met	O'Malley et al. (2015)
31	Ho Chi Minh City, Vietnam	0.03–0.07	28	0.2	0.1	Simulation	Envi-Met	Chau Huynh and Ronald Eckert (2012)
32	Agia Paraskevi, Athens, Greece	0.05–0.15	29	0.8	0.2	Simulation	Envi-Met	GRBES (2016)
33	Ilioupolis, Athens, Greece	0.15–0.25	37.5	1.4	0.4	Simulation	CFD	GRBES (2016)
34	Suburban Thessaloniki, Greece	0.15–0.25	36	1.3	0.34	Simulation and experimental	Envi-Met	Papadopoulos (2016)
35	Marousi, Greece	0.2–0.3	34	1.9	0.7	Simulation and experimental	EEnvi-Met	GRBES (2016)
36	Veneto, Italy	0.15–0.25	34	2.2	0.6	Simulation	EEnvi-Met	Campostrini (2013)
37	Madrid, Spain	0.05–0.15	29	1	0.3	Simulation	EEnvi-Met	Tumini (2014)
38	Agia Paraskevi2, Athens, Athens	0.05–0.1	34	1.8	0.36	Simulation and experimental	CFD	GRBES (2016)
39	Florina, Greece	0.05–0.15	35	1.4	0.3	Simulation	CFD	Zoras et al. (2014)

TABLE 8.11 Characteristics and Reported Performance For All Projects Using Reflective Materials

No	Location	Increase of the Albedo in the Study Area (0–1)	Design Temperature (C)	Maximum Temperature Reduction (C)	Average Temperature Reduction (C)	Experimental/ Si.	Simulation Tool Used	Reference
Cool roofs *and cool pavements*								
40	Los Angeles, USA	0.05–0.15	33	1	0.46	Simulation	MMS	Taha (2008b)
41	Los Ange les, USA	0.15–0.25	33	2.1	0.9	Simulation	MMS	Taha (2008b)
42	Pomona, USA	0.05–0.15	34	1.1	0.45	Simulation	MMS	Taha (2008b)
43	Pomona, USA	0.15–0.25	34	2.1	0.75	Simulation	MMS	Taha (2008b)
44	San Fernando, USA	0.15–0.25	34	2	0.9	Simulation	MMS	Taha (2008b)
45	San Fernando, USA	0.15–0.25	34	1.5	0.7	Simulation	MMS	Taha (2008b)
46	Atlanta, USA	0.15–0.25	32		0.23	Simulation	WRF	Stone et al. (2014)
47	Philadelphia, USA	0.15–0.25	30		0.35	Simulation	WRF	Stone et al. (2014)

48	Phoenix, USA	0.15–0.25	36		0.16	Simulation	WRF	Stone et al. (2014)
49	Colombo, Sri Lanka	0.05–0.l5	30	0.2	0.16	Simulation	Envi-Met	Emmanuel et al. (2007)
50	Campinas. Brazil	0.10–0.20	33	1.3	0.45	Simulation	Envi-Met	Alchapar et al. (2016)
51	Mentoza. Argentina	0.10–0.20	36	0.7	0.3	Simulation	Envi-Met	Alchapar et al. (2016)
52	Grenoble France	0.05–0.15	34	0.3	0.2	Simulation	Envi-Met	Pisello et al. (2016)
53	New York, USA	0.10–0.20	31	0.62	0.4	Simulation	MMS	Savio et al. (2006)
54	Los Angeles, USA	0.05–0.10	32	1.4	0.6	Simulation	Colorado State University Mesoscale Model	Sailor (1995)
55	Los Angeles, USA	0.10–0.20	32	3	0.5	Simulation	Colorado State University Mesoscale Model	Rosenfeld et al. (1995)
56	Los Angeles, USA	0.10–0.20	32	1.5	0.3	Simulation	Colorado State University Mesoscale Model	Rosenfeld et al. (1998)

(Continued)

TABLE 8.11 (Continued)

No Location	Increase of the Albedo in the Study Area (0–1)	Design Temperature (C)	Maximum Temperature Reduction (C)	Average Temperature Reduction (C)	Experimental/ Si.	Simulation Tool Used	Reference	
Cool roofs *and cool pavements*								
57	Colombus, USA	0.01–0.05	26		0.1	Simulation	WRF	Millstein and Menon (2011)
58	San Antonio, USA	0.01–0.05	33		0.14	Simulation	WRF	Millstein and Menon (2011)
59	San Diego, USA	0.01–0.05	37		0.13	Simulation	WRF	Millstein and Menon (2011)
60	Jacksonville, USA	0.01–0.05	30		0.12	Simulation	WRF	Millstein and Menon (2011)
6l	San Jose USA	0.01–0.05	33		0.23	Simulation	WRF	Millstein and Menon (2011)
62	Dallas, USA	0.03–0.07	34		0.20	Simulation	WRF	Millstein and Menon (2011)

63	Phoenix, USA	0.03–0.07	35		0.16	Simulation	WRF	Millstein and Menon (2011)
64	Miami USA	0.04–0.08	34		0.20	Simulation	WRF	Millstein and Menon (2011)
65	Chicago, USA	0.05–0.l	26		0.27	Simulation	WRF	Millstein and Menon (2011)
66	Atlanta, USA	0.0 5–0. L	30		0.14	Simulation	WRF	Millstein and Menon (2011)
67	Philadelphia, USA	0.07–0.12	28		0.24	Simulation	WRF	Millstein and Menon (2011)
68	Houston, USA	0.05–0.15	30		0.19	Simulation	WRF	Millstein and Menon (2011)
69	New York, USA	0.05–0.15	28		0.35	Simulation	WRF	Millstein and Menon (2011)

(*Continued*)

TABLE 8.11 (Continued)

No	Location	Increase of the Albedo in the Study Area (0–1)	Design Temperature (C)	Maximum Temperature Reduction (C)	Average Temperature Reduction (C)	Experimental/ Si.	Simulation Tool Used	Reference
Cool roofs *and cool pavements*								
70	Dedroid, USA	0.1–0.15	25		0.39	Simulation	WRF	Millstein and Menon (2011)
71	Los Angeles, USA	0.1–0.15	33		0.53	Simulation	WRF	Millstein and Menon (2011)
72	Philadelphia, USA	0.05–0.15	28	0.5	0.35	Simulation	MMS	Sailor et al. (2002)
73	Atlanta, USA	0.25–0.35	30	2.5		Simulation	WRF	Zhou et al. (2010)
74	Huston, USA	0.1–0.2	30	3.5	0.4	Simulation	MMS	Taha (2008a)
75	New York, USA	0.3–0.4	28	2	1.1	Simulation	MMS	Lynn et al. (2009)

Adapted from Santamouris et al. (2017).

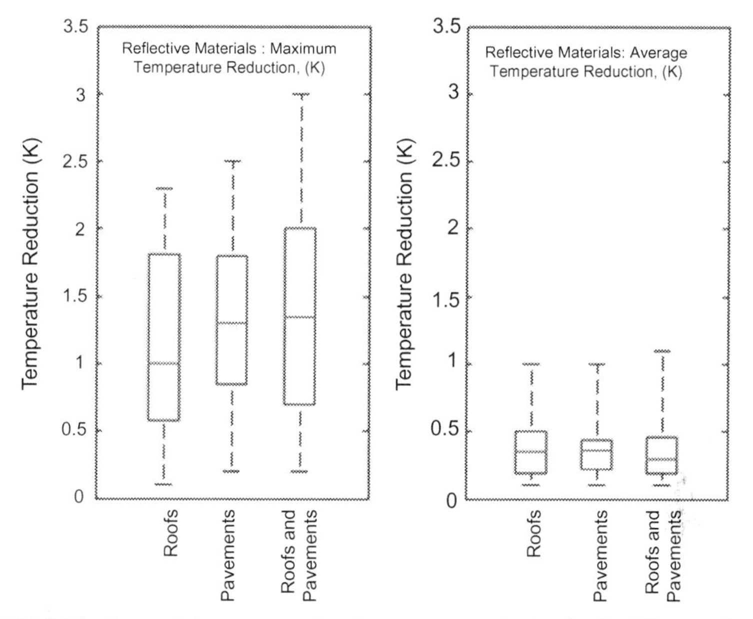

FIGURE 8.34 Range of the average and peak temperature reduction for the different reflective technologies. Source: *Santamouris et al., 2017.*

ground level. Studies performed for high rise buildings in Hong Kong and Tokyo show that the cooling potential of cool roofs is quite low.

- The highest mitigation potential is observed when a combination of cool roofs and pavements is applied. However, the combined cooling potential is substantially lower than the sum of the cooling capacity of each individual system.
- An analysis of the correlation between the albedo increase and the potential maximum and average temperature drop was performed for each individual technology and their combination (Figs. 8.35 and 8.36). The calculated decrease of the average and maximum temperature drop when the albedo increases by 10% is found to be close to 0.27 K and 0.78 K. It is also concluded that for cool roof technology, the average and maximum temperature decrease per 10% increase of the albedo is 0.23 K and 0.62 K, for the cool pavements it is 0.27 K and 0.94 K, while the corresponding figures for their combination are 0.35 K and 0.91 K.
- When the mitigation performance of cool materials is estimated using simulation techniques and real experimental data, it is always concluded that simulated results overestimate the performance of reflective technologies. A comparison of the experimental against the simulated performance data is given by Santamouris et al. (2012a). The range of the simulation and experimental performance is given below in Fig. 8.37. It is found that the experimental performance is about 16% lower than the

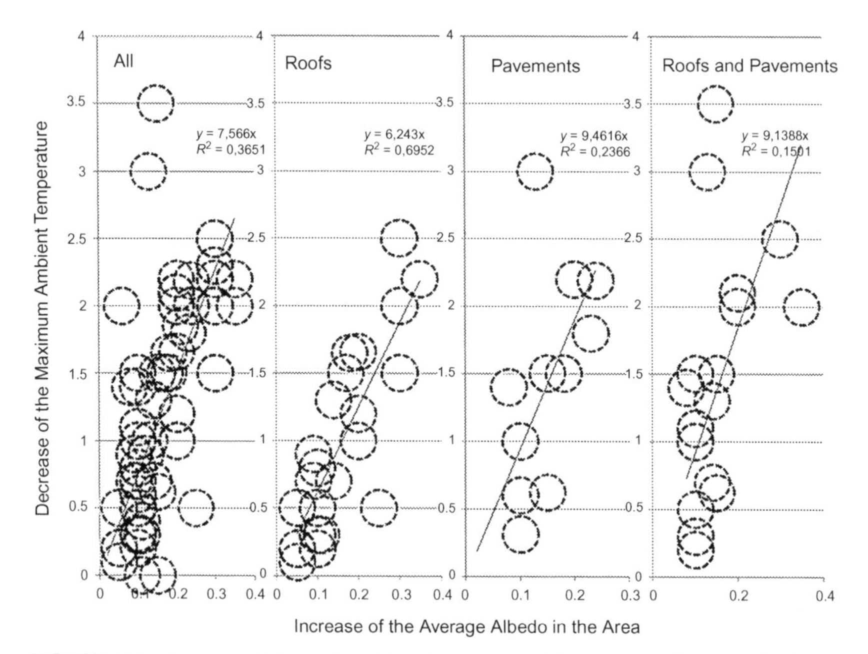

FIGURE 8.35 Decrease of the peak ambient temperature of the various reflective technologies as a function of the albedo increase. *Adapted from Santamouris et al. (2017).*

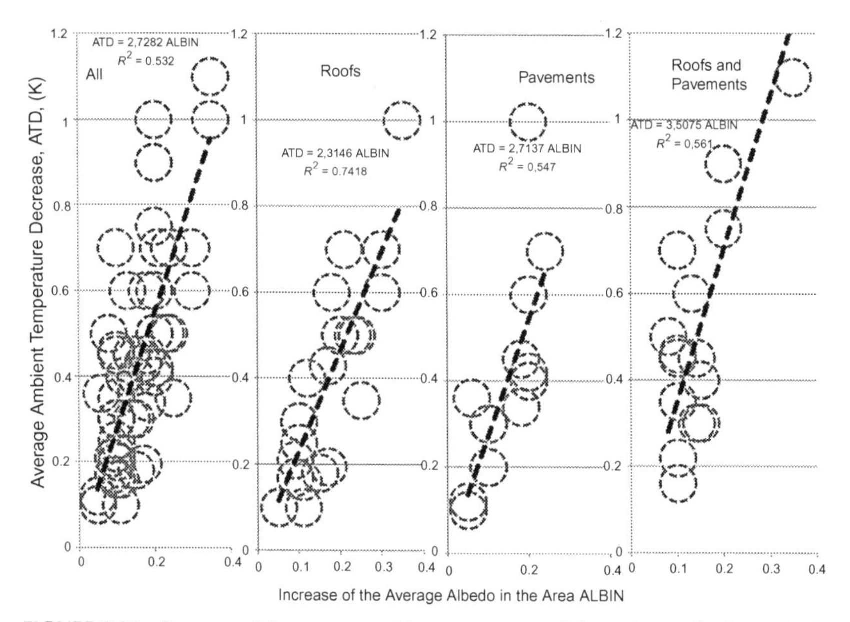

FIGURE 8.36 Decrease of the average ambient temperature of the various reflective technologies as a function of albedo increase. *Adapted from Santamouris et al. (2017).*

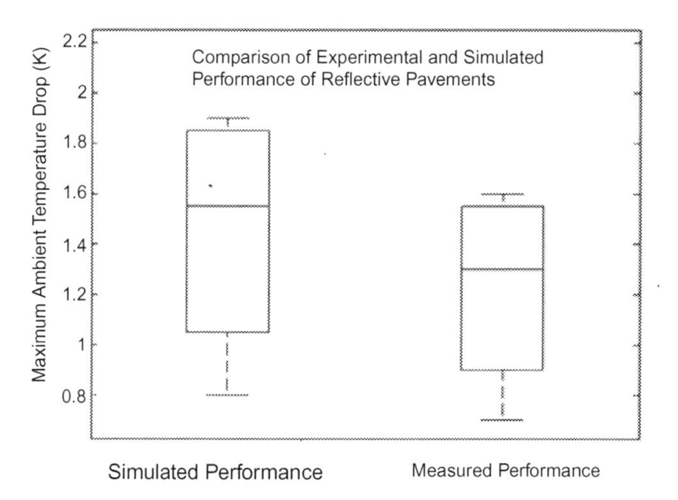

FIGURE 8.37 Comparison of the predicted and experimental temperature drop for four projects using cool pavements. *Adapted from Santamouris et al. (2017).*

simulated one. This can be attributed mainly to the aging caused by the depositing of rubber from the tires of cars on the surface of cool bitumen materials just after their installation (Synnefa et al., 2011). An second factor that decreases the real performance of reflective technologies has to do with the deposition of several atmospheric pollutants on their surface (Mastrapostoli et al., 2014).

Combined Projects Involving the Use of Reflective Pavements with Greenery, Shading and Evaporative Technologies

In most large scale mitigation projects, a combination of different mitigation technologies is used. Santamouris et al. have reported the results of 25 different projects using a combination of greenery and reflective technologies. The characteristics of the specific case studies and the reported mitigation results are given in Table 8.12 below. The range of the maximum and average temperature decrease for all of the projects under consideration is given in Fig. 8.38.

As shown, the peak temperature decrease varies between 0.4 K and 5 K, with an average value close to 1.95 K. In parallel, the corresponding decrease of the mean temperature ranges between 0.1 K and 1.4 K, with a median value close to 0.7 K. The combined use of greenery and reflective materials is found to increase the peak and average temperature decrease by 0.95 K, and 0.3 K respectively.

When reflective and greenery technologies are used in combination, the resulting peak temperature drop per 10% increase of the albedo is higher by 0.4 K, compared to a corresponding decrease when solely reflective materials are used. In parallel, the average temperature drop increases by 0.17 K per

TABLE 8.12 Characteristics and Reported Performance of Projects Involving Reflective Materials and Urban Greenery Technologies

Reflective materials and Greenery								
No	**Location**	**Increase of the Albedo in the Study Area, (0–1)**	**Design Temperature (C)**	**Maximum Temperature Reduction (C)**	**Average Temperature Reduction, (C)**	**Experimental/ Simulation**	**Simulation Tool Used**	**Reference**
1	Grenoble, France	0.05–0.15	34	0.4	0.2	Simulation	Envi-Met	Pisello et al (2016)
2	Athens Suburb, Greece	0.15–0.25	37.2	2.1	1.0	Simulation	CFD	GRBES (2016)
3	Phoenix, USA	0.15–0.25	36		0.1	Simulation	WRF	Stone et al. (2014)
4	Athens, Suburb, Greece	0.1–0.2	33	1.5	0.6	Simulation	CFD	GRBES (2016)
5	Atlanta, USA	0.15–0.25	32		0.3	Simulation	WRF	Stone et al. (2014)
6	Athens, Suburb, Greece	0.05–0.15	26	0.5	0.2	Simulation	CFD	GRBES (2016)
7	Athens, Suburb, Greece	0.15–0.25	32	1.8	0.9	Simulation	CFD	GRBES (2016)
8	Toronto, Canada, Greece	0.05–0.15	33	1	0.3	Simulation	Envi-Met	Wang et al. (2016)
9	Flisvos, Athens, Greece	0.20–0.25	32	2,2	0.9	Simulation	CFD	Santamouris et al. (2012a)

10	Ho Chi Minh City, Vietnam	0.03–0.07	28	0.6	0.15	Simulation	Envi-Met	Chau Huynh and Ronald Eckert (2012)
11	Central Athens, Greece	0,15–0,25	32	1.5	0.6	Simulation	CFD	Gaitani et al. (2011)
12	Agios Demetrios, Athens, Greece	0.15–0.20	34	1.5	0.55	Simulation	CFD	GRBES (2016)
13	Philadelphia, USA	0.15–0.25	30		0.3	Simulation	WRF	Stone et al. (2014)
14	Mentoza, Argentina	0.10–0.20	36	3.5	1.0	Simulation	Envi-Met	Alchapar et al. (2016)
15	Katerini, Greece	0.3–0.4	32	3	1.4	Simulation	CFD	GRBES (2016)
16	Campinas, Brazil	0.10–0.20	33	5	1.4	Simulation	Envi-Met	Alchapar et al. (2016)
17	Rethimnon, Greece	0.15–0.25	31	1.7	0.7	Simulation	Envi-Met	Tsitoura et al. (2016)
18	Veneto, Italy	0.15–0.25	34	3	0.9	Simulation	Envi-Met	Campostrini (2013)
19	Athinas Avenue, Athens, Greece	0.3–0.4	31	3	1.4	Simulation	CFD	GRBES (2016)
20	Los Angeles, USA	0.15–0.25	33	3.6	1	Simulation	MM5	Taha (2008a)
21	Salonica, Greece	0.10–0.15	40.9	0.9	0.25	Simulation	Envi-Met	Chadzidimitriou et al. (2013)
22	Pomona, USA	0.15–0.25	34	3.4	0.9	Simulation	MM5	Taha (2008a)
23	Katerini, Greece	0.25–0.35	33	3	1.4	Simulation	CFD	GRBES (2016)

(Continued)

TABLE 8.12 (Continued)

No	Location	Increase of the Albedo in the Study Area, (0–1)	Design Temperature (C)	Maximum Temperature Reduction (C)	Average Temperature Reduction, (C)	Experimental/ Simulation	Simulation Tool Used	Reference
Reflective materials and Greenery								
24	San Fernando, USA	0.15–0.25	34	3.4	1	Simulation	MM5	Taha (2008a)
25	Dafni, Athens	0.10–0.20	36	1.1	0.4	Simulation	CFD	GRBES (2016)
Reflective Materials + Greenery and Shading								
1	Olympic Village, Athens, Greece	0.20–0.25	39	3.5	1.3	Simulation	CFD	GRBES (2016)
2	Kesariani Square, Athens, Greece	0.15–0.20	35	1.9	0.8	Simulation	CFD	GRBES (2016)
3	North Athens, Greece	0.15–0.20	33	1.8	0.7	Simulation	CFD	GRBES (2016)
4	University Campus, Athens, 2009	0.20–0.25	32	2.3	1.0	Simulation	CFD	GRBES (2016)
5	Athinas Avenue, Athens, Greece	0.3–0.4	31	4	1.9	Simulation	CFD	GRBES (2016)

Reflective Materials + Greenery + Shading + Water								
1	Salonica, Greece	0.10–0.15	40.9	2.4	1.1	Simulation	Envi-Met	Chadzidimitriou et al. (2013)
2	Panepistimiou Avenue, Athens	0.10–0.15	33	1.4	0.6	Simulation	Envi-Met	
3	Eastern Macedonia, Greece	0.25–0.35	36	4.4	2.1	Simulation	Envi-Met	Papadopoulos (2016)
4	Suburb Salonica, Greece	0.30–0.40	34	5.8	2.4	Simulation	Envi-Met	Papadopoulos (2016)
Reflective Materials + Greenery + Water								
1	Ptolemaida, Greece	0.15–0.25	32	2.7	1.3	Simulation	CFD	Zoras et al. (2015)
2	Salonica, Greece	0.10–0.15	40.9	2.3	0.9	Simulation	Envi-Met	Chadzidimitriou et al. (2013)
3	Agia Varvara, Athens, Greece	0.20–0.25	36	3.1	1.2	Simulation	CFD	GRBES (2016)
4	Immitos, Athens, Greece	0.05–0.15	33	1.4	0.8	Simulation	Envi-Met	Lontorfos et al. (2018)

Adapted from Santamouris et al. (2017).

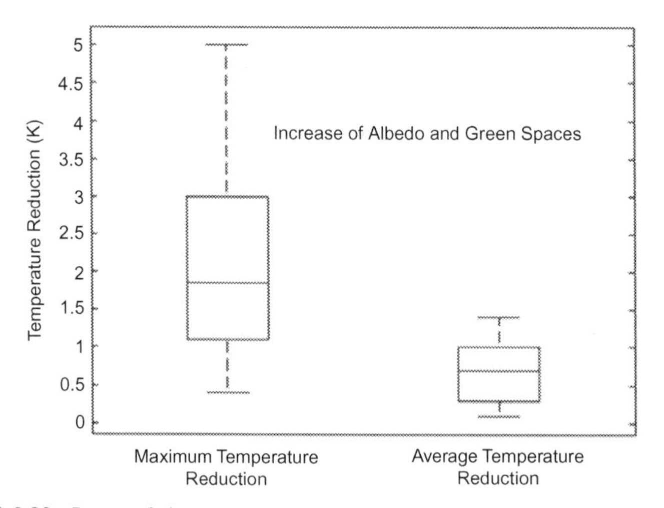

FIGURE 8.38 Range of the maximum and average temperature drop in projects involving reflective materials and greenery technologies. *Adapted from Santamouris et al. (2017).*

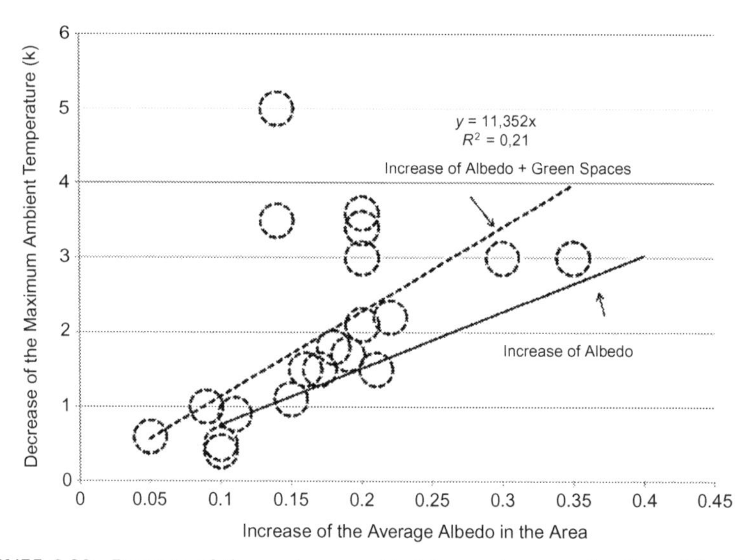

FIGURE 8.39 Decrease of the maximum ambient temperature as a function of the albedo increase for projects involving reflective materials and greenery. *Adapted from Santamouris et al. (2015).*

10% albedo increase. The corresponding results are reported in Figs. 8.39 and 8.40.

Cool roofs and cool pavements may be combined with greenery, evaporative and solar control systems. Four urban mitigation project combining

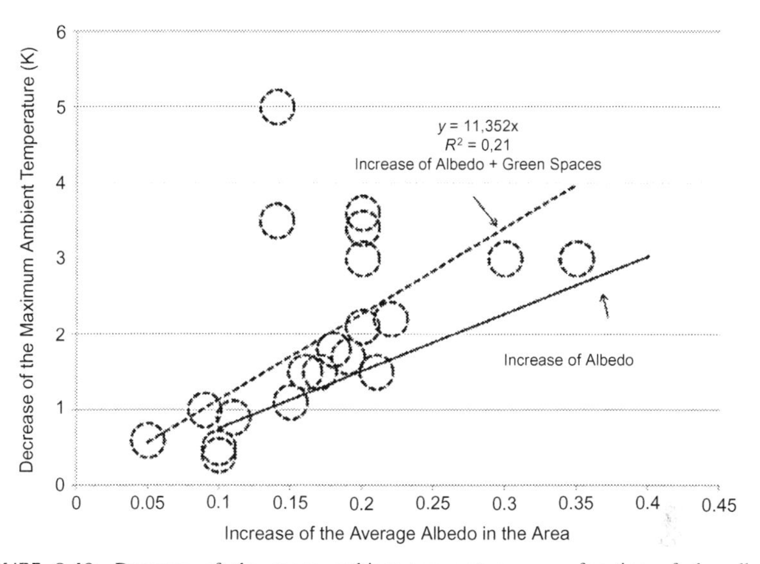

FIGURE 8.40 Decrease of the mean ambient temperature as a function of the albedo increase for projects involving reflective materials and greenery. *Adapted from Santamouris et al. (2015).*

reflective pavements, greenery and water for evaporation are reported in Table 8.12. It is found that the decrease of the peak and average temperature varies from 1.4–3.1 K and 0.8–1.3 K respectively. This corresponds to a peak temperature decrease per 10% of the albedo value close 1.44 K. This is about 0.3 K higher than the corresponding value for the combination of reflective materials and greenery (Fig. 8.41).

Parallel to the above, the characteristics and performance data from five urban mitigation projects combining reflective pavements, greenery and shading devices are reported in Table 8.12. As shown, the peak and average temperature decrease varies between 1.8–4 K and 0.7–1.9 K respectively. The corresponding decrease of the peak temperature per 10% increase in the albedo value is found to be close to 1.29 K. This is about 0.16 K higher compared to the case where greenery and reflective materials are used (Fig. 8.42). The increase of the mitigation potential can be attributed to the impact of urban shading devices (Paolini et al., 2014).

Finally, four projects combining reflective pavements, greenery, shading and water are fully reported in Table 8.12. The calculated peak and average temperature decrease vary between 1.4–5.8 K and 0.6–2.4 K respectively. The corresponding peak temperature drop per 10% increase of the albedo is close to 1.51 K. This is about 0.22 K higher than in the previous case.

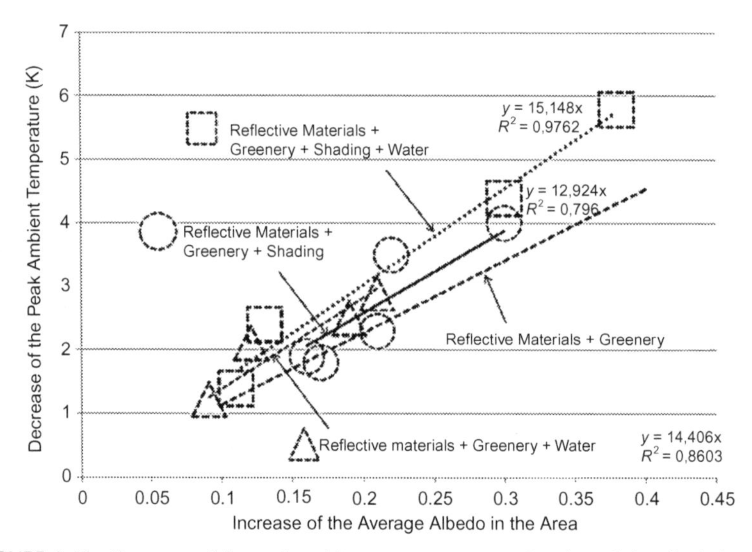

FIGURE 8.41 Decrease of the peak ambient temperature as a function of the albedo increase for projects involving a combination of reflective materials, greenery, water and shading. *Adapted from Santamouris et al. (2017).*

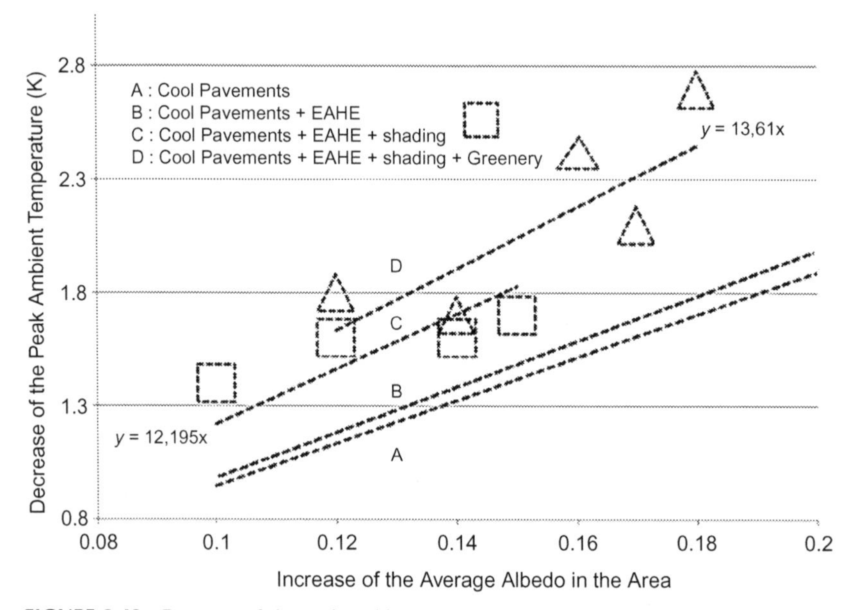

FIGURE 8.42 Decrease of the peak ambient temperature as a function of the albedo increase for projects involving a combination of earth to air heat exchangers and other mitigation techniques. *Adapted from Santamouris et al. (2017).*

Mitigation Projects Using Earth to Air Heat Exchangers Combined with Other Technologies

Santamouris et al. have reported the characteristics and the mitigation potential of nine projects combining reflective technologies and earth to air heat exchangers. The corresponding characteristics and the results are shown in Table 8.13. The number of buried pipes used in the considered projects varied from 3–74, and they delivered cooling between 0.6–10.2 W/m^2 of area. The maximum mitigation potential of the projects varied between 1.7 K and 2.3 K with an average value close to 1.95 K. In parallel, the average temperature decrease is found to vary between 0.65 K and 0.9 K, with an average value close to 0.7 K. Fig. 8.43 reports the calculated peak temperature drop of the considered projects against the corresponding increase of the albedo (Fig. 8.43).

It is estimated that the peak and average temperature reduction per 10% increase of the albedo are close to 1.0 K and 0.37 K respectively. Given that the peak and average temperature decrease when only cool pavements are used is 0.94 K and 0.27 K, it can be concluded that when ground cooling systems are used in combination with reflective pavements, their average mitigation potential may vary between 0.04 K and 0.1 K per 10% increase of the albedo.

Four projects reported in Table 8.13 employed reflective pavements, earth to air heat exchangers and solar shading systems to mitigate urban heat. The decrease of the peak and average ambient temperature was found to vary from 0.14–1.7 K and 0.4–0.7 K respectively. In parallel, the temperature decrease per 10% increase of albedo was found to be around 1.22 K, about 0.23 K higher than when reflective materials and ground cooling systems are used (Fig. 8.43).

Four other projects reported in Table 8.13 employed a ground cooling system (earth to air heat exchangers) combined with reflective pavements, solar control systems and urban. The peak and average temperature drop varies between 1.7–2.7 K and 0.7–1.1 K respectively. The maximum temperature decrease per 10% of albedo increase is found to be around 1.36 K, which is about 0.12 K higher than when earth to air heat exchangers are combined with reflective materials and shading.

Concluding Remarks About the Real Cooling Potential of the Considered Mitigation Technologies

The analysis presented in the previous chapters revealed that existing mitigation technologies can efficiently decrease the ambient temperature and counterbalance the impacts of the UHI and local climate change. Based on the characteristics of the considered projects and the reported mitigation potential, the following conclusions may be drawn:

TABLE 8.13 Characteristics and Reported Performance of Projects Involving the Use of Earth to Air Heat Exchangers in Combination with Other Mitigation Technologies

Use of Reflective Materials in Combination with EAHE								
No	Location	Increase of the Albedo in the Study Area, (0-1)	Design Temperature (C)	Number of Heat Exchangers	Average Flow Rate (m^3/h)	Peak Temperature Drop, (C)	Average Temperature Drop, (C)	Reference
1	Marousi, Athens	0.05–0.15	36	3	1450	1.05	0.35	Santamouris et al. (2012b)
2	Western Greece	0.15–0.25	32.5	54	565	1.9	0.7	Kolokotsa (2011)
3	Western Athens, Greece	0.20–0.30	39	20	565	2,3	0.9	Kolokotsa (2011)
4	Western Athens2, Greece	0.15–0.25	39	20	565	1.95	0.8	Kolokotsa (2011)
5	Thessaly, Greece	0.15–-0.20	40	40	600	1.7	0.85	Kolokotsa (2011)
6	South Peloponese, Greece	0.15–0.20	33	74	600	1.85	0.65	Kolokotsa (2011)
7	Central Greece	0.15–0.25	33	4	1100	2.15	0.6	GRBES (2016)

8	Northern Athens, Greece	0.2–0.25	37	12	565	2.15	0.8	Kolokotsa (2011)
9	Ioanian Island, Greece	0.2–0.30	39	4	565	2	0.65	Kolokotsa (2011)
Reflective Materials in Combination with EAHE and Shading								
1	Central Athens, Greece	0.15–0.25	32	6	1420	1.7	0.7	Gaitani et al. (2011)
2	Central Athens, Greece	0.10–0.15	32	3	1230	1.6	0.6	GRBES (2016)
3	South Athens, Greece	0.10–0.15	32	4	1150	1.6	0.45	GRBES (2016)
4	Northern Athens	0.05–0.15	33	5	1200	1.4	0.4	Gaitani et al. (2014)
Reflective Materials in Combination with EAHE, Shading and Greenery								
1	Central Western Athens, Greece	0.10–0.15	35	6	1150	1.8	0.8	GRBES (2016)
2	Central Northern Athens	0.15–0.20	33	4	1200	2.7	1.1	GRBES (2016)

(*Continued*)

TABLE 8.13 (Continued)

Use of Reflective Materials in Combination with EAHE								
No	Location	Increase of the Albedo in the Study Area, (0-1)	Design Temperature (C)	Number of Heat Exchangers	Average Flow Rate (m^3/h)	Peak Temperature Drop, (C)	Average Temperature Drop, (C)	Reference
3	Western Athens	0.10–0.20	33	4	1300	1.7	0.75	GRBES (2016)
4	Tirana, Albania	0.15–0.20	33	15	1800	2.1	1	Fintikakis et al (2011)

Adapted from Santamouris et al. (2017).

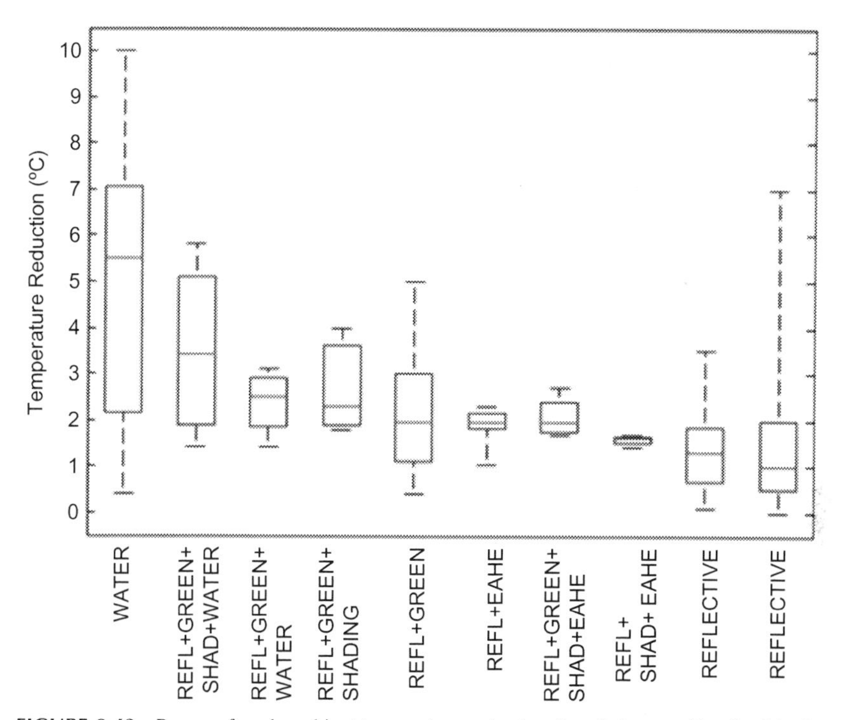

FIGURE 8.43 Range of peak ambient temperature reduction for all the considered mitigation systems and technologies. *Adapted from Santamouris et al. (2017).*

a. The average peak temperature decrease calculated for all the considered projects is close to 2 K. In parallel, the corresponding drop of the average ambient temperature is around 0.7–0.8 K. The analysis shows that about 31% of the considered projects resulted in a peak temperature decrease below 1 K, 62% below 2 K (0–2 K), 82% below 3 K (0–3 K), and finally for almost 90% of the projects it was below 4 K.

b. When various mitigation technologies are combined, the expected mitigation potential increases considerably. It must be pointed out that on average the peak temperature drop of all projects based on the use of just one technology is around 1.9 K and rises to 2.26 K when two or more technologies are employed together. Just 10% of the considered projects based on more than one mitigation technologies resulted in a peak temperature reduction below 1 K, compared with 45% of the case studies using only one mitigation technology.

c. The range of mitigation potential of the various technologies and their combinations varies substantially. Fig. 8.43 shows the values obtained for all technologies and combinations.

d. The average peak temperature decrease of all the considered greenery projects is around 1.7 K. Almost 50% of the urban greenery case studies present a peak temperature drop below 1 K, 78% below 2 K, and almost 90% below 3 K. Among the various greenery types, trees present the highest cooling potential, followed by grass and green roofs.

e. Cool roofs and cool pavements present a very significant mitigation potential. It is estimated that the mean peak temperature decrease for all considered projects is near to 1.3 K. In parallel, almost 45% of the case studies presented a peak temperature reduction below 1 K and 81% below 2 K. Cool pavements presented a higher cooling potential than cool roofs, mainly because of the proximity to the ground level. It is calculated that the peak temperature is reduced by almost 0.8 K per 10% increase of the urban albedo.

f. Evaporative techniques present a considerable mitigation potential and may significantly decrease the ambient temperature at the local level. The average peak temperature decreases of all the considered evaporative case studies was close to 4.5 K. Almost 28% of the projects presented a peak temperature decrease lower than 2 K, 45% below 4 K, and 65% below 6 K. Most of the data were collected in dry climatic zones and the performance of the evaporative systems is expected to decrease substantially in humid climates. The use of evaporative coolers and ambient misting systems is an excellent solution for small to medium size dry urban zones.

g. The mitigation potential of the proposed reflective technologies increases considerably when combined with other mitigation technologies. Table 8.14 presents the estimated equivalent peak temperature drop per 10% increase of the albedo. It is found that reflective materials may contribute to a temperature reduction of between 0.6 K and 0.95 K per 10% increase of the albedo. In parallel, the equivalent contribution of the combined mitigation technologies may increase by up to 1.5 K.

h. To estimate the additional mitigation contribution of the systems combined with reflective technologies, the slope of the curves between the peak temperature decrease and the albedo for all the combination of technologies can be used. The analysis has shown that the contribution of urban greenery and shading technologies is equal to 0.2 K, while the corresponding decrease caused by the use of water and EAHE systems are between 0.3–0.4 K and 0.05–0.1 K respectively.

i. It is concluded that the more the technologies are combined, the less the absolute contribution of each individual mitigation technology.

It is evident that the existing mitigation technologies are able to counterbalance the UHI and the impact of local climate change. The developed and proposed mitigation technologies applied at the city level can decrease the peak ambient temperature by up 2–3 K on average.

TABLE 8.14 Calculated Decrease of the Peak Ambient Temperature of the Various Mitigation Technologies as a Function of Albedo Increase

Mitigation Technology	Decrease of Peak Temperature per 10% Increase of Albedo, (K)
Cool Roofs	0.60
All Reflective Technologies	0.70
Cool Pavements	0.95
Cool Pavements and Earth to air Heat Exchangers	1.00
All Reflective Technologies and Greenery	1.10
Reflective Pavements and EAHE and Shading	1.20
Reflective Pavements and Greenery and Shading	1.30
Reflective Pavements and Greenery and Shading and EAHE	1.40
Reflective Pavements and Greenery and Water	1.45
Reflective Pavements, Shading, Greenery and Water	1.50

Source: Santamouris et al., 2017.

EXISTING AND FUTURE POLICIES TO FIGHT LOCAL CLIMATIC CHANGE AND URBAN VULNERABILITY - ANALYSIS OF THE POTENTIAL IMPACT OF A COMPLETE MITIGATION OF LOCAL CLIMATE CHANGE

The presentation and analysis of the existing mitigation techniques to counterbalance the impact of the local climate change shows that reflective materials can significantly decrease the ambient temperature in cities (Akbari et al., 2016). Reflective materials may be used in the roofs of urban buildings (cool roofs) or in pavements (cool pavements) and can be white or colored. Cool roofs are mature mitigation technologies that are used in hundreds of thousands of applications around the world and their performance is very well documented (Santamouris, 2014). Cool pavement technologies are increasingly used in large scale projects presenting a very significant mitigation potential (Santamouris, 2013). The development of new, advanced reflective materials for pavements further enhances the cooling potential of reflective pavements (Kyriakodis et al., 2016; Lontorfos et al., 2018).

To investigate the investments required to implement reflective technologies that will fully mitigate the UHI in 41 the major European cities, and also assess the associated economic and social benefits, the following methodology was followed (Santamouris, 2016b). In all the cities considered. the population exceeds 500,000 citizens, and UHI is a major problem (as discussed in Chapter 3 of this book). The selected cities as well as the identified UHI intensity is given in Figs. 8.44–8.46.

For all the selected cities, the total land area as well the local population is taken from Demographia (2015), The global population in all the cities considered exceeds 96 million. Based on the estimations given by Akbari et al. (2009), almost 60% of the urban areas are covered by pavements and roofs, in which 30%–45% are pavements and 20%–25% are roofs. We have considered that roofs occupy 20% and pavements 35%, and the total surface of roofs and pavements is calculated for each city. Shading roofs and pavements is important in cities. Existing studies concerning several European cities have concluded that the non-obstructed area is between 54%–64% of the total surface (Defaix et al., 2012). About the 58% of the total roof and pavement urban areas can support cool roofs and pavements.

The total calculated suitable surface was around 2780 km^2 and 4873 km^2 respectively. Actual cost data for cool roofs and pavements are used to

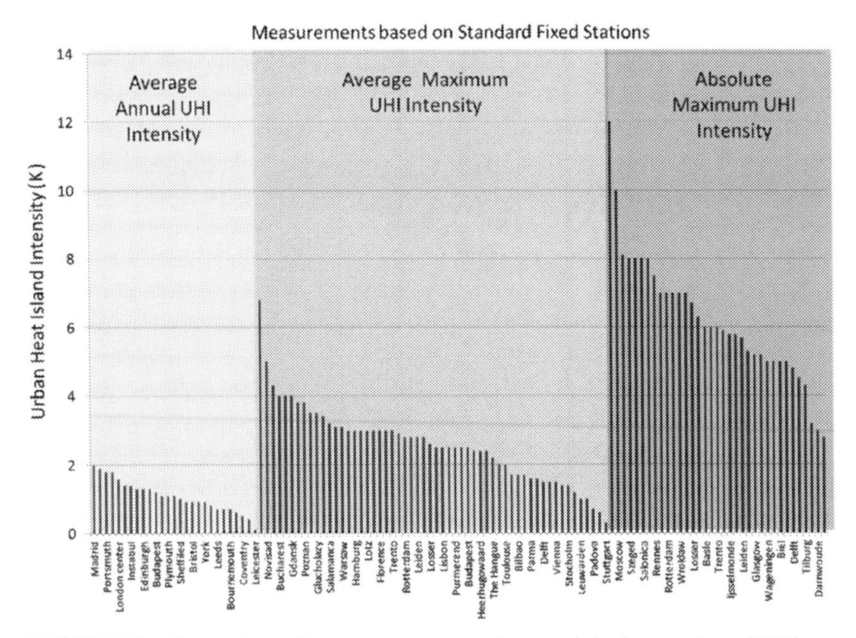

FIGURE 8.44 Measured annual average, average maximum and absolute maximum UHI intensity for all experiments based on the use of standard fixed meteorological stations. *Adapted from Santamouris (2016b).*

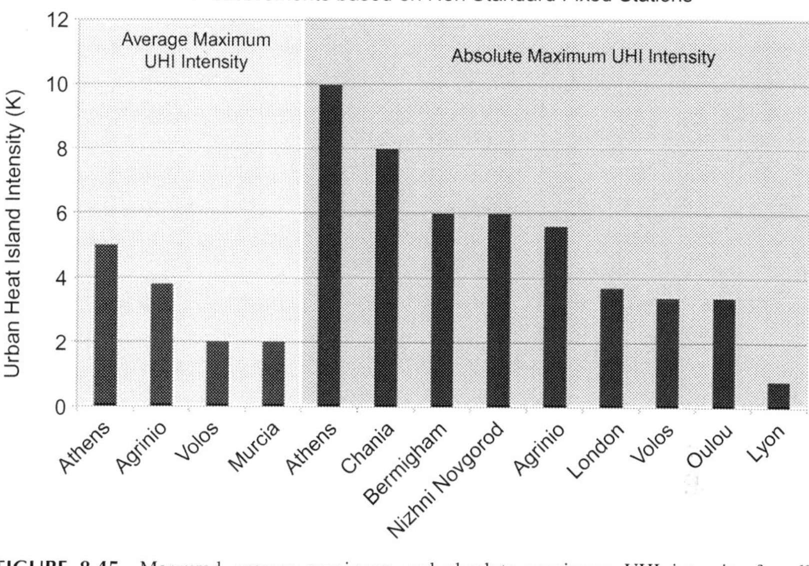

FIGURE 8.45 Measured average maximum and absolute maximum UHI intensity for all experiments based on the use of non-standard fixed meteorological stations. *Adapted from Santamouris (2016b).*

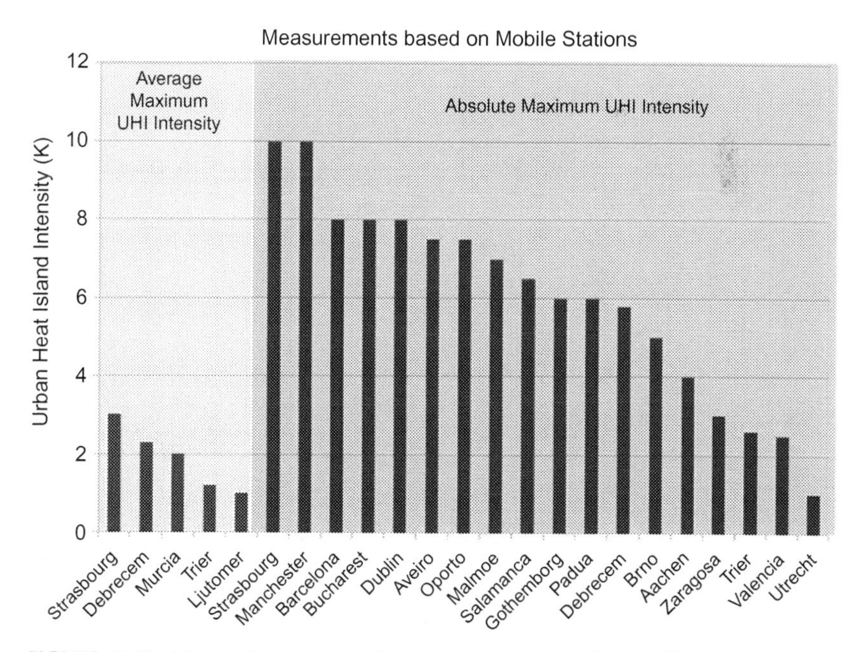

FIGURE 8.46 Measured average maximum and absolute maximum UHI intensity for all experiments based on the use of mobile traverses. *Adapted from Santamouris (2016b).*

calculate the necessary investments (Gaitani et al., 2011; Synnefa and Santamouris, 2012). It is assumed that the cost of installing a cool roof is close to 8 euros/m^2. Based on the existing data, this is a quite representative average value in Europe. Prices may vary by $\pm 20\%$ depending on the quality of the system used. Given the important variability of the cost of cool pavements, two scenarios have been developed. The high cost scenario considers a price close to 26 euros/m^2, while for the low cost one the price was taken as close to 16 euros/m^2. The above values are close to the cost thresholds in USA as proposed by the Cool Cities Alliance (2012) and the EPA (2008), and do not include the contractor's profit.

The estimated financial investments to implement cool roofs and pavements in each European city are given in Fig. 8.47.

The total estimated cost to implement cool roofs in the selected European cities is close to 22.2 billion euros. In parallel, the corresponding financial investments for cool pavements varies between 78 billion euros and 131.5 billion euros for the low and high cost scenarios respectively. About half of the budget corresponds to the investments for Barcelona, Rome, Madrid, Istanbul, London, Milan and Paris. The estimated investment per person for both the cool roofs and pavements are given in Fig. 8.48. It is estimated that the mean cost of cool roofs per citizen is close to 277 euros, while the

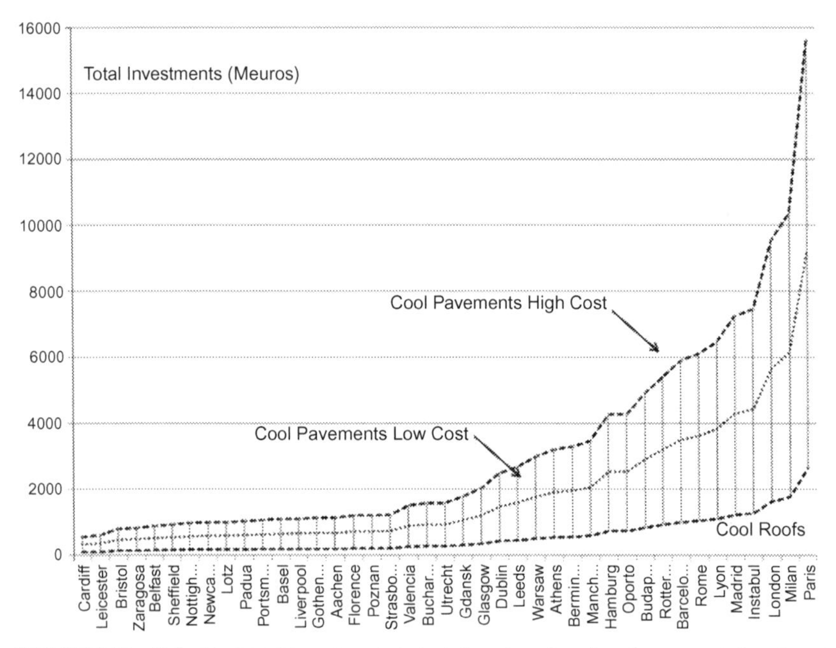

FIGURE 8.47 Calculated total investments to install cool roofs and cool pavements in selected European cities. *Adapted from Santamouris (2016b).*

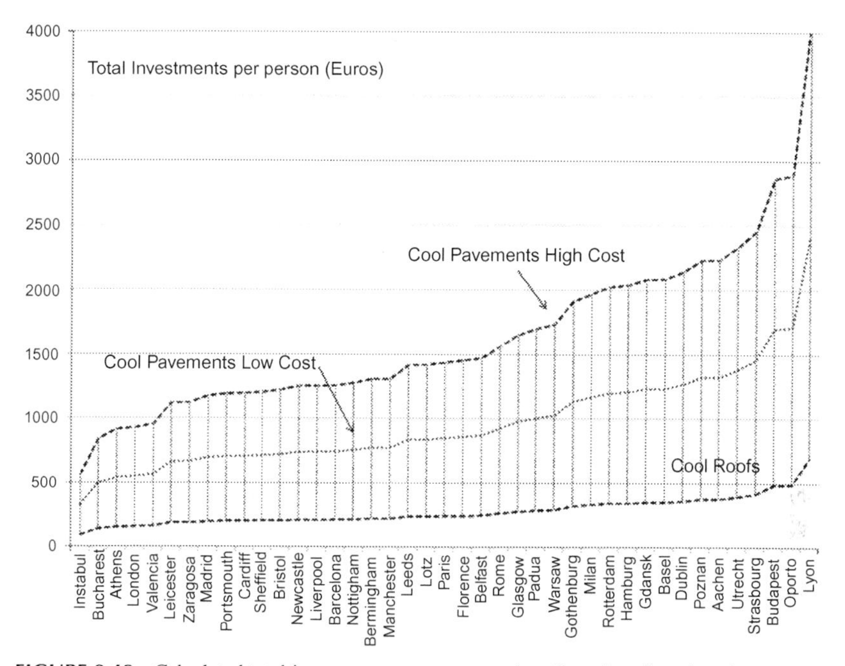

FIGURE 8.48 Calculated total investments per person to install cool roofs and cool pavements in selected European cities. *Adapted from Santamouris (2016b).*

average cost per person for cool pavements varies between 972 and 1640 euros for the low and high cost scenario respectively.

The possible energy benefits associated with the implementation of cool roofs and pavements in the cities are calculated using the energy penalty data given by Santamouris (2014). It is found that the mean energy penalty per person induced by local climate is close to 237 ($\pm$ 130) kWh/person. Simple calculations have shown that the associated energy gains from the mitigation of the UHI vary between 0.022 PWh and 0.035 PWh. It is also estimated that the implementation of cool roofs and pavements in European cities will decrease carbon dioxide emissions. Given this, Akbari et al. (2009) estimate that the use of reflective technologies may offset 64 kg and 38 kg CO_2/m^2 respectively, it is estimated that the expected decrease of CO_2 emissions induced by the proposed installation of cool roofs and pavements in the selected cities is close to 0.178 and 0.185 $GtCO_2$ equivalent respectively. Such a quantity corresponds to almost 9.6% of the total emissions in the European Union in 2013. Considering a price close to 23 euros per tonne of CO_2, the total financial value of the reduced emissions reaches 8.3 billion euros.

The proposed investments in reflective technologies may have a significant impact on the labor market. Considering that the proposed reflective technologies will be implemented by 2050, and assuming a multiplication factor for labor close to 2 jobs per million euros, it is estimated that the total additional jobs for the cool roofs will be 44,500, or 1270 per year. In parallel, the corresponding values for the cool pavements varies between 156,000 and 263,000 or 4450 and 7520 per year. When the average multiplication factor for labor is considered (European Commission, 2015) (6 jobs/million of investments), then the estimated values are 133,600 additional jobs for cool roofs and between 468,000 and 789,000 for cool pavements. These values correspond to about 3820 new jobs annually for cool roofs and 13,400 and 22,600 for cool pavements.

REFERENCES

Akbari, H., Rosenfeld, A., Menon, S., 2009. Global cooling: Increasing world-wide urban albedos to offset CO_2. Clim. Change 94 (3–4), 275–286.

Akbari, H., Cartalis, C., Kolokotsa, D., Muscio, A., Pisello, A.L., Rossi, F., et al., 2016. Local climate change and urban heat island mitigation techniques – the state of the art. J. Civil Eng. Manage. 22 (1), 1–16.

Alchapar, N., Pezzuto, Claudia Cotrim, Correa, Erica Norma, Labaki, Lucila Chebel, 2016. The impact of different cooling strategies on urban air temperatures: The cases of Campinas, Brazil and Mendoza, Argentina. Theor. Appl. Climatol. Available from: https://doi.org/10.1007/s00704-016-1851-5.

Alcoforado, M., 1996. Comparaison des ambiances bioclimatiques estivales d'espaces verts de Lisbonne [Comparison of the bioclimatic environment of green spaces in Lisbon during summer], 9. Publications de l' Association Internationale de Climatologie, pp. 273–280.

Alexandri, E., Jones, P., 2007. Developing a one-dimensional heat and mass transfer algorithm for describing the effect of green roofs on the built environment: Comparison with experimental results. Building and Environment 42 (8), 2835–2849.

Ambrosini, D., Galli, Giorgio, Mancini, Biagio, Nardi, Iole, Sfarra, Stefano, 2014. Evaluating mitigation effects of urban heat islands in a historical small center with the ENVI-Met climate model. Sustainability 6, 7013–7029. Available from: https://doi.org/10.3390/su6107013.

Amor, B., Lacheheb, Dhia Eddine Zakaria, Bouchahm, Yasmina, 2015. Improvement of thermal comfort conditions in an urban space (case study: The square of independence, Sétif, Algeria). Eur. J. Sustain. Dev. 4 (2), 407–416.

Andrade, H., Vieira, R., 2007. A climatic study of an urban green space: The Gulbenkian Park in Lisbon (Portugal). Finisterra - Revista Portuguesa de Geografia XLII (84), 27–46.

Bacci, L., Morabito, M., Raschi, A., & Ugolini, F. (2003). Thermohygrometric conditions of some urban parks of Florence (Italy) and their effects on human well-being. In: The Fifth International Conference on Urban Climate Lodtz, (pp. 245–248).

Barradas, V.L., 1991. Air temperature and humidity and human comfort index of some city parks of Mexico city. International Journal of Biometeorology 35, 24–28.

Bass, B., Krayenhoff, S., Martilli, A., 2002. Mitigating the Urban Heat Island With Green Roof Infrastructure. Urban Heat Island Summit, Toronto.

Bencheikh, H., Ameur, R., 2012. The effects of green spaces (palme trees) on the microclimate in arid zones, case study: Ghardaia. Algeria. Architecture Research 2 (5), 60–67.

Berdahl, Paul, Chen, Sharon S., Destaillats, Hugo, Kirchstetter, Thomas W., Levinson, Ronnen M., Zalich, Michael A., 2016. Fluorescent cooling of objects exposed to sunlight – The ruby example. Solar Energy Materials & Solar Cells 157, 312–317.

Bowler, D.E., Buyung-Ali, L., Knight, T.M., Pullin, A.S., 2010. Urban greening to cool towns and cities: A systematic review of the empirical evidence. Landscape and Urban Planning 97 (3), 147–155.

Bruce, M., Skinner, C., 1999. Rooftop Greening and Local Climate: A Case Study in Melbourne Available through: <http://www.envi-met.com/documents/papers/Rooftop1999.pdf>.

Ca, T., Asaeda, V.T., Abu, E.M., 1998. Reductions in air conditioning energy caused by a nearby park. Energy and Buildings 29 (1), 83–92.

Campostrini, P., 2013: UHI – urban heat island and its mitigation in the Veneto region, Available through: <http://www.iprpraha.cz/uploads/ssets/soubory/data/projekty/UHI/UHI_v_regionu_Veneto_a_jeho_zmirovani.pdf>.

Castleton, H.F., Stovin, V., Beck, S.B.M., Davison, J.B., 2010. Green roofs: Building energy savings and the potential for retrofit. Energy Build 42, 1582–1591.

Chadzidimitriou, A., Liveris, P., Bruse, M., Topli, L., 2013. Urban redevelopment and microclimate improvement: A design project in Thessaloniki, Greece. In: PLEA2013 – 29th Conference, Sustainable Architecture for a Renewable Future, Munich, Germany, 10–12 September 2013.

Chang, C.-R., Li, M.-H., Chang, S.-D., 2007. A preliminary study on the local cool-island intensity of Taipei city parks. Landscape and Urban Planning 80 (4), 386–395.

Chen, H., Ooka, Ryozo, Huang, Hong, Tsuchiya, Takashi, 2009. Study on mitigation measures for outdoor thermal environment on present urban blocks in Tokyo using coupled simulation. Build. Environ 44, 2290–2299.

Chow, W.T.L., Pope, R.L., Martin, C.A., Brazel, A.J., 2010. Observing and modeling the nocturnal park cool island of an arid city: Horizontal and vertical impacts. Theoretical and Applied Climatology 103 (1–2), 197–211. Available from: https://doi.org/10.1007/s00704-010-0293-8.

Cohen, P., Potchter, O., Matzarakis, A., 2012. Daily and seasonal climatic conditions of green urban open spaces in the Mediterranean climate and their impact on human comfort. Building and Environment 51, 285–295.

Cool Cities Alliance, 2012. A Practical Guide to Cool Roofs and Cool Pavements. <http://www.coolrooftoolkit.org/wp-content/pdfs/CoolRoofToolkit_Full.pdf>.

Defaix, P.R., van Sark, W.G.J.H.M., Worrell, E., de Visser, E., 2012. Technical potential for photovoltaics on buildings in the EU-27. Sol. Energy 86, 2644–2653.

Demographia: World Urban Areas, 11th Edition, 2015. <http://www.demographia.com/db-worldua.pdf>.

Djukic, A., 2015. Retrofitting of communal open spaces towards climate comfort: Case study mega block in New Belgrade. J. Sustain. Archit. Civ. Eng. 2015/1/10.

Dominnguez, S. Alvarez, de la Flor, F.J. Sanchez, 2016. The effect of evaporative techniques on reducing urban heat. In: Santamouris, M., Kolokotsa, D. (Eds.), Urban Climate Mitigation echniques. Routledge, London.

Eliasson, I., 1996. Urban nocturnal temperatures, street geometry and land use. Atmospheric Environment 30 (3), 379–392.

Emmanuel, R., Rosenlund, H., Johansson, E., 2007. Urban shading – A design option for the tropics? A study in Colombo, Sri Lanka. Int. J. Climatol. 27, 1995–2004.

Environmental Protection Agency, EPA, 2008. Reducing Urban Heat Islands: Compendium of Strategies. <http://www2.epa.gov/heatislands/heat-island-compendium>.

European Commission, 2015. Energy Renovation: The Trump Card for the New Start for Europe, JRC Science and policy Reports. <http://iet.jrc.ec.europa.eu/energyefficiency/system/tdf/eur26888_buildingreport_online.pdf?file = 1&type = node&id = 9069>.

Feng, C., Meng, Q., Zhang, Y., 2010. Theoretical and experimental analysis of the energy balance of extensive green roofs. Energy and Buildings 42, 959–965.

Ferrante, A., 2016. Towards Nearly Zero Energy: Urban Settings in the Mediterranean Climate. Elsevier.

Fintikakis, N., Gaitani, N., Santamouris, M., Assimakopoulos, M., Assimakopoulos, D.N., Fintikaki, M., et al., February 2011. Bioclimatic design of pen public spaces in the historic centre of Tirana. Albania Original Research Article Sustainable Cities and Society 1 (1), 54–62.

Ford, B., Francis, E., Alvarez, S., Thomas, P., Schiano-Phan, R., 2010. The Architecture and engineering of Downdraught Cooling: A Design Sourcebook. PHDC Press.

Gaffin, S.R., Rosenzweig, C., Khanbilvardi, R., Parshall, L., Mahani, S., Glickman, H., et al., 2008. Variations in New York city's urban heat island strength over time and space. Theoretical and Applied Climatology 94 (1–2), 1–11. Available from: https://doi.org/10.1007/s00704-007-0368-3.

Gaitani, N., Spanou, A., Saliari, M., Synnefa, A., Vassilakopoulou, K., Papadopoulou, K., et al., 2011. Improving the microclimate in urban areas. A case study in the centre of Athens. J. Build. Serv. Eng. 32 (1), 53–71.

Georgakis, Ch, Zoras, S., Santamouris, M., 2014. Studying the effect of "cool" coatings in street urban canyons and its potential as a heat island mitigation technique. Sustain. Cities Soc 13, 20–31.

Georgescu, M., Mahalov, A., Moustaoui, M., 2012. Seasonal hydroclimatic impacts of sun corridorexpansion. Environ. Res. Lett. 7, 034026.

Georgescu, M., Morefieldb, Philip, E., Bierwagen, Britta, G., Weave, et al., 2014. Urban adaptation can roll back warming of emerging megapolitan regions. PNAS 111 (8), 2909–2914.

Gill, S.E., Handley, J.F., Ennos, A.R., Pauleit, S., 2007. Adapting cities for climate change: The role of green infrastructure. Built Environment 33, 115–133.

Gober, P., Brazel, Anthony, Quay, Ray, Myint, Soe, Grossman-Clarke, Susanne, Miller, Adam, et al., 2009. Using watered landscapes to manipulate urban heat island effects: How much water will it take to cool phoenix? J. Am. Plan. Assoc. 76 (1), 109–121. Available from: https://doi.org/10.1080/01944360903433113.

Gober, P., Brazel, A., Quay, R., Myint, S., Grossman-Clarke, S., Miller, A., et al., 2010. Using watered landscapes to manipulate urban heat island effects: how much water will it take to cool phoenix? J. Am. Plan. Assoc. 76 (1). Available from: https://doi.org/10.1080/01944360903433113.

Gomez, F., Gaja, E., Reig, A., 1998. Vegetation and climatic changes in a city. Ecological engineering 10 (4), 355–360. Available from: https://doi.org/10.1016/S0925-8574(98)00002-0.

Hamada, S., Ohta, T., 2010. Seasonal variations in the cooling effect of urban green areas on surrounding urban areas. Urban Forestry Urban Greening 9 (1), 15–24.

Hodo-Abalo, Samah, Banna, Magolmèèna, Zeghmati, Belkacem, 2012. Performance analysis of a planted roof as a passive cooling technique in hot-humid tropics. Renewable Energy 39, 140–148.

Huang, J., Ooka, Ryozo, Okada, Akiko, Omori, Toshiaki, Huang, Hong, 2009. The effects of urban heat island and mitigation strategies on the outdoor thermal environment in central Tokyo – a numerical simulation. In: The Seventh AsiaPacific Conference on wind Engineering, November 8–12, 2009, Taipei, Taiwan.

Hull, R., 1992. Brief encounters with urban forests produce moods that matter. Journal of Arboriculture 118, 322–324.

Huynh, Chau, Eckert, Ronald, 2012. Reducing heat and improving thermal comfort through urban design – A case study in Ho Chi Minh City. Int. J. Environ. Sci. Dev. 3 (5).

Jansson, C., Jansson, P.E., Gustafsson, D., 2006. Near surface climate in an urban vegetated park and its surroundings. Theoretical and Applied Climatology 89 (3–4), 185–193.

Jauregui, E., 1990. Influence of a large urban park on temperature and convective precipitation in a tropical city. Energy and Buildings 16 (3–4), 457–463. Available from: https://doi.org/10.1016/0378-7788(90)90021-A.

Jim, C.Y., He, Hongming, 2010. Coupling heat flux dynamics with meteorological conditions in the green roof ecosystem. Ecological Engineering 36, 1052–1063.

Jonsson, P., 2004. Vegetation as an urban climate control in the subtropical city of Gaborone, Botswana. International Journal of Climatology 24 (10), 1307–1322.

Jusuf, S. Kardinal, Hien, Wong Nyuk, La Win, Aung Aung, Thu, Htun Kyaw, Negara, To Syatia, Xuchao, Wu, 2006. Study on effect of greenery in campus area. In: PLEA2006 – The 23rd Conference on Passive and Low Energy Architecture, Geneva, Switzerland, 6–8 September 2006.

Jusuf, S.K., Wong, N., Hagen, E., Anggoro, R., Hong, Y., 2007. The influence of land use on the urban heat island in Singapore. Habitat International 31 (2), 232–242. Available from: https://doi.org/10.1016/j.habitatint.2007.02.006.

Karachaliou, P., Santamouris, Mat, Pangalou, Helli, 2016. Experimental and numerical analysis of the energy performance of a large scale intensive green roof system installed on an office building in Athens. Energy Build. 114 (February), 15.

Karlessi, Theoni, Santamouris, Mat, 2013. Improving the performance of thermochromic coatings with the use of UV and optical filters tested under accelerated aging conditions. J. low Carbon Technologies 2015. Available from: https://doi.org/10.1093/ijlct/ctt027. Frst published online.

Karlessi, T., Santamouris, M., Apostolakis, K., Synnefa, A., Livada, I., April 2009. Development and testing of thermochromic coatings for buildings and urban structures. Solar Energy 83 (4), 538–551.

Karlessi, T., Santamouris, M., Synnefa, A., Assimakopoulos, D., Didaskalopoulos, P., Apostolakis, K., March 2011. Development and testing of PCM doped cool colored coatings to mitigate urban heat island and cool buildings. Building and Environment 46 (3), 570–576.

Kjelgren, R.K., Clark, J.R., 1992. Microclimates and tree growth in three urban spaces. Journal of Environmental Horticulture 10 (3), 139–145.

Kleerekoper, L., van Escha, M., Salcedob, T.B., 2012. How to make a city climate proof, addressing the urban heat island effect. Resour. Conserv. Recycl. 64, 30–38.

Kolokotsa, D., 2011. Urban Bioclimatic Rehabilitation Studies. Technical University of Crete, Chania, Greece. Available through the Author.

Kolokotsa, D., Santamouris, M., Zerefos, S.C., 2013. Green and cool roofs' urban heat island mitigation potential in European climates for office buildings under free floating conditions. Sol. Energy 95 (September), 118–130.

Kyriakodis, G., Mastrapostoli, E., Santamouris, M., 2016. Experimental and numerical assessment of bioclimatic rehabilitation of a large urban area in Western Athens using reflective and photocatalytic materials. In: Proc. 4th International Conference on Countermeasures to UHI National University of Singapore, Singapore.

Lahme, E., & Bruse, M. (2003). Microclimatic effects of a small urban park in a densely build up area: Measurements and model simulations. In: The Fifth International Conference on Urban Climate Lodtz.

Lazzarin, Renato M., Castellotti, Francesco, Busato, Filippo, 2005. Experimental measurements and numerical modelling of a green roof. Energy and Buildings 37, 1260–1267.

Lee, S.-H., Lee, K.-S., Jin, W.-C., Song, H.-K., 2009. Effect of an urban park on air temperature differences in a central business district area. Landscape Ecological Engineering 5 (2), 183–191.

Levinson, R., Berdahl, P., Akbari, H., 2005. Spectral solar optical properties of pigments Part II: Survey of common colorants. Solar Energy Materials and Solar Cells 89, 351–389.

Lewis, J.E., Nicholas, F.W., Scales, S.M., Woolum, C.A., 1971. Some effects of urban morphology on street level temperatures at Washington, DC. Journal of the Washington Academy of Science 61, 258–265.

Lontorfos, V., Efthymiou, C., Santamouris, M., January 2018. On the time varying mitigation performance of reflective geoengineering technologies in cities. Renewable Energy 115, 926–930.

Lu, J., Li, C.-D., Yang, Y.-C., Zhang, X.-H., Jin, M., 2012. Quantitative evaluation of urban park cool island factors in mountain city. Journal of Central South University of Technology (English Edition) 19 (6), 1657–1662.

Lynn, B.H., Carlson, T.N., Rosenzweig, C., Goldberg, R., Druyan, L., Cox, J., et al., 2009. A modification to the NOAH LSM to simulate heat mitigation strategies in the New York city metropolitan area. J. Appl. Meteorol. Climatol. 48, 200–216.

Maleki, A., Mahdavi, A., 2016. Evaluation of urban heat islands mitigation strategies using 3Dimentional urban microclimate model ENVI-MET. Asian J. Civ. Eng. (BHRC) 17 (3), 357–371.

Manteghi, G., bin Limit, Hasanuddin, Remaz, Dilshan, 2015. Water bodies an urban microclimate: A review. Mod. Appl. Sci. 9 (6), 10.

Martins, T., Adolphe, Luc, Bonhomme, Marion, Bonneaud, Frédéric, Faraut, Serge, Ginestet, Stéphane, et al., 2016. Impact of urban cool island measures on outdoor climate and pedestrian comfort: Simulations for a new district of Toulouse, France. Sustain. Cities Soc 26, 9–26.

Mastrapostoli, E., Karlessi, Theoni, Pantazaras, Alexandros, Gobakis, Kostas, Kolokotsa, Dionysia, Santamouris, Mattheos, 2014. On the cooling potential of cool roofs in cold climates: Use of cool fluorocarbon coatings to enhance the optical properties and the energy performance of industrial buildings. Energy Build. 69 (February), 417–425.

Middel, A., Chhetri, Nalini., 2014. City of phoenix cool urban spaces project urban forestry and cool roofs: Assessment of heat mitigation strategies in phoenix. Report prepared by Center for Integrated Solutions to Climate Challenges at Arizona State University July 2014 in collaboration with Climate Assessment for the Southwest at University of Arizona.

Mihalakakou, G., Santamouris, M., Asimakopoulos, D., 1994a. Modelling the thermal performance of the earth to air heat exchangers. Sol. Energy 53 (3), 301–305.

Mihalakakou, G., Santamouris, M., Asimakopoulos, D., 1994b. On the cooling potential of earth to air heat exchangers. J. Energy Convers. Manage. 35 (5), 395–402.

Millstein, D., Menon, Surabi, 2011. Regional climate consequences of large-scale cool roof and photovoltaic array deployment. Environ. Res. Lett. 6, 034001.

Mitchell, V.G., Cleugh, H.A., Grimmond, C.S.B., Xu, J., 2008. Linking urban water balance and energy balance models to analyse urban design options. Hydrol. Process. 22, 2891–2900.

Ng, E., Chen, Liang, Wang, Yingna, Yuan, Chao, 2012. A study on the cooling effects of greening in a high-density city: An experience from Hong Kong. Build. Environ. 47, 256–271.

Nishimura, N., Nomura, T., Iyota, H., Kimoto, S., 1998. Novel water facilities for creation of comfortable urban micrometeorology. Sol. Energy 64 (4), 197–207.

Nunes, J., Zoilo, I., Jacinto, N., Nunes, A., Associated Prof., Campos, T., et al., 2016. Misting-cooling systems for microclimatic control in public space. Available through: <http://www.proap.pt/847/misting-cooling-systemsfor-microclimatic-control-in-public-space/>.

Oke, T.R., Johnson, G.T., Steyn, D.G., Watson, I.D., 1991. Simulation of surface urban heat islands under ideal conditions at night. 2. Diagnosis of Causation. Boundary Layer Meteorology 56 (4), 339–358.

Oliveira, S., Andrade, H., Vaz, T., 2011. The cooling effect of green spaces as a contribution to the mitigation of urban heat: A case study in Lisbon. Building and Environment 46 (11), 2186–2194.

O'Malley, C., Piroozfar, Poorang, Farr, Eric R.P., Pomponi, Francesco, 2015. Urban Heat Island (UHI) mitigating strategies: A case-based comparative analysis. Sustain. Cities Soc 19, 222–235.

Padmanabhamurty, B., 1990. Microclimates in tropical urban complexes. Energy and Buildings 15 (C), 83–92.

Paolini, R., Mainini, A.G., Poli, T., Vercesi, L., 2014. Assessment of thermal stress in a street canyon in pedestrian area with or without canopy shading. Energy Procedia 48, 1570–1575. Available from: https://doi.org/10.1016/j.egypro.2014.02.177.

Papadopoulos, A., 2016. Aristotle University Thessaloniki. Personal Communication.

Parizotto, S., Lamberts, R., 2011. Investigation of green roof thermal performance in temperate climate: A case study of an experimental building in Florianopoliscity. Southern Brazil, Energy and Buildings 43, 1712–1722.

Pisello, A.L., Castaldo, Veronica Lucia, Piselli, Cristina, Fabiani, Claudia, Colleluori, Marco, Pignatta, Gloria, et al., 2016. Elaboration of the simulation tools and models of the NZE settlements with integrated the proposed systems and analysis of the simulated results. Report Zero Plus Project. European Commission.

Polyzos, V., Istapolou, A., 2014. Bioclimatic Design of Open Spaces Using Envimet. National Technical University of Athens.

Potchter, O., Cohen, P., Bitan, A., 2006. Climatic behavior of various urban parks during hot and humid summer in the Mediterranean city of Tel Aviv, Israel. International Journal of Climatology 26 (12), 1695–1711. Available from: https://doi.org/10.1002/joc.

Rajabi, T., Abu-Hijleh, B., 2014. The Study of Vegetation Effects on Reduction of Urban Heat Island in Dubai. SB Conference, Barcelona.

Rezaei, F., 2005. Evapotranspiration Rates from Extensive Green Roof Plant Species, M.S. Thesis. Agricultural and Biological Engineering, The Pennsylvania State University, State College, PA.

Rosenfeld, A.H., Akbari, Hashem, Bretz, Sarah, Fishman, Beth L., Kurn, Dan M., Sailor, David, et al., 1995. Mitigation of urban heat islands: Materials, utility programs, updates. Energy Build. 22, 255–265.

Rosenfeld, A.H., Akbari, Hashem, Romm, Joseph J., Pomerantz, Melvin, 1998. Cool communities: Strategies for heat island mitigation and smog reduction. Energy Build. 28, 51–62.

Sailor, D., 1995. Simulated urban climate response to modifications in surface albedo and vegetative cover. J. Appl. Meteorol. 34, 1694–1700.

Sailor, D.J., Kalkstein, L.S., Wong, E., 2002. The potential of urban heat island mitigation to alleviate heat related mortality: Methodological overview and preliminary modeling results for Philadelphia. In: Proceedings of the 4th Symposium on the Urban Environment, May 2002, Norfolk, VA, vol. 4, pp. 8– 69.

Sani, S., 1990. Urban climatology in Malaysia: An overview. Energy and Buildings 15 (C), 105–117.

Santamouris, M., 2001. In: Santamouris, M. (Ed.), Energy and Climate in the Urban Built Environment. James and James Science Publishers, London.

Santamouris, M., 2013. Using cool pavements as a mitigation strategy to fight urban heat island—A review of the actual developments. Renew. Sustain. Energy Rev. 26, 224–240.

Santamouris, M., 2014. Cooling the cities – A review of reflective and green roof mitigation technologies to fight heat Island and improve comfort in urban environments. Solar Energy 103, 682–703. 2014.

Santamouris, M., 2016b. Cooling of buildings, past, present and future. Energy Build. 128, 617–638.

Santamouris, M., April 2016b. Innovating to zero the building sector in Europe: Minimising the energy consumption, eradication of the energy poverty and mitigating the local climate change. Solar Energy 128, 61–94.

Santamouris, M., Assimakopoulos, D.N. (Eds.), 1997. Passive Cooling of Buildings. James and James Science Publishers, London.

Santamouris, M., Kolokotsa, D., 2016. Urban Climate Mitigation Techniques. Routledge.

Santamouris, M., Cartalis, C., Synnefa, A., Kolokotsa, D., 2015. On the impact of urban heat island and global warming on the power demand and electricity consumption of buildings—a review. Energy Build. 98 (1), 119–124. Available from: https://doi.org/10.1016/j.enbuild.2014.09.052.

Santamouris, M., Mihalakakou, G., Argiriou, A., Asimakopoulos, D.N., 1995. On the performance of buildings coupled with earth to air heat exchangers. Solar Energy 54 (6), 375–380.

Santamouris, M., Mihalakakou, G., Balaras, C.A., Lewis, J.O., Vallindras, M., Argiriou, A., 1996. Energy conservation in greenhouses with buried pipes. Journal of Energy 21 (5), 353–360.

Santamouris, M., Gaitani, N., Spanou, A., Saliari, M., Gianopoulou, K., Vasilakopoulou, K., 2012a. Using cool paving materials to improve microclimate of urban areas – design realisation and results of the Flisvos project. Build. Environ. 53 (July), 128–136.

Santamouris, M., Xirafi, F., Gaitani, N., Spanou, A., Saliari, M., Vassilakopoulou, K., 2012b. Improving the microclimate in a dense urban area using experimental and theoretical techniques – The case of Marousi, Athens. Int. J. Ventil. 11 (1), 1–16.

Savio, P., Rosenzweig, Cynthia, Solecki, William, D., Slosberg, Ronald, B., 2006. Mitigating New York City's Heat Island with Urban Forestry, Living Roofs, and Light Surfaces. New York City Regional Heat Island Initiative. The New York State Energy Research and Development Authority, Albany, NY.

Sharma, A., Conry, P., Fernando, H.J.S., Hamlet, Alan F., Hellmann, J.J., Chen, F., 2016. Green and cool roofs to mitigate urban heat island effects in the Chicago metropolitan area: Evaluation with a regional climate model. Environ. Res. Lett. 11, 064004.

Skelhorn, C., Lindley, Sarah, Levermore, Geoff, 2014. The impact of vegetation types on air and surface temperatures in a temperate city: A fine scale assessment in Manchester, UK. Landsc. Urban Plan. 121, 129–140.

Skoulika, F., Santamouris, M., Boemi, N., Kolokotsa, D., 2014. On the thermal characteristics and the mitigation potential of a medium size urban park in Athens, Greece. Landscape and Urban Planning 123, 73–86.

Smith, K., Roeber, P., 2011. Green roof mitigation potential for a proxy future climate scenario in Chicago, Illinois. J. Appl. Meteorol. Climatol. 50, 507–522.

Song, B., Park, Kyunghun, 2015. Contribution of greening and high-albedo coatings to improvements in the thermal environment in complex urban areas. Adv. Meteorol. 14, 792172. 2015.

Soutullo Castro, S., Guaita, Cristina Sanjuan, Egido, Maria Nuria S.ánchez, Zarzalejo, Luis F., Miranda, Ricardo Enríquez, Celemín, Maria del Rosario Heras, 2012. Comfort evaluation in an urban boulevard by means of evaporative wind towers. Energy Procedia 30, 1226–1232.

Spronken-Smith, R.A., Oke, T.R., 1998. The thermal regime of urban parks in two cities with different summer climates. International Journal of Remote Sensing 19 (11), 2085–2104.

Stone, Jr.B., Vargo, Jason, Liu, Peng, Habeeb, Dana, DeLucia, Anthony, Trail, Marcus, et al., 2014. Avoided heat-related mortality through climate adaptation strategies in three US cities. PLoS ONE 9 (6), e100852.

Sugawara, H., Narita, K., Mikami, T., Honjo, T., Ishii, K., 2006. Cool island intensity in a large Urban green: Seasonal variation and relationship to atmospheric conditions. Tenki (in Japanese, with English summary) 53 (5), 3–14.

Sun, R., Chen, L., 2012. How can urban bodies be designed for climate adaptation? Landsc. Urban Plan. 105, 27–33.

Sun, R., Chen, Ailian, Chen, Liding, Lü, Yihe, 2012. Cooling effects of wetlands in an urban region: The case of Beijing. Ecol. Ind 20, 57–64.

Sundersingh, S.D., 1991. Effect of heat islands over urban madras and measures for its mitigation. Energy and Buildings 15 (1–2), 245–252.

Synnefa, Santamouris, M., 2012. Advances on technical, policy and market aspects of cool roof technology in Europe: The Cool Roofs project. Energy Build. 55, 35–41.

Synnefa, A., Santamouris, M., Livada, I., August 2006. A study of the thermal performance of reflective coatings for the urban environment. Solar Energy 80 (8), 968–981.

Synnefa, A., Santamouris, M., Apostolakis, K., 2007. On the development, optical properties and thermal performance of cool colored coatings for the urban environment. Solar Energy 81, 488–497.

Synnefa, A., Dandou, A., Santamouris, M., Tombrou, M., 2008. On the use of cool materials as a heat island mitigation strategy. J. Appl. Meteorol. Climatol. 47, 2846–2856.

Synnefa, A., Karlessi, T., Gaitani, N., Santamouris, M., Assimakopoulos, D.N., Papakatsikas, C., 2011. Experimental testing of cool colored thin layer asphalt and estimation of its potential to improve the urban microclimate. Build. Environ. 46 (1), 38–44.

Taha, H., 2008a. Meso-urban meteorological and photochemical modeling of heat island mitigation. Atmos. Environ. 42 (38), 8795–8809.

Taha, H., 2008b. Urban surface modification as a potential ozone air quality improvement strategy in California: A mesoscale modelling study. Boundary layer Meteorol. 127, 219–239. Available from: https://doi.org/10.1007/s10546-007-9259-5.

Takebayashi, H., Moriyama, Masakazu, 2007. Surface heat budget on green roof and high reflection roof for mitigation of urban heat island. Building and Environment 42, 2971–2979.

Takebayashi, H., Moriyama, Masakazu, 2009. Study on the urban heat island mitigation effect achieved by converting to grass-covered parking. Sol. Energy 83, 1211–1223.

Taleghani, M., Sailor, David J., Tenpierik, Martin, van den Dobbelsteen, Andy, 2014a. Thermal assessment of heat mitigation strategies: The case of Portland, State University, Oregon, USA. Build. Environ 73, 138–150.

Taleghani, M., Tenpierik, Martin, van den Dobbelsteen, Andy, Sailor, David J., 2014b. Heat mitigation strategies in winter and summer: Field measurements in temperate climates. Build. Environ 81, e309–e319.

Theeuwes, N.E., Solcerová, A., Steeneveld, G.J., 2013. Modeling the influence of open water surfaces on the summertime temperature and thermal comfort in the city. J. Geophys. Res.: Atmos. 118, 8881–8896. Available from: https://doi.org/10.1002/jgrd.50704.

Thorsson, S., Honjo, T., Lindberg, F., Eliasson, I., Lim, E.M., 2007. Thermal comfort and outdoor activity in Japanese urban public places. Environment and Behavior 39 (5), 660–684.

Tsilini, V., Papantoniou, Sotiris, Kolokotsa, Dionysia-Denia, Maria, Efpraxia-Aithra, 2014. Urban gardens as a solution to energy poverty and urban heat island. Sustain. Cities Soc 14, 323–333.

Tsitoura, M., Michailidou, Marina, Tsoutsos, Theocharis, 2016. Achieving sustainability through the management of microclimate parameters in mediterranean urban environments during summer. Sustain. Cities Soc 26, 48–64.

Tumini, I., 2014. The Urban Microclimate in Open Space. Case Studies in Madrid Ph. D. Thesis. Escuela Técnica Superior de Arquitectura de Madrid, Spain.

Upmanis, H., Eliasson, I., Lindqvist, S., 1998. The influence of green areas on nocturnal temperatures in a high latitude city (Goteborg, Sweden). International Journal of Climatology 18 (6), 681–700.

Velazquez, R., Alvarez, Servando, Guerra, Jose, 1992. Control Climatico de Los Espacios Abiertos en Expo'92, DYNA – NQ 2 – Marzo.

Wang, Y., Berardi, Umberto, Akbari, Hashem, 2016. Comparing the effects of urban heat island mitigation strategies for Toronto, Canada. Energy Build. 114, 2–19.

Watkins, R., Palmer, J., Kolokotroni, M., Littlefair, P., 2002. The London heat island – Surface and air temperature measurements in a park and street gorges. ASHRAE Transactions 108 (1), 419–427.

Xu, J., Wei, Q., Huang, X., Zhu, X., Li, G., 2009. Evaluation of human thermal comfort near urban water body during summer. Build. Environ. 45, 1072–1080.

Yamada, H., Yoon, G., 2008. Study of cooling system with water mist sprayers: Fundamental examination of particle size distribution and cooling effects. In: The 10th International Building Performance Simulation Association Conference and Exhibition. Tsinghua Press and Springer-Verlag, Beijing, pp. 1389–1394.

Yoon, G.H.Y., 2008. Study on a cooling system using water mist sprayers; system control considering outdoor environment. In: Korea-Japan Joint Symposium on Human-Environment System, Cheju, Korea, p. 4.

Zhou, Yan, Shepherd, Z.J., Marshall, 2010. Atlanta's urban heat island under extreme heat conditions and potential mitigation strategies. Nat. Hazards 52, 639–668. Available from: https://doi.org/10.1007/s11069-009-9406-z.

Zoras, S., Tsermentselis, A., Kosmopoulos, P., Dimoudi, A., 2014. Evaluation of the application of cool materials in urban spaces: a case study in the center of Florina. Sustain. Cities Soc 13, 223–229.

Zoras, S., Dimoudi, Argyro, Evagelopoulos, Vasilis, Lyssoudis, Spyros, Dimoudi, Sofia, Tamiolaki, Anna-Maria, et al., 2015. Bioclimatic rehabilitation of an open market place by a computational fluid dynamics simulation assessment. Future Cities Environ 1, 6.

Zoulia, I., Santamouris, M., Dimoudi, A., 2009. Monitoring the effect of urban green areas on the heat island in Athens. Environmental Monitoring and Assessment 156 (1–4), 275–292.

FURTHER READING

Group Building Environmental Research, University of Athens, 2016. Bioclimatic studies on the application of mitigation technologies in the built environment. Available through the author.

Jauregui, E. (1990). Infl.

G.E. Kyriakodis and M. Santamouris : Using Reflective Pavements to Mitigate Urban Heat Island in Warm Climates - Results from a large scale urban mitigation project., Urban Climate, In Press, 2017.

Santamouris, M., Ding, L., Fiorito, F., Oldfield, P., Paul Osmond, Paolini, R., et al., 2016. Passive and active cooling for the outdoor built environment – Analysis and assessment of the cooling potential of mitigation technologies using performance data from 220 large scale projects. Solar Energy In Press, Corrected Proof, Available online 19 December.

Chapter 9

Eradicating Energy Poverty in the Developed World

PRESENTATION AND ASSESSMENT OF THE EXISTING PROGRAMS TO FIGHT ENERGY POVERTY IN DEVELOPED COUNTRIES

Policies to eradicate energy poverty usually comprise measures of an economic, social and technological nature. In general, policies aiming to fight energy poverty are classified either as short-term relief measures and/or structural ones (Dheret and Giuli, 2017). Short-term measures involve financial interventions aiming to provide financial relief to low income households, while structural measures include more radical measures, such as additional protection, investments to improve the energy efficiency of buildings, and measures to increase information and awareness (Dheret and Giuli, 2017). A full list of the potential good practices to eradicate energy poverty is proposed and managed by the European Union and is given in Table 9.1 (EC, 2015a,b). The list includes policies aiming to protect consumers and also control and regulate the relevant markets.

Financial instruments usually focus on the provision of subsidies to cover part of the energy cost of the low income households, while other economic tools and measures may include tax reductions, labor bonuses, and so on. According to Dheret and Giuli (2017), several financial interventions of different natures are being implemented in European countries. These include social support (36%), direct payments (39%), social tariffs (20%), and negotiated tariffs (5%). Other measures to protect low income households include disconnection safeguards, debt protection measures and utility codes of conduct (Dheret and Giuli, 2017). Financial measures to support low income households may provide temporal assistance, but are highly criticized by experts for two main reasons: (1) they do not offer any structural interventions and do not provide incentives towards behavioural change, while at the same time they do not contribute towards a reduced and a more efficient energy consumption, and (2) as they are identified through a social welfare system, they suffer from poor targeting and may not support the population that needs them.

Minimizing Energy Consumption, Energy Poverty and Global and Local Climate Change in the Built Environment: Innovating to Zero. DOI: https://doi.org/10.1016/B978-0-12-811417-9.00009-X

TABLE 9.1 Good Practices to Mitigate and Tackle Energy Poverty

Category	Description
Financial measures	Earmarked subsides i.e partial or full payments of market priced energy bills
	General (non-earmarked) social support
	Targeted Social tariffs i.e.: regulated prices limited to a well-defined group
Consumer protection	Safeguards against disconnections (registry of vulnerable consumers which cannot be disconnected, disconnection moratorium during winter; active customer engagement by the energy provider prior to disconnection; possibility of debt-restructuring; provision of Pay As You Go meters for the provision of a minimum amount of energy)
	Consumer complaints (alternative dispute resolution e.g. by the National Regulatory Authority, an Ombudsman or a consumer body)
	A wareness campaigns and information dissemination through a single point of contact (e.g. consumer inforamtion point, advice centre)
	Supplier of last resort designated according to competitive procedures
Market-centred	Impartial, verified price comparison tools
	Codes of good conduct for utilities agreed with the Energy Regulator
	Retrofit grants, loans, or tax incentives
Energy efficiency measures for buildings	Minimum energy efficiency standards in the social, private and private rented sector
	Energy Company obligations to invest a share of energy efficiency investments in low-income and vulnerable households or deprived areas
	Grants for appliances

Source: *Adapted from European Commision. 2015a. Energy Renovation: The Trump Card for the New Start for Europe, JRC Science and policy Reports. <http://iet.jrc.ec.europa.eu/energy efficiency/system/tdf/eur26888_buildingreport_online.pdf?file = 1&type = node&id = 9069>; European Commission. 2015b. Working paper on Energy Poverty – Vulnerable Consumers Working Group, available at: <https://ec.europa.eu/energy/sites/ener/files/documents/Working%20Paper%20on%20Energy%20Poverty.pdf>.*

It is widely accepted that technological measures, and in particular the retrofitting of low income houses or the construction of new energy efficient houses, are among the more efficient instruments to eradicate energy poverty and protect the vulnerable population. Programs to eradicate energy poverty, as well as many relevant studies, have shown that there are many economic, social and environmental advantages associated with the deep retrofitting of low income houses and the construction of new energy efficient residential buildings.

Low income households live in houses with poor energy performance. As shown in the previous chapters, most low income houses do not provide the required climatic and environmental protection to the tenants. Improving the energy efficiency of low income residential buildings requires a substantial financial investment. The necessary investments to eliminate energy poverty depend on three major factors: (1) the number of households under energy poverty status, (2) the existing housing conditions, and (3) the cost of the required investments per house.

Given that energy poverty is not accounted for everywhere using the same metrics, there is considerable uncertainty regarding the total number of energy poor in the developed nations of the world. In parallel, housing conditions vary substantially in different parts of the world and specific energy and environmental needs are completely different, while the cost of the retrofitting depends on the local financial conditions and characteristics, and the specific technologies required to be employed.

According to EPEE (2009), the number of energy poor in Europe is between 50–125 million. Bird et al. (2010) estimate a much higher number, approaching almost 150 million people. In parallel, the EU Survey on Income and Living Conditions (EU SILC) states that in 2012 approximately 54 million European citizens, which correspond to about 10.8% of the EU population, were unable to satisfy their basic energy needs. Considering that in Europe the average size of households is 2.74 people (Doll and Haffner, 2010), it can be reasonably concluded that the total number of energy poor households in Europe varies between 18.3 and 54.8 million. Although existing information on the conditions and the qualitative characteristics of energy poor residential buildings in Europe is very limited, several national projects aiming to retrofit low income houses provide significant information on the required investments and potential energy and environmental benefits arising from the retrofitting activities.

Santamouris (2016) has identified and summarized fifteen different sources of information reporting data about the financial and technical characteristics of relevant energy retrofitting activities in low income residential buildings in Europe. The main characteristics of the identified programs are reported in Table 9.2 (Santamouris, 2016).

All data reported in Table 9.2, refer to either specific studies or to real projects mainly carried out in the UK. As shown, to achieve a reduction in energy needs of at least 60%, investments of between EUR€20,000 and EUR€45,000 are required. The specific magnitude of the cost depends on the financial characteristics of each country and the actual conditions of the building stock under consideration. Based on the above figures, it can be

TABLE 9.2 Presentation of Different Financial Scenarios and Policies Regarding the Refurbishment of Low Income Houses in Europe

Scenario No.	Investments per Household, (Euros)	Country	Remarks	Reference
1	257	UK	This is the calculated minimum retrofitting cost to decrease the energy cost just below the threshold of 10% of the household income.	Preston et al. (2008).
2	400	France	The European Regional and Development Fund allocated 320 million Euros to renovate 800,000 social housing by 2020	USH (2013)
3	462	UK	Installation of Insulation in 70,645 buildings between 2007–2010 using 32,62 million Euros	Butterworth et al. (2011)
4	863	Ireland	From 2000 to 2013 95,000 homes were supported by the Warmer Homes Scheme spending 82 million euros	SEAI (2013)
5	1712	UK	Warm Front Scheme spent 3,98 billion Euros for 2324,500 households from 2000 to 2012	WFS (2013)
6	2691	Romania	Refurbishment of 65,000 apartments with a total budget of 16 million Euros	
7	3974	France	Almost 233,7 million euros were used to renovate 58,800 low income houses.	USH (2013)
8	4200	UK	The Warm Front Scheme gives a grant of 4200 Euros for professional energy efficiency measures.	WFS (2013)
9	6832	UK	In Wales 51,250,000 million Euros spent for the refurbishment of 7500 dwellings	WSA (2012)

10	8400	UK	The Warm Front Scheme gives a grant of 8400 Euros for advanced energy efficiency measures.	WFS (2013)
11	9800	UK	Estimated Cost of Refurbishment of the EST Anaysis, (60% CO_2 Reduction-Low Cost)	EST (2008)
11	20,927	Hungary	Cost of the Low income housing Refurbishment in Hungary. Final Energy Consumption below 40 kWh/m^2	Hermelink (2006)
12	28,000	UK	Estimated Cost of Refurbishment of the EST Anaysis, (60% CO_2 Reduction-High Cost)	EST (2008)
12	29,820	UK	Estimated Cost of Refurbishment of the Tarbase Anaysis, (46% CO_2 Reduction-Low Cost)	Jenkings (2010)
13	33,484	Europe	Cost of deep refurbishment 300–400 $Euros/m^2$	EC (2015)
14	35,000	UK	Estimated Cost of Refurbishment of the Green Features Anaysis, (60% CO_2 Reduction)	Green Futures, 2008
15	44,660	UK	Estimated Cost of Refurbishment of the Tarbase Anaysis, (57% CO_2 Reduction-High Cost)	Jenkings, 2010

Source: *Adapted from Santamouris et al. (2015).*

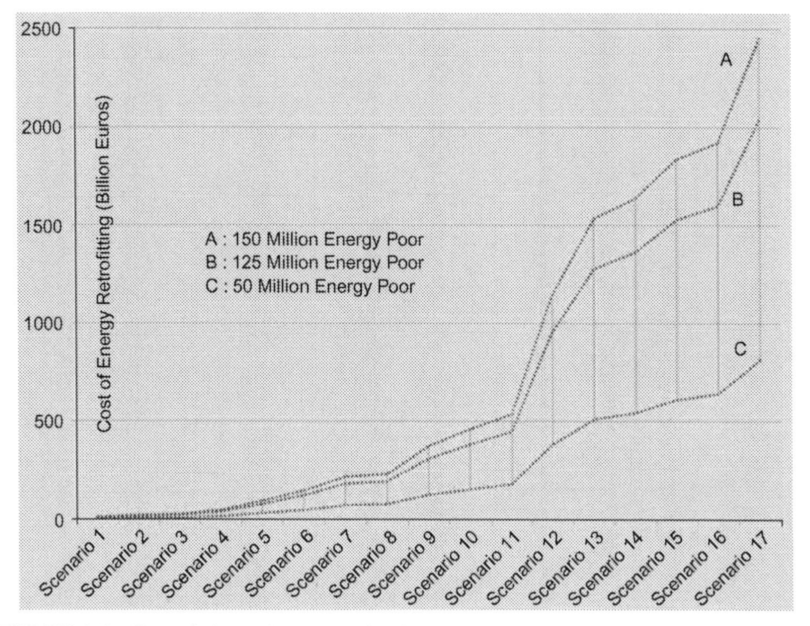

FIGURE 9.1 Cost of the various scenarios for the rehabilitation of low income houses in Europe. *Adapted from Santamouris (2016).*

estimated that the necessary investments to eradicate energy poverty in the countries of the European Union may vary between EUR€1.1 to EUR€2.5 trillion (Santamouris, 2016). Fig. 9.1 presents the results of the calculations performed by Santamouris (2016) regarding the necessary investment to fully eradicate energy poverty in Europe. Three scenarios are considered: for 50, 125 and 150 million energy poor people. Calculations are performed for the 15 different costs given in Table 9.2.

To fight energy poverty in Europe, several projects have been designed and financed by the European Union. The projects aim to either introduce low cost energy efficiency measures or to plan and optimize the refurbishment and the construction of new low income residential buildings. Projects incorporating low cost and high energy efficiency measures aim to provide information and advice to consumers, train so-called energy ambassadors, employ teachers and students as energy advisors, and develop tools and practices leading to energy saving behavior.

Smart-Up is an EU project (Smart-UP, 2018) that "will encourage vulnerable customers in those Member States that have embarked on the roll-out of Smart Meters, to actively use their Smart Meters and In-House Displays to achieve energy savings." The project aims to train 225 smart meter installers, social workers and other frontline staff on the best way to advise vulnerable households. Almost 5000 home visits are foreseen within the project. As mentioned, several studies have indicated that the use of smart metering technologies do not result in substantial energy savings in the residential sector unless the tenants use them in an active way that modifies their energy-use behaviour. The project aims to provide advice on the best way to use smart meters and in-home displays, to maximise the use and performance of the energy systems. It also aims to provide guidance towards more energy efficient behaviour, and advise people how to choose new appliances.

The TRIME Project (2018), aims 'to explore how social housing residents can be helped to reduce their energy bills and cost of energy'. It endeavours to provide training to 60 energy ambassadors who will try to help social housing residents decrease their energy consumption. Almost 1600 home visits by the so-called energy ambassadors to social housing are foreseen during the project.

The REACH project (2018), aims to 'to contribute to energy poverty abatement at practical and structural level [sic]. To empower energy poor households to take actions to save energy and change their habits, and to establish energy poverty as an issue that demands structural solutions at local, national and EU level.' The project aims to train students and teachers in vocational training schools to provide energy advice to low income households. Almost 1600 visits in low income houses are foreseen to provide tailor-made advice, guidebooks and energy saving device kits. It is expected that the project will result in energy savings of close to 300 toe/year.

The ACHIEVE project (2018), aims 'to contribute to practical (energy uses and behaviours) and structural (retrofitting buildings) solutions for reduction of fuel poverty in Europe.' The project is developing tools and practices to address energy poverty in Europe. About 150 energy experts have been trained and have performed more than 3000 visits to install low energy consumption devices in low income residential buildings to save energy and water. It is foreseen that the project will contribute to overall energy savings exceeding 250 toe/year.

European projects related to the refurbishment and construction of new low income houses aim to engage social housing companies in building renovations, and also to share best practices and guidance. Social housing companies engaged in the renovation or the construction of new social houses receive financial assistance from the European Union to perform the necessary feasibility and market studies, business plans, structuring of projects, preparation of tendering procedures and energy audits. The LEMON project (2018), aims 'to increase saving opportunities, improve housing space quality and create value for all the actors involved in the energy refurbishment of buildings: from people living in the dwellings to the technicians who manage and keep high building's [sic[performances, from innovators to investors in the construction sector.' The project will invest almost EUR€15 million to improve the energy conditions of about 600 private and public social houses in Italy. Energy savings of up to 40% are guaranteed by the involved

ESCOs. In a similar way, the project EnerSHIFT (2018) focuses on 'the energy renovation of public housing in Liguria, Italy, promoting and applying innovative financing models. In line with the Regional Environmental Energy Plan, EnerSHIFT seeks to reduce energy consumption, improve tenants' quality of life and boost the local economy. As part of this process, EnerSHIFT will launch a public tender process reserved for ESCo (Energy Service Companies) after which energy performance contracts (EPC) will be signed.' The project will invest almost EUR€14.6 million to retrofit 43 social houses in Liguria, Italy.

The AFTER and SHELTER projects of the European Union involve social housing organisations from several European countries aiming to develop and implement energy saving measures and integration methods in their asset and facility management. Energy efficiency measures were implemented in about 820 dwellings, resulting in energy savings close to 500 toe/year.

In the United States of America, the Low Income Home Energy Assistance Program, (LIHEAP) assists low income households in managing energy costs and facing energy crises, while it provide assistance to improve the energy efficiency of low income houses. In parallel, in 2015, the Federal Low Income Home Energy Assistance program invested more than USD$3 billion to help the low income population. However, as stated by the Solargaines block (2018), only 22% of the families needing assistance received funding. According to LIHEAP (2018), in 2015 almost USD$1.7 billion was spent on heating assistance, USD$202 million on cooling assistance, and USD$336 million was spent to improve the energy performance of low income houses.

TECHNOLOGICAL SOLUTIONS TO IMPROVE THE ENERGY AND ENVIRONMENTAL QUALITY OF LOW INCOME HOUSING

Building technologies provide a plethora of solutions to improve the energy and environmental quality of low income houses. The selection of appropriate technological solutions depends on several economic, climatic and technological criteria. Building technologies to be employed in the retrofitting of low income buildings should satisfy a number of these, among which the more important are:

a. Technologies have to respond to the specific energy and environmental needs of houses. If, for example, houses present very low temperatures during the winter period, then technologies that are able to reduce heating losses and increase heating and solar gains must be used. These techniques involve, among others, the use of additional insulation in the walls and the roof of the building, increasing in solar gains, decreasing unnecessary ventilation and infiltration, the use of double windows, and the implementation of high efficiency heating systems. On the contrary, when the main energy and environmental problem of the houses are related to summer overheating, then the technologies to be employed should maximize energy losses, minimize internal heat as well as all energy and solar gains, and make use of any available atmospheric sinks where the excess heat of the building may be dissipated.

b. The employed technologies should present the maximum possible energy benefits with the minimum possible financial cost. The investments performed should be optimized to achieve the maximum possible improvements for the household and the community.

c. Selected technologies should be simple and should be managed and controlled easily. Sophisticated and quite complicated energy systems and techniques requiring complex treatment may not operate at optimum conditions and may not contribute their maximum energy potential, mainly because of sub-optimal use by the tenants.

d. Energy metering techniques and smart indoor environmental monitoring should be installed in houses and should be used by the tenants to control and regulate their energy consumption. Knowledge of the magnitude of energy consumption helps foster a more conscious management by the tenants. The energy rebound affect that results in an increase in energy consumption after the energy retrofitting of the houses should be minimized or fully avoided. This can be achieved through the implementation of proper metering technologies and the appropriate training of tenants on the optimum ways to use energy efficiently.

e. The selected technologies should have a quite long lifespan and should not present a high maintenance cost.

f. The operational cost of the installed technologies should be the minimum possible to avoid an excessive running energy cost for the low income households. The integrated capital, maintenance and operational costs should be optimized using life cycle techniques.

g. The technologies to be selected should provide the maximum possible added value. In particular, they should create employment and income in the region and the country, and should promote the local and global economy.

h. Retrofitting of low income houses should be always associated with a substantial improvement in the local microclimate. It is well known that low income and vulnerable populations live in poor urban zones presenting serious environmental problems. Atmospheric pollution and urban heat islands (UHIs) are among the more serious problems in devaluated urban zones. UHIs cause serious environmental problems in dense and poor urban areas that present significantly higher ambient temperatures than the less dense and well developed zones. The intensity of local overheating may exceed 5–10°C, and result in a tremendous increase in the cooling load of the buildings, as well as in serious indoor and outdoor environmental problems (Santamouris 2–14, 2015; Santamouris et al., 2015). Advanced mitigation technologies have been developed in recent years to counter-balance the negative effects of urban overheating on energy demand and the environment quality of the urban built environment (Santamouris et al., 2007). Among the various available mitigation technologies, those that are of most interest are: (1) the employment of advanced materials that are able to amortize, dissipate and reflect heat and solar radiation, (2) the use of greenery as well as the implementation of cool and green roofs, (3) solar control and evaporation systems, (4) techniques aiming to dissipate excess heat in environmental heat sinks with a temperature that is lower than the ambient temperature, such as the ground, the water and the ambient air, (5) the integration of low and zero carbon systems such as renewables energy and micro generation technologies at the city level (Santamouris and Kolokotsa, 2013; Santamouris, 2014a,b).

The use of cool roofs in cooling dominated climates may contribute significantly to improving the energy performance and the total environmental quality of low income houses. Cool roofs are roofing systems and techniques that deliver a very high reflectance together with a very high thermal emmitance. Their application helps to decrease the surface temperature of the roofs and consequently the cooling energy needs of the buildings, while also helping to improve indoor comfort conditions (Figs. 9.2 and 9.3).

Recent research and numerous applications have shown that the use of cool roofs in buildings may reduce their cooling load by up to 45%, with an average contribution close to 20% (Synnefa et al., 2006; Kolokotsa et al., 2011). An analysis of the cooling potential of cool roofs, together with an estimation of the potential heating penalty, is provided by Synnefa et al. (2007). Fig. 9.4 reports the calculated decrease of the cooling load and the corresponding increase of the heating demand for a representative residential building for various parts of the world.

As shown, in cooling dominated climates, cool roofs can significantly decrease the cooling load of low income houses, while the corresponding decrease in the heating demand is very low.

Parallel to the important energy and environmental benefits provided by cool roofs to individual buildings, cool roofs help to decrease the ambient temperature and thus further decrease the cooling needs of the buildings. Recent research shows that the use of cool materials in the built environment may decrease the maximum ambient temperature in cities by up to 1°C on average (Figure 8.5). Such a temperature reduction has a serious impact on the cooling energy consumption of buildings. Estimations have shown that when cool materials are used in the urban fabric, the resulting decrease in the ambient temperature may reduce the cooling load of the buildings by up to 20% (Santamouris et al., 2018) (Fig. 9.5).

The use of appropriate shading and solar control devices in buildings and open spaces is a significant technique to improve the microclimate of cities and protect buildings from solar radiation. Shading open spaces reduces the

FIGURE 9.2 Visible and infrared images of the roof surface depicting the differences in the surface temperature prior and after the cool roof application. *Adapted from Synnefa, A., Santamouris, M. December 2012. Advances on technical, policy and market aspects of cool roof technology in Europe: The Cool Roofs project. Energy Build. 55, 35–41.*

absorption of solar radiation and decreases the mean radiant temperature of urban areas, contributing to better thermal comfort conditions. Trees can protect open spaces and the façade of buildings, and thereby decrease the cooling load. As mentioned by Papadakis et al. (2001), when trees are planted close to the façade of the buildings, the incident solar radiation may be reduced by up to 85%. In parallel, trees may significantly reduce the ambient temperature as a result of evapotranspiration processes (Skoulika et al., 2014). Analysis of several projects aiming to mitigate urban overheating by using additional greenery show that it can decrease peak ambient urban temperatures by up to 2.5°C (Fig. 9.6).

Adapted from Santamouris, M., Ding, L., Fiorito, F., Oldfield, P., Osmond, P., Paolini, R., et al. September 2017. Passive and active cooling for the outdoor built environment – Analysis and assessment of the cooling potential of mitigation technologies using performance data from 220 large scale projects. Sol. Energy, 154, 14–33.

Ground cooling techniques can reduce both the cooling load of buildings and decrease the ambient temperatures in open spaces (Santamouris and Assimakopoulos, 2017). The use of earth to air heat exchangers to mitigate

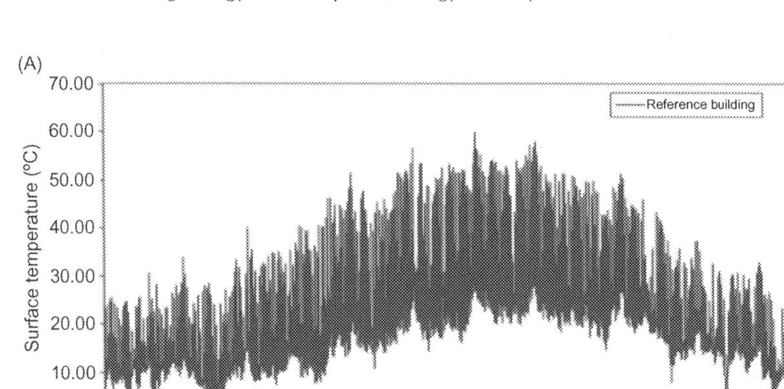

FIGURE 9.3 Hourly values of surface temperatures for both the reference (A) and cool (B) roof buildings. *Adapted from Synnefa, A., Santamouris, M. December 2012. Advances on technical, policy and market aspects of cool roof technology in Europe: The Cool Roofs project. Energy Build. 55, 35–41.*

the UHI in Tirana, Albania (Fintikakis et al., 2011) shows that it is possible to decrease the maximum local air temperature by up to 0.7 K. In parallel, earth to air heat exchangers can contribute highly to decreasing the cooling needs of low income houses (Santamouris and Kolokotsa, 2015).

ECONOMIC AND SOCIAL TOOLS AND POLICIES TO FIGHT ENERGY POVERTY

The magnitude of energy prices is a decisive parameter that plays a significant role in the level of energy poverty in a country. Although most effort to reduce energy poverty focuses on improving the energy performance of low income houses through a deep retrofitting process or the construction of new

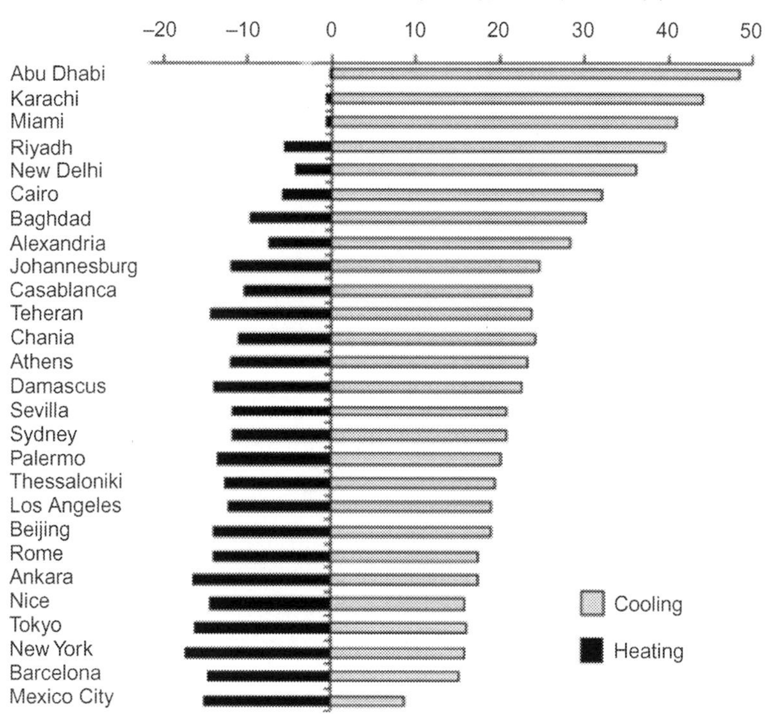

FIGURE 9.4 Climate effect on cooling and heating load changes for a change in roof solar reflectance of 0.65. *Adapted from Synnefa, A., Santamouris, M., Livada, I. 2006. A study of the thermal performance of reflective coatings for the urban environment. Sol. Energy 80(8), 968–981.*

energy efficient houses, future energy prices will greatly determine the balance of household economics and, in turn, the levels and the characteristics of energy poverty.

Most the future energy scenarios foresee a very significant reduction in energy consumption within the building sector by 2050. However, most of the scenarios simultaneously predict a very significant increase in energy prices. As mentioned by Santamouris (2016), the expected increase in electricity prices by 2050 may vary from 35%–40% compared to 2010 prices.

This increase in energy prices may place a substantial future burden on the low-income population and may contribute significantly to increasing energy poverty levels. It is characteristic that the European Commission (2014) reports that based on the results of the future scenarios EE35 and EE40, the average cost of energy for the European households in 2050 may reach 13.2% of household income, while under current conditions it does not exceed 7.5%.

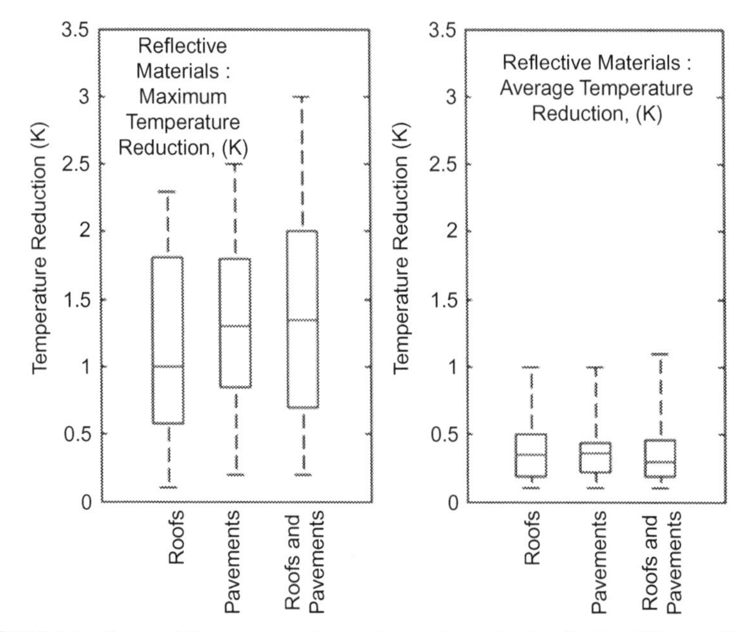

FIGURE 9.5 Range of the average and peak temperature reduction for the different reflective technologies. *Adapted from Santamouris, M., Ding, L., Fiorito, F., Oldfield, P., Osmond, P., Paolini, R., et al. September 2017. Passive and active cooling for the outdoor built environment – Analysis and assessment of the cooling potential of mitigation technologies using performance data from 220 large scale projects. Sol. Energy 154, 14–33.*

ANALYSIS OF THE POTENTIAL IMPACT OF A COMPLETE ERADICATION OF ENERGY POVERTY IN THE DEVELOPED WORLD

A potential complete eradication of energy poverty in the developed world is associated with significant economic, energy, environmental and social benefits.

It is estimated that the complete eradication of energy poverty in Europe by 2050 will create 197,000 to 420,000 additional jobs per year (Santamouris, 2016). The specific figures are estimated by considering that the total number of energy poor in Europe is currently approaching 150 million people, the refurbishment of their houses will be performed by 2050, and that each million euros of investment in refurbishing the dwellings of energy poor households creates almost 6 new additional jobs. In the case where higher labor multiplication values are considered, then the above figures must be adjusted accordingly.

Investments aiming to improve the energy performance of low income housing may boost the local economy and increase the global impact on the society through the creation of additional capital ventures. As mentioned by Santamouris (2016), investments of about EUR€234 million aiming to

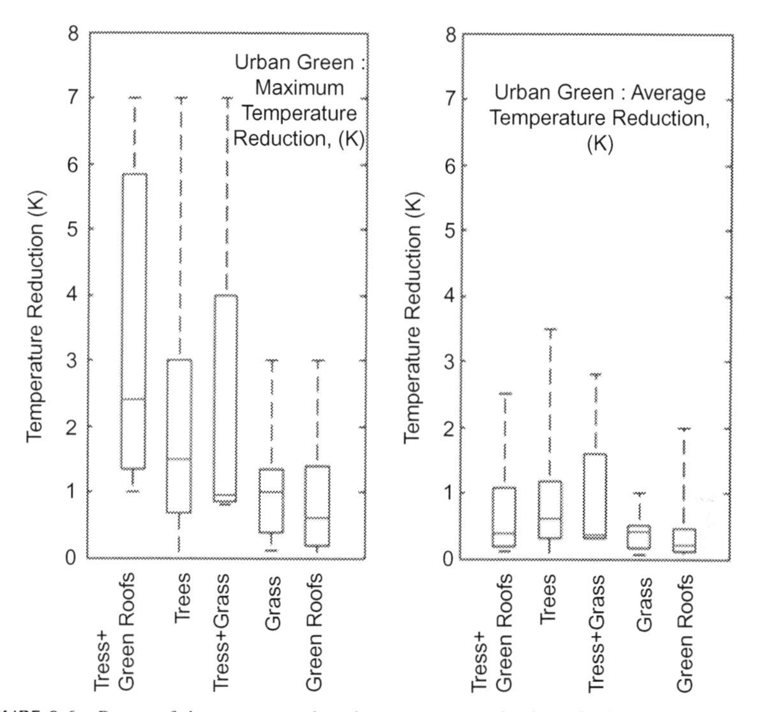

FIGURE 9.6 Range of the average and peak temperature reduction of urban greenery mitigation techniques.

improve the energy performance of low income housing, performed in France by the European Development Fund from 2009–2013, generated a global investment of almost EUR€1.22 billion and almost 17,225 additional jobs. This corresponds to a labor multiplication factor of close to 18.5 additional jobs per million euros of investment in energy performance improvements.

An additional benefit arising from potential investments in improving the energy performance of low income housing is the significant reduction of the emitted greenhouses gases. Given the previously estimated energy benefit per unit of energy efficiency investments (1.8 PWH per trillion euros invested), it is calculated that the reduction in energy consumption associated with the deep retrofitting of low income housing, and the corresponding improvement of their energy performance, may vary between 0.61 PWH and 1.39 PWH (Santamouris, 2016). This may decrease greenhouse gas emissions from low income houses by up to 60%, (Santamouris et al., 2016).

Apart from the considerable energy and global environmental benefits associated with the deep retrofitting of low income houses, significant improvements in indoor environmental quality and health are reported (Kolokotsa and Santamouris, 2015; Santamouris and Kolokotsa, 2015). As reported by SHU (2008), the refurbishment of many low income houses performed by the Warm

Front Scheme in the UK has decreased the levels of anxiety and depression experienced by the beneficiaries by almost 50%. In a similar manner, the Sustainable Energy Authority of Ireland (2009) found that after the energy retrofitting of low income houses performed by the Warmer Homes Schemes, the number of beneficiaries suffering from disorders and long term illness decreased by almost 88%. In parallel, is reported that headaches were reduced by 83.3%, heart attacks by 70%, circulatory problems by 64%, asthma by 62.5% and hypertension problems by 56%.

The reduction of health problems among low income and vulnerable households caused by the deep retrofitting of their housing has a serious impact on the health care expenses and the total budget of the National Health System. As reported by Butterworth et al. (2011), who provide estimations performed in the frame of the UK based project Kirklees, the global financial benefits to the UK National health system were close to the 20.6% of the budget invested to retrofit and improve the energy performance of a large number of low income houses.

It is evident that appropriate measures to fight or even eradicate energy poverty may have significant energy, social, environmental and employment benefits. It is also clear that significant investments are required to improve the living conditions of the vulnerable population. While expensive, such a level of investment contributes greatly to generating growth and additional employment, while it substantially reduces greenhouse emissions and improves the social status of the low income population.

REFERENCES

ACHIEVE EU project: <https://ec.europa.eu/energy/intelligent/projects/en/projects/achieve>, 2018.

Bird, J., Campbell, R., Lawton, K., 2010. The long cold winter: Beating fuel poverty, institute for public policy research and national energy action, March 2010, <www.vhscotland.org.uk/library/misc/The_Long_Cold_Winter.pdf>.

Butterworth, N., Southernwood, J., Dunham, C., 2011. Kirklees Warm Zone Economic Impact Assessment. <https://www.kirklees.gov.uk/you-kmc/partners/other/pdf/warmZone_EconomicImpactAssessment.pdf>.

Dheret, C., Giuli, M., June 2017. The long journey to end energy poverty in Europe, Policy Brief, European Policy Center. Available through: <http://www.epc.eu/documents/uploads/pub_7789_ljtoendenergypovertyineurope.pdf?doc_id = 1871>.

Doll, K., Marietta Haffner, 2010. Housing Statistics in the European Union. The Hague: Ministry of the Interior and Kingdom Relations. <http://www.bmwfw.gv.at/Wirtschaftspolitik/Wohnungspolitik/Documents/housing_statistics_in_the_european_union_2010.pdf>.

Energy Saving Trust (EST), 2008. In: Roadmap to 60%: Eco-refurbishment of 1960's flats. EST, London. <http://tools.energysavingtrust.org.uk/Publications2/Housing-professionals/Refurbishment/Roadmap-to-60-eco-refurbishment-of-1960s-flats-2008-edition>.

ENERShift EU project: <https://enershift.eu/en/il-progetto/>, 2018.

EPEE, 2009. Tackling Fuel Poverty in Europe, Recommendations Guide for Policy Makers. Report produced in the framework of the Intelligent Energy Europe project EPEE- European Fuel Poverty and Energy Efficiency. <http://www.fuel-poverty.org/files/WP5_D15_EN.pdf>.

EU Statistics on Income and Living Conditions: Available through: <https://www.eui.eu/Research/Library/ResearchGuides/Economics/Statistics/DataPortal/EU-SILC>.

European Commission, 2014. Commission Staff Working Document Impact Assessment Accompanying the document Communication from the Commission to the European Parliament and the Council Energy Efficiency and its contribution to energy security and the 2030 Framework for climate and energy policy, COM(2014) 520 final} {SWD(2014) 256 final.

European Commision, 2015a. Energy Renovation: The Trump Card for the New Start for Europe, JRC Science and policy Reports. <http://iet.jrc.ec.europa.eu/energyefficiency/system/tdf/eur26888_buildingreport_online.pdf?file = 1&type = node&id = 9069>.

European Commission (2015b), Working paper on Energy Poverty – Vulnerable Consumers Working Group, available at: <https://ec.europa.eu/energy/sites/ener/files/documents/Working%20Paper%20on%20Energy%20Poverty.pdf>.

Fintikakis, N., Gaitani, N., Santamouris, M., Assimakopoulos, M., Assimakopoulos, D.N., Fintikaki, M., et al., 2011. Bioclimatic design of open public spaces in the historic centre of Tirana, Albania. Sustain. Cities Soc. 1 (1), 54–62.

Green Futures, 2008. The Future is Retro-fit: Bringing the Housing Stock up to Scratch. GreenFutures, London, <http://www.ourfutureplanet.org/newsletters/resources/Built%20environment/THE%20FUTURE%20IS%20RETRO-FIT%20-Bringing%20the%20housing%20stock.pdf>.

Hermelink, A., 2006. SOLANOVA, Proc European Conference and Cooperation Exchange. <http://www.solanova.org/3.html>.

Jenkings, D.P., 2010. The value of retrofitting carbon-saving measures into fuel poor social housing. Energy Policy 38, 832–839.

Kolokotsa, D., Santamouris, M., 2015. Review of the indoor environmental quality and energy consumption studies for low income households in Europe. Sci. Tot. Environ. 536, 316–330.

Kolokotsa, D., Diakaki, C., Papantoniou, S., Vlissidis, A., 2011. Numerical and experimental analysis of cool roofs application on a laboratory building in Iraklion, Crete, Greece. Energy Build. 55, 85–93.

LEMON EU Project: <http://www.lemon-project.eu/>, 2018.

LIHEAP: Available through: <https://www.acf.hhs.gov/ocs/programs/liheap/about>, 2018.

Papadakis, G., Tsamis, P., Kyritsis, S., 2001. An experimental investigation of the effect of shading with plants for solar control of buildings. Energy Build. 33 (8), 831–836.

Preston, I., Moore, R., Guertler, P., 2008. How Much? The Cost of Alleviating Fuel Poverty, Report to the EAGA Partnership Charitable Trust. CSE, Bristol.

REACH EU Project: <http://reach-energy.eu/> 2018.

Santamouris, M., 2014a. Cooling the Cities – A review of reflective and green roof mitigation technologies to fight heat Island and improve comfort in Urban environments. Sol. Energy 103, 682–703.

Santamouris, M., 2014b. On the energy impact of Urban Heat Island and global warming on buildings. Energy Build. 82, 100–113.

Santamouris, M., 2015. Regulating the damaged thermostat of the Cities – Status, Impacts and Mitigation Strategies. Energy Build. Energy Build. 91, 43–56.

Santamouris, M., April 2016. Innovating to zero the building sector in Europe: Minimising the energy consumption, eradication of the energy poverty and mitigating the local climate change. Sol. Energy 128, 61–94.

Santamouris, Mattheos, Kolokotsa, Dionysia, February 2013. Passive cooling dissipation techniques for buildings and other structures: The state of the art review article. Energy Build. 57, 74–94.

Santamouris, M., Kolokotsa, D., 2015. On the impact of Urban overheating and extreme climatic conditions on housing energy comfort and environmental quality of vulnerable population in Europe. Energy Build. 98, 125–133. Available from: https://doi.org/10.1016/j.enbuild.2014.08.050.

Santamouris, M., Assimakopoulos, D.N. (Eds.), 2017. Passive Cooling of Buildings. Earthscan, London.

Santamouris, M., Pavlou, K., Synnefa, A., Niachou, K., Kolokotsa, D., 2007. Recent progress on passive cooling techniques. Advanced technological developments to improve survivability levels in low-income households. Energy Build. 39 (7), 859–866.

Santamouris, M., Cartalis, C., Synnefa, A., Kolokotsa, D., July 2015. On the impact of Urban Heat Island and global warming on the power demand and electricity consumption of buildings—A review. Energy Build. 98, 119–124. Available from: https://doi.org/10.1016/j.enbuild.2014.09.052.

Santamouris, M., Ding, L., Fiorito, F., Oldfield, P., Paul Osmond, Paolini, R., et al., September 2017. Passive and active cooling for the outdoor built environment – Analysis and assessment of the cooling potential of mitigation technologies using performance data from 220 large scale projects. Sol. Energy 154, 14–33.

Santamouris, Mattheos, Haddad, Shamila, Saliari, Maria, Vasilakopoulou, Konstantina, Synnefa, Afroditi, Paolini, Riccardo, et al., 2018. On the energy impact of urban heat island in Sydney: Climate and energy potential of mitigation technologies. Energy Build. 166, 154–164.

Sheffield Hallam University, 2008. Health Impact Evaluation of the Warm Front Scheme.

Skoulika, F., Santamouris, M., Boemi, N., Kolokotsa, D., 2014. On the thermal characteristics and the mitigation potential of a medium size Urban Park in Athens, Greece. Landsc. Urban Plan. 123, 73–86.

Smart – Up EU Project: Assessed through: <https://www.smartup-project.eu/>, 2018.

Solargaines: Available through: <http://solargaines.com/10-stats-about-energy-poverty-in-the-u-s-that-will-shock-you/2018>.

Sustainable Energy Authority of Ireland, 2013. Energy Efficiency in the Residential Sector.

Synnefa, A., Santamouris, M., December 2012. Advances on technical, policy and market aspects of cool roof technology in Europe: The cool roofs project. Energy Build. 55, 35–41.

Synnefa, A., Santamouris, M., Livada, I., 2006. A study of the thermal performance of reflective coatings for the urban environment. Sol. Energy 80 (8), 968–981.

Synnefa, A., Santamouris, M., Akbari, H., 2007. Estimating the effect of using cool coatings on energy loads and thermal comfort in residential buildings in various climatic conditions. Energy Build. 39 (11), 1167–1174.

TRIME EU Project: <http://www.trime-eu.org/> 2018.

Union Sociale pour l'Habitat, 2013. L'Europe investit dans le logement social, Plan Europeen pour la relance economique, Evaluation 2009–2013-France".

Warm Front Scheme (2013)– Commons Library Standard Note, <www.parliament.uk>.

Welsh School of Architecture, Cardiff University, (2012): Arbed 1 Scheme, Evaluation of the Warm Wales Programme.

Chapter 10

Concluding Remarks and Policy Proposals

DESIGNING OUR FUTURE

The building sector is one of the more important economic sectors. It fuels the global economy, offers employment to millions of people, provides shelter, and protects the health and the well-being of dwellers. At the same time, it is the source of serious economic technological, environmental, and social problems. It is highly responsible for local and global climate change and global overheating problems, while it consumes tremendous quantities of energy and raw materials. In parallel, because of the significant cost of housing and associated services to supply them, a significant portion of the world's population cannot afford to satisfy their needs, and as a result live in a condition of poverty. It is by far the highest consumer of energy and materials, among all major economic sectors, while about 15% of the world's population live in unhealthy and inadequate shelters, placing the health and even the life of the poor population in danger. A major objective of the sector is to achieve the sustainability goals in the built environment; however, the specific targets set by the global sustainability agenda, mainly defined at the beginning of the present century, are not fully sufficient to face the current and future problems and challenges of the built environment. Recent urban growth, and the tremendous increase in the population expected until 2050, call for a more radical and efficient agenda to adapt the built environment to changing conditions. Greening the built environment is not enough any more. Humanity must fight existing poverty levels and at the same time must shelter more than 3–4 billion new people by 2050 and provide them with the basic health, sanitary and education services and networks. Most of the new population will live in cities within low income nations. It is therefore reasonable to consider that the additional stress placed on the urban built environment and infrastructure will be considerable. To satisfy such a requirement is a major challenge for the broad building industry, which must develop and apply innovative economic, political, social, and technological solutions and tools to the problems.

Inevitably, the various creators of an agenda for the future built environment have identified and documented all the above problems. International

Minimizing Energy Consumption, Energy Poverty and Global and Local Climate Change in the Built Environment: Innovating to Zero. DOI: https://doi.org/10.1016/B978-0-12-811417-9.00010-6

institutions have prepared scenarios and studies forecasting future conditions in the building sector. Unfortunately, knowledge and even identification of the issues are not enough to drive change, and may not suffice to provide credible solutions. Quantitative and qualitative technological, environmental, political, social, and economic targets and objectives concerning our future must be set and agreed. Once such an agreement is achieved, the most important thing is that humanity decides on an executive road map to follow until the objectives are met and acceptable solutions are implemented. Given the history of the world and the conditions under which we are living, such a scenario seems to be unrealistic or even a naïve expectation. Reaching such an international agreement is very difficult; some people may think it is impossible. Conflicting interests and the indifference of powerful groups and lobbies seem to present insurountable obstacles. Everyone agrees that moral principles and philanthropy are not strong enough motives to design and provide solutions. It is more realistic to search, define and implement alternative, efficient and powerful policies.

It seems that a potential translation of the identified challenges into corresponding opportunities may offer some perspectives and a light at the end of the tunnel. But how this may be realised ? The answer is clear through the economic development associated with the greening of the built environment in the developed countries and the drastic improvement of the quality of the built environment in the same countries, including the satisfaction of the needs of the expected new population. It is estimated that the budget required just to decrease the energy consumption of buildings by 2050 to meet the climatic targets of Paris, and decarbonize the energy infrastructures, exceeds USD$100 trillion. On top of this, the budget to buid at least 1.5 billion additional homes for the new population, extend the cities, and finally provide the necessary new urban and energy infrastructures multiplies the above figure by a factor of two or even three. In parallel, the budget to fight urban overheating and completely eradicate energy poverty all around the world may exceeds USD$100 trillion. Such a huge capital investment by 2050 is the most serious incentive and financial opportunity to attract and recruit everyone in a major world project aiming to improve the environment we are living . . .

Such a large available budget will offer opportunities for all the involved parties and all players in the market. Industry can produce and market new high-performance energy and environmental products and systems. Construction companies can benefit from the additional production of new residential and commercial building units and urban infrastructures. Employees in the building sector, including engineers, technicians and simple workers, will profit from the increase employment and market opportunities. Citizens will enjoy healthier buildings that consume fewer resources, and a cleaner, healthier and more comfortable outdoor environment. Service providers, including the operators of major networks, will further penetrate the industry and increase their market share. Policy makers will realize their plans to retrofit the buildings stock, improve the quality of the urban environment and fight energy poverty. The associated industrial sectors will advance their sales, since the budget of the building sector will increase and the buying capacity of the main players will rise.

The main question is whether this new market will be open and free to everyone. The answer is "yes," but under two conditions. First, all products, systems, buildings and new urban infrastructures offered should present a superior energy and environmental performance, satisfying the requirements for low energy consumption and carbon emissions, and in parallel a higher indoor and outdoor environmental quality. The second condition is related to the accessibility of the newly developed technologies. It is crucial that most of the energy and environmentally efficient technologies are financially accessible to the highest possible fraction of the world's population. If these requirements are met, new win-win conditions for almost all players will be created. Energy consumption and the carbon emissions will decrease significantly, the amplitude of the global and local climate change will de-escalate, vulnerable and low-income populations will benefit from healthier dwellings, and industry and the global commercial sector will see their profits increase tremendously. Besides these direct benefits, indirect social and economic advantages will be enormously important. The new economic activities will boost the fight against poverty, especially in the developing countries. They will increase the consumption capacity of local populations and significantly reduce immigration flows to developed countries. They will boost education and culture, and will help to weaken local conflicts. They will promote social and economic equity, protect the health of the population, increase life expectancy (especially in the developing countries), and will reduce vulnerability. Such an objective, although it seems very ambitious, is an unequivocal choice that will create substantial opportunities for future growth, alleviate the population from the consequences of the specific problems, and create short, medium and long-term benefits and opportunities

Is the world ready to implement such a policy in the built environment? It is evident that it is not. During the last 10–15 years, the development of building energy-related technologies has not exhibited significant progress, and the average increase in energy efficiency in the building sector is very slow. Although energy consumption for heating and lighting purposes is decreasing considerably in developed countries, energy consumption by the building sector globally is increasing on a continuous basis, even in the developed world. Needs for cooling are increasing and are expected to be almost double that of the corresponding heating requirements by the end of the century. Climatic change and urban overheating, the growth of the world's population, the rise in buying capacity in the developing world, and the increase in space occupied per person are skyrocketing cooling energy consumption in the building environment. As mentioned in the previous chapters, the cooling energy consumption of just one Indian city will be equal to the consumption of the whole of the United States by 2050. In addition, the thermal conditions in cities are continuously worsening. The reported intensity of the urban heat island (UHI) is increasing, and heat-related vulnerability, heat related mortality and morbidity are also expected to increase considerably. The number of people living in informal settlements and slums is foreseen to increase at least by 50 by 2050. Moreover, a very high portion of the world's population does not have access to clean energy and electricity, and are essentially forced to consume polluting, low-cost fuels in detriment to their health.

Future forecasts of progress in energy efficient building technologies are not optimistic either. The forecasted increase in energy efficiency cannot counterbalance even a small fraction of the energy increase caused by the other energy drivers. In the developing countries, the expected increase in the necessary energy infrastructures and the absolute energy consumption are tremendous and ask for giant future investments. Under the current conditions, the building industry seems to be satisfied and quite happy as the existing technological products sell well in the market and the profits are quite high. For most building-related industrial sectors annual turnover is increasing continuously. Competition in the market is also high, but not for the very efficient energy products, which remain very expensive and are not prominent in the market. Exemplary demonstration projects for advanced zero or positive energy buildings or settlements do not present a high replication potential, as they are not easily adopted by the mainstream building market. New legislation in the developed countries aims to boost technologies to decrease the energy consumption of buildings, despite the fact that the prices of the very advanced technologies remain high and are not accessible to most of the population, even in the developed world. If energy legislation is not combined with the widespread adoption of energy efficient and environmentally efficient products, it may put low income population under further stress and increase social inequalities. Increased energy prices, and the recurrence of economic crises, combined with the poor environmental and energy quality of low income housing stock, continuously shifts more social groups towards energy poverty conditions.

There is a serious need to further advance building energy-related research and innovation. New innovative and cutting edge energy and environmental technological products must be developed and must be available on the market. It is not enough to continue to slowly improvement energy efficiency. Humanity needs a technological revolution in building technologies, similar to the progress achieved recently in the electronics, telecommunication and pharmaceutical sectors. The countries and institutions that invest in advanced building research over the next few years will be the future technological leaders and will virtually monopolize the future market. The rest of the countries will follow the technological leaders and will simply purchase products.

These new, efficient technologies should available a reasonable cost. Profits should come from the extensive penetration of the products in the market and their widespread use by households and the commercial companies, not because of the high cost of the product per se. A marketing policy like the one followed by the electronics market must be designed and implemented. In addition, new economic and policy tools must be made available in the market. Investments by SMEs must be encouraged and promoted. Building-related research in the developing countries should try to develop and apply advanced energy technologies based on local knowledge and offering high added value for the country.

Innovative new products that are accessible to everyone will create substantial opportunities for businesses. Tremendous capital will be invested and will change hands. Here there is risk. Investments should result in the significant improvement of the economic status of society as a whole, and should be distributed fairly to all income groups. Otherwise, very few will profit and the social and economic inequalities in society will increase enormously. In this case, energy and environmental conditions may improve, but overall economic conditions will be worsened for the benefit of the few people who manage and own the capital. The price of energy may increase considerably and despite the significant reduction of energy consumption per household, the final economic burden on families may increase. This is clear in almost all future projections performed for developed countries. The cost of buildings may also increase, creating prohibitive conditions for most of the population, who will be required to pay higher costs to cover mortgages or pay expensive rents. In addition, the cost of living in urban zones with a high environmental quality will increase significantly, and the average and low income populations will move to areas of lower urban quality. Urban stratification resulting from the significant environmental and social differences between the various urban zones will inevitably create social tensions in cities, while critical problems of criminality and social diversity will be generated.

In conclusion, the proposed new urban agenda is a wonderful opportunity for development. If planned properly and implemented in a fair way, it can result in much better living conditions for most of the population in both the developed and developing countries, and it will generate wealth, employment and social equality. On the contrary, if investments are manipulated and used for the benefits of a small group of people, then the improvements achieved in the energy and environmental status will fail to have a positive impact. Instead, economic and social sustainability and equity will be damaged seriously, and social stratification and discrimination will increase tremendously and the additional social stress will be considerable. Investments should be targeted at diminishing economic disparities and discrimination. The efficiency of models of aggregate economic growth must be reconsidered in order to increase as much as possible the incremental benefits of global growth to the whole population and improve upon past poverty-reduction efforts, where in order for EUR€1 to reach the poorest population, almost EUR€166 had to be produced and consumed in the global economy (Sustainable Energy Security, 2015).

Such ambitious goals will be achieved gradually and will be associated with major technological breakthroughs. It is evident that the definition and the implementation of a clear, ambitious, holistic, and multifaceted vision for the future of the built environment is an imperative (Santamouris, 2016).

REFERENCES

Santamouris, M., April 2016. Innovating to zero the building sector in Europe: Minimising the energy consumption, eradication of the energy poverty and mitigating the local climate change. Solar Energy 128, 61–94.

Sustainable Energy Security, 2015. Sustainable Growth or Growth vs. Sustainability Examining the relationships between resource use, climate emissions and economic growth. <https://qceablog.files.wordpress.com/2011/07/qcea_sustainable_energy_security_briefing_paper_2.pdf>

Index

Note: Page numbers followed by "*f*" and "*t*" refer to figures and tables, respectively.

D

E

F

G

V

W

Z

Printed in the United States
By Bookmasters